Genetic
Principles

*Human
and Social
Consequences*

Genetic Principles

Human and Social Consequences

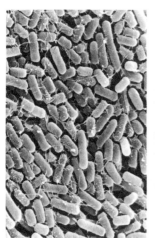   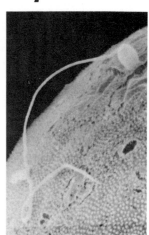

Gordon Edlin

UNIVERSITY OF CALIFORNIA, DAVIS

Jones and Bartlett Publishers, Inc.

BOSTON AND PORTOLA VALLEY

Editorial offices: 30 Granada Court, Portola Valley, CA 94025

Sales and customer service offices: 20 Park Plaza, Boston, MA 02116

Library of Congress Cataloging in Publication Data

Edlin, Gordon, 1932–
    Genetic principles.

    Includes index.
    1. Genetics.   I. Title.   [DNLM: 1. Genetics.   QH 431 E23g]
QH430.E34   1983   575.1   83-16257

ISBN 0-86720-016-2

Production: Del Mar Associates
Designers: Louis Neiheisel and Paul Slick
Manuscript editor: Jackie Estrada
Technical art director: Richard Carter
Illustrators: Kim Fraley and Linda McVay
Technical artists: Kim Fraley and Pam Posey
Compositor: Boyer and Brass
Printer and Binder: Alpine Press

Printed in the United States of America

Printing number (last digit) 10 9 8 7 6 5 4 3 2

for Merrill

Preface

The past few decades have witnessed an explosion in our knowledge of the principles of heredity and in our understanding of how genes function in organisms. From the pioneering pea experiments of Gregor Mendel in 1865, the science of genetics has blossomed into a discipline that impinges on the lives of everyone and that affects the organization of society as a whole. Genetics is playing an increasingly important role in the applications of modern medicine, the practices of modern agriculture, and the manufacture of drugs and chemicals.

In the past decade genetic discoveries have spawned a vast new biotechnology industry commonly referred to as genetic engineering. The applications of genetic engineering techniques to human and social problems have also raised complex ethical and moral questions that are not easily resolved. It is scarcely possible to pick up a newspaper or magazine today without finding an article on the potential profits of genetic engineering companies or people's concerns about fertilizing human eggs in test tubes or implanting human embryos in surrogate mothers. Another controversial topic is the teaching of creationism versus evolution in our public schools, which also involves complex moral and legal issues.

More than ever, applications of genetics affect how we live our lives, whether or not we decide to have children, and even how our children are educated. Whether we like it or not, the principles and uses of genetics are changing not only the society and world in which we live but the one that we will bequeath to future generations. Some understanding of the basic principles of genetics is thus essential if individuals are to function successfully and intelligently in modern societies.

Despite this need for basic knowledge, a recent poll showed widespread ignorance of one of the most basic principles of genetics. In 1983 this nationwide survey found that almost two-thirds of all adults could not explain what DNA is; 2 percent of those polled even believed DNA to be some kind of poison! Another survey conducted in 1982 showed that almost half of all Americans believe that humans similar to ourselves first appeared on earth only a few thousand years ago, despite overwhelming scientific evidence that human ancestors were walking about more than 3 million years ago. The fact that so many Americans are unaware of the basic concepts of genetics shows that our educational system has not been overly successful in conveying to students the principles of the biological world. Nor are students generally informed about the serious social consequences that occurred in the past as the result of misunderstanding and misuse of genetic principles. As with all scientific information, genetics can be used to benefit or to harm individuals and society.

In 1982 David S. Saxon, then president of the University of

California, wrote: "It should be possible to convey to students both the power and the limits of general scientific laws and why we can, in the light of both, draw reliable conclusions from those laws . . . If the ability to distinguish sense from nonsense is an indispensable aspect of a liberal education, . . . then in a technological society science is an indispensable part of the liberal arts curriculum." This book was written in the hope that students who read it will subsequently be able to distinguish sense from nonsense in the area of genetics—especially as it applies to their own lives and to the society they construct.

This book describes for readers with little or no scientific background how DNA functions as the carrier of genetic information in cells of all living organisms. It explains how genes are expressed and regulated in cells and how genes are cloned in the laboratory. And it describes how mutations change the genetic information and how mutations have accumulated over billions of years of biological evolution, giving rise to new organisms and new species. The first ten chapters are devoted to the principles of molecular genetics, that branch of genetics dealing with the chemistry of the molecules that carry and express genetic information in all living cells. Included in this section are two chapters of particular relevance to the health of the individual—one on the genetic aspects of cancer and the other on how genetics is used in biotechnology and recombinant DNA experiments.

Chapters 11 and 12 explain the fundamental patterns of inheritance discovered by Gregor Mendel, the determination of sex in animals, and the genetic principles of human reproduction. Chapters 13 and 14 discuss the causes of human hereditary diseases and developmental abnormalities and the use of modern medical techniques to detect and screen for hereditary diseases. Chapters 15 and 16 describe how genetic experiments can be carried out using human cells grown in test tubes and how the millions of genetically different cells of the human immune system are generated during embryonic development. Chapters 17 and 18 explain the genetics of populations and the principles of evolutionary biology. The final chapter discusses in depth one of the most important aspects of genetics: the controversy over how much of human behavior (such as intelligence) is determined by nature (heredity) and how much by nurture (environment). Readers who desire to understand the human and social consequences of genetics—past and present—may wish to read Chapter 19 first.

Understanding genetics, as is the case for all sciences, depends on understanding a new language of scientific terms. Thus, in each chapter key words are defined as they are introduced, and definitions are also provided at the end of the chapter as well as in the glossary at the end of the book. Review questions are included at the end of each chapter to help students assess their understanding of the material. Answers to these questions as well as essay topics are provided at the back of the book.

Many persons contributed valuable advice and helped to eliminate errors of fact and of interpretation in the manuscript. Any flaws or inaccuracies that remain are my responsibility. I would like to express my deep appreciation to those who critically read part or all of various drafts of the manuscript: Willard Centerwall, David Freifelder, Jon Beckwith, Jon Gallant, Rachel Freifelder, Barton Slatko, Gerard O'Donovan, Ric Davern, A. John Clark, Eric Golanty, Les Gottlieb, and Ram Bhagavan. I am especially grateful to my editors, John Hendry and Jackie Estrada, for their expertise and advice; to my publisher, Art Bartlett, Jones and Bartlett; to Nancy Sjoberg and all her associates who designed and produced the book; to Connie Wilson, who typed more drafts than either of us wants to recall; and to Morton Mandel, who provided me with an ideal environment for writing. Finally, I would like to acknowledge my debt to the thousands of students whose interest in genetics and whose probing questions are largely responsible for my own understanding of genetics and for my writing this book.

Gordon Edlin
DAVIS, CALIFORNIA

Contents

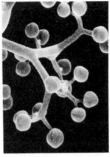

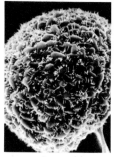

4
GENETIC INFORMATION
How Genes Are Exchanged 69

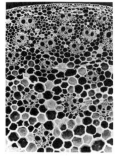

5
GENETIC INFORMATION
How Genes Are Expressed 97

6
THE GENETIC CODE
The Dictionary of Life 117

7
MUTATIONS
Changes in DNA   *137*

8
CANCER
Not a Hereditary Disease   *159*

9
GENE REGULATION
Switching Genes Off and On   *179*

10
RECOMBINANT DNA TECHNOLOGY
Opening Pandora's Box   *203*

14
GENETIC DISEASES
Screening and Counseling 293

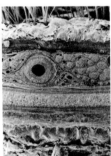

15
SOMATIC CELL GENETICS
Manipulating Genes in the Laboratory 315

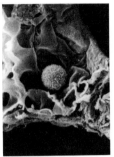

16
IMMUNOGENETICS AND ANTIBODIES
Protection Against Disease 333

17
EVOLUTION
Natural Selection and Populations 355

18
EVOLUTION
Biological History in Fossils and Molecules 377

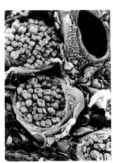

19
NATURE VERSUS NURTURE
The Social Consequences of Biological Determinism 405

DIATOMS, A FORM OF ALGAE, HAVE INTRICATE PATTERNS OF STONY CELL WALLS. (2,340 ×)

1

PREFACE
TO
LIFE

*Atoms
and
Molecules*

*Science is nothing else than the search to
discover unity in the wild variety of
nature–or more exactly, in the variety of
our experience.*
JACOB BRONOWSKI

1
PREFACE TO LIFE:
ATOMS AND MOLECULES

All living organisms are composed of cells. The simplest organisms, such as bacteria and algae, are single cells; more complex organisms, such as plants and animals, are composed of millions, billions, even trillions of cells. The human brain alone is estimated to have about 20 billion nerve cells that regulate all our mental processes and many of our body's physiological functions as well.

Two essential properties of living cells are growth and reproduction. Bacteria growing in a liquid medium in a test tube make identical copies of themselves every half hour or so. Bacteria grow and reproduce exponentially: one cell produces two, two produce four, those four give rise to eight, and so on. If the growth of bacteria in cultures were not limited, by nutrients or space, within two days a single bacterium, which weighs about 10^{-12} grams, would produce so many cells that their combined weight would exceed the weight of the earth, which is about 10^{24} grams. Obviously, the growth and reproduction of all cells—not only bacteria—is limited by various factors.

A natural forest contains trees that are of certain sizes, numbers, and kinds. These variations, along with the geographic extent of the forest, are limited by ecological factors. Within a forest there is a dynamic flux, albeit a leisurely one, as individual trees grow, die, and are replaced by new seedlings and even new species of trees. The actual status of the forest at any moment depends on the age, size, and diversity of the trees themselves, on nutrients in the soil, on diseases that attack the trees, on foraging animals, on weather conditions, and on many other factors.

People grow to a certain size and then stop growing, because each organ in the human body grows to a certain size and then stops. If certain tissues in plants or animals are destroyed or damaged, cells are stimulated to grow, replacing the damaged tissue. What controls the growth, reproduction, and diversity of the various kinds of cells in bacteria, trees, and humans? By some means, the properties of all living cells and of multicellular organisms are controlled by their *genetic information.* Therefore, understanding the chemical composition, physical structure, and biological functions of the genetic material is essential to understanding the biology and behaviors of living organisms.

Despite all efforts to find something uniquely different about living cells as compared to inanimate objects, science has consistently found that the properties of all cells—their similarities, differences, and functions—derive from the chemical and physical properties of the atoms and molecules of which they are composed. Atoms combine chemically to form simple molecules, these molecules may combine to produce even more complex molecules, many thousands of different molecules give structure and functions to cells, and cells aggregate to form tissues and organisms. Understanding the mechanisms that govern the chemistry of molecules is thus the first step toward understanding how cells grow and reproduce.

An organism can be thought of as something like a house. A house is a complicated structure of wood, metals, concrete, and other materials. It has plumbing to carry water, wiring for electricity, mechanical devices to heat or cool the rooms, paint and other materials to protect its surfaces, and devices to warn of fire or unwelcome intruders. If you were to see all the pieces of a house in a random pile—the boards, nails, wires, pipes, shingles, and so on—you would have no concept of the structure it could become or of how the pieces should be put together. Organizing all these materials requires a blueprint: a diagram that provides the assembly instructions. Similarly, the genetic information of every living cell provides a blueprint for all the components of that cell. *Genes* direct the assembly of all kinds of molecules that, together, produce that most wondrous and unique property of cells: life.

In order to understand how single cells and organisms grow, reproduce, evolve, and age, one must first understand the properties of the basic building materials of all cells, the atoms and molecules. This chapter describes the few basic principles that govern the chemical and physical interactions of atoms and molecules and provides a first look at how cells might have arisen and at how the genetic information is organized.

Elements

All matter is composed of **elements**, which are substances that cannot be broken down further or transformed into other substances by ordinary chemical or physical means. For example, metals such as iron, copper, and gold do not yield simpler substances when subjected to any chemical treatment. In all of the matter on earth, and presumably elsewhere in the universe, only ninety-two elements occur naturally (although heavier ones have been created by physicists). The elements can be ordered in a small number of groups so that those elements sharing particular chemical properties reside in the same group in the periodic table of elements (Figure 1-1). How all of the elements came into being is discussed in Box 1-1.

Most of the elements that are present in large quantities in living cells are found at the lighter end of the periodic table. About 70 percent of a living cell is water and another 10 percent consists of the element carbon. A variety of other elements are also found in cells in very small (trace) amounts; all together the trace elements contribute about 2–3 percent to a cell's weight (Table 1-1).

Carbon plays a key role in cells because of its unique chem-

PERIODIC TABLE OF ELEMENTS

The periodic table:

Group / Transition Elements

	I	II	Transition Elements										III	IV	V	VI	VII	0
1	**H** 1 1.0080 Hydrogen																		**He** 2 4.0026 Helium
2	**Li** 3 6.94 Lithium	**Be** 4 9.012 Beryllium											**B** 5 10.81 Boron	**C** 6 12.011 Carbon	**N** 7 14.007 Nitrogen	**O** 8 15.999 Oxygen	**F** 9 18.998 Fluorine	**Ne** 10 20.18 Neon
3	**Na** 11 22.98 Sodium	**Mg** 12 24.31 Magnesium											**Al** 13 26.98 Aluminum	**Si** 14 28.09 Silicon	**P** 15 30.97 Phosphorus	**S** 16 32.06 Sulfur	**Cl** 17 35.453 Chlorine	**Ar** 18 39.948 Argon
4	**K** 19 39.102 Potassium	**Ca** 20 40.08 Calcium	**Sc** 21 44.96 Scandium	**Ti** 22 47.90 Titanium	**V** 23 50.94 Vanadium	**Cr** 24 51.996 Chromium	**Mn** 25 54.94 Manganese	**Fe** 26 55.85 Iron	**Co** 27 58.93 Cobalt	**Ni** 28 58.71 Nickel	**Cu** 29 63.54 Copper	**Zn** 30 65.37 Zinc	**Ga** 31 69.72 Gallium	**Ge** 32 72.59 Germanium	**As** 33 74.92 Arsenic	**Se** 34 78.96 Selenium	**Br** 35 79.91 Bromine	**Kr** 36 83.80 Krypton
5	**Rb** 37 85.47 Rubidium	**Sr** 38 87.62 Strontium	**Y** 39 88.906 Yttrium	**Zr** 40 91.22 Zirconium	**Nb** 41 92.91 Niobium	**Mo** 42 95.94 Molybdenum	**Tc** 43 (99) Technetium	**Ru** 44 101.1 Ruthenium	**Rh** 45 102.91 Rhodium	**Pd** 46 106.4 Palladium	**Ag** 47 107.87 Silver	**Cd** 48 112.40 Cadmium	**In** 49 114.82 Indium	**Sn** 50 118.69 Tin	**Sb** 51 121.75 Antimony	**Te** 52 127.60 Tellurium	**I** 53 126.90 Iodine	**Xe** 54 131.30 Xenon
6	**Cs** 55 132.91 Cesium	**Ba** 56 137.34 Barium	57-71 Lanthanide series (rare earth)	**Hf** 72 178.49 Hafnium	**Ta** 73 180.95 Tantalum	**W** 74 183.85 Tungsten	**Re** 75 186.2 Rhenium	**Os** 76 190.2 Osmium	**Ir** 77 192.2 Iridium	**Pt** 78 195.09 Platinum	**Au** 79 196.97 Gold	**Hg** 80 200.59 Mercury	**Tl** 81 204.37 Thallium	**Pb** 82 207.2 Lead	**Bi** 83 208.98 Bismuth	**Po** 84 (210) Polonium	**At** 85 (218) Astatine	**Rn** 86 (222) Radon
7	**Fr** 87 (223) Francium	**Ra** 88 (226) Radium	89-103 Actinide series (uranium)	**Rf** 104 (261) Rutherfordium	**Ha** 105 (262) Hahnium	106 (263)	107 (261)											

Key:
Symbol — **Cl** 17 — Atomic number
Atomic mass — 35.453
Element name — Chlorine

Period (left vertical label)

Figure 1-1 Simplified periodic table of the elements. Elements in a *group* (vertical column) have similar chemical properties. The elements in group 0, for example, are all chemically inactive gases, such as helium and neon. Elements in a *period* (horizontal row) have different chemical properties. Generally speaking, across each period from left to right there is a transition from a chemically active metal, through less active metals, through highly active nonmetals, to an inert (inactive) gas. The first ninety-two elements occur naturally; the rest are experimental creations of physicists.

Table 1-1. Elements Present in Human Cells

Abundant elements in biological molecules	Carbon (C)
	Hydrogen (H)
	Nitrogen (N)
	Oxygen (O)
	Phosphorus (P)
	Sulfur (S)
Trace elements required by all cells	Chlorine (Cl)
	Sodium (Na)
	Potassium (K)
	Magnesium (Mg)
	Calcium (Ca)
	Manganese (Mn)
	Iron (Fe)
	Cobalt (Co)
	Copper (Cu)
	Zinc (Zn)
Trace elements required by certain cells	Boron (B)
	Fluorine (F)
	Silicon (Si)
	Vanadium (V)
	Chromium (Cr)
	Selenium (Se)
	Molybdenum (Mo)
	Tin (Sn)
	Iodine (I)

Where Did the Elements Come From?

BOX 1-1

A quite convincing body of physical and astronomical evidence points to the fact that the universe and all matter that now exists was created 10 to 20 billion years ago in an explosion of unimaginable intensity that is referred to as the "Big Bang." Prior to that event, there was no universe; neither time, nor matter, nor energy, nor even space itself existed before the Big Bang. We cannot even hazard a scientific guess as to where the matter and energy contained in that primordial explosion came from. However, from the moment of that singular event, all the laws of physics as they are understood today have applied, and they can be used to predict how the universe expanded, cooled, and evolved to give rise to the atoms, stars, and galaxies that we now observe.

At the moment of the Big Bang there was only energy; atoms could not have existed at the elevated temperature of the explosion. However, the enormous amount of energy in the fireball of the Big Bang instantly began to radiate outward, creating space and time. Within billionths of a second, the temperature of the fireball began to drop and particles of matter began to form. (Energy and matter are interconvertible, according to Einstein's famous law $E = mc^2$: energy is equal to mass multiplied by the square of the speed of light in a vacuum.) Within seconds after the explosion, energy was converted into protons, which are hydrogen atoms without an electron. Hydrogen, then, was the first and the simplest element. The energy in this dense soup of protons was enough to fuse some protons together, creating an isotope of helium, the next element. Protons are still being fused into helium in nuclear reactions deep within our sun; these reactions provide the energy that sustains all life on earth.

As the cloud of hydrogen nuclei expanded and cooled further, the force of gravity began to pull portions of it together to form local concentrations of denser material; the formation of stars and galaxies had begun. Billions of years ago, a galaxy known as the Milky Way was formed, and within that galaxy, about 5 billion years ago, a unique solar system consisting of an average-size star (our sun) and its nine planets was formed.

The many different elements that are now found on the earth and other planets in our solar system were—except for hydrogen and some of the helium—formed in a complex series of nuclear reactions in the hot interior of stars formed much earlier in the history of the universe. The heavier elements, such as carbon, oxygen, nitrogen, iron, and lead, were formed in nuclear reactions within stars and were spread through space after the stars exploded in spectacular explosions called *supernovas*. Even now some stars are being formed and some are disintegrating; this continuing process allows us to decipher the history and evolution of the universe.

The elements on other planets and elsewhere in the universe are identical to the elements on earth. We know this because each star gives off a characteristic pattern of light called a *spectrum*, which, when analyzed, tells us what elements are present in the star. Because our planet contains almost all of the elements, it could not have been formed during the early history of the universe, when only the lighter elements were available. (Even today most of the matter in the universe is still in the form of hydrogen.) Billions of years from now, our sun will eventually explode, again spewing its atoms and those of earth and the other planets back into space.

Reading: Steven Weinberg, *The First Three Minutes*. New York: Bantam Books, 1977.

ical properties: every carbon atom can form chemical bonds or links with other carbon atoms or with as many as four other atoms of one or more other elements. **Organic** molecules contain one or many carbon atoms; for example, methane (CH_4) has one carbon, ethyl alcohol (CH_3CH_2OH) has two carbons, and the milk protein ($C_{1864} H_{3012} O_{576} N_{468} S_2$) has nearly two thousand.

Where Did the First Molecules Come From?

BOX 1-2

Organic molecules are synthesized in living organisms, from bacteria to humans. But since cells themselves are composed of many millions of biological molecules, which in turn can only be synthesized by biological processes in living cells, the dilemma sounds like the well-known chicken-and-egg puzzle: Which came first, molecules or cells?

As the solar system was being formed, simple inorganic molecules, such as minerals, salts, and gases, were synthesized as a consequence of the high temperatures and pressures within the earth and on its surface. But what was the source of the organic molecules necessary for the development of primitive cells and ultimately of complex living organisms? How were the essential small biological molecules such as amino acids and sugars first created? And what is the origin of the important macromolecules such as proteins and nucleic acids?

To answer these questions, researchers have created laboratory conditions thought to reflect the conditions that existed on the primitive earth. While the earth was still quite hot and fluidlike, there was very little oxygen in the atmosphere and little liquid water on the surface. Gigantic electrical storms raged in the sky, and enormous volcanoes erupted on the surface, spewing forth matter from the molten interior. Elements on the earth's surface were exposed to ionizing radiation from space and from radioactivity on earth itself; thus, energy necessary for chemical reactions to occur was readily available. To simulate these conditions, scientists have exposed simple gas mixtures such as ammonia (NH_3), hydrogen (H_2), methane (CH_4), and water vapor (H_2O) to electrical discharges or have subjected them to high heat and pressure. Under these conditions, many of the complex organic molecules that serve as building blocks for biologically active molecules have been found to form spontaneously; thus, these molecules are presumed to have also formed in the chemical and physical environment of the primitive earth.

The geological development of the earth was apparently favorable to the formation of increasingly complex organic molecules. It need not have been so, and we are fortunate that earth is situated as it is relative to the sun. Although the other planets of our solar system were formed at the same time as earth and contain the same elements, they evolved differently and—as far as we know—have no living cells or organisms on them. Planets that are a little too far from the sun (which most of them are) are frozen and barren. Those that are a little too close to the sun (which two of them are) are too hot for life. Indeed, if the earth's orbit had been just slightly different, it is quite likely that the molecules of life would have failed to form and that the earth, too, would be without living things.

To date, experiments support the hypothesis that complex organic molecules—the building blocks of primitive cells—were formed billions of years ago on earth and that these molecules had the chemical properties that enabled them to evolve into the biologically active molecules that eventually combined to form living cells. (For a different, highly controversial view of how the all-important molecules of DNA came to be on earth, see Box 4-3.)

Reading: Linda Gormon, "As It Was in the Beginning." *Science News*, January 31, 1981.

Inorganic molecules do not contain carbon; for example, water (H_2O), ammonia (NH_3), and sulfuric acid (H_2SO_4) are inorganic molecules. (By convention carbon dioxide, CO_2, and carbon monoxide, CO, are classified as inorganic molecules.) How the first organic molecules arose on earth is discussed in Box 1-2.

Atoms

All of the elements are composed of **atoms**, which are the smallest units of any substance. Each atom, in turn, is composed of three kinds of fundamental particles: protons, neutrons, and electrons. **Protons** and **neutrons** are dense, heavy particles located in the nucleus of the atom. Protons and neutrons are identical except that protons carry a positive electrical charge ($+1$) whereas neutrons have no electrical charge. **Electrons** are much lighter particles that carry a negative electrical charge (-1) and circle about the nucleus in designated orbital shells. All elements, from the simplest to the most complex, are composed of different numbers of protons, neutrons, and electrons.

The number of positive charges (one per proton) in an atom's nucleus determines the atom's atomic number, chemical properties, and position in the periodic table. Normally, atoms do not have either a positive or negative charge, since they contain equal numbers of protons and electrons. If an atom's number of electrons is greater or less than the number of protons in the atom, the atom is called an **ion**; ions may be either positively or negatively charged.

An electron does not actually exist as a discrete particle. It is more appropriately thought of as a cloud of negative electrical charge that is distributed in the space immediately surrounding the atom's nucleus (Figure 1-2). The chemical properties of any element are due solely—and this is an important point to keep in mind—to the electrons that orbit the nuclei of its atoms, and in particular to the electrons in the outermost orbital shell, which are the ones freest to interact with the electrons in the outermost orbital shells of other elements. Protons and neutrons do not participate in chemical reactions or in the formation of chemical bonds to create molecules. Chemistry, therefore, is governed by the behaviors of electrons.

The weight of an atom is the sum of all the protons and neutrons in the nucleus (the contribution of electrons is negligible); the atomic weights of most elements are listed in the periodic table in Figure 1-1. For our purpose, the terms *atomic weight* and *atomic mass* are used interchangeably, although physicists define them somewhat differently. The same element may have different numbers of neutrons in the nucleus, and therefore different atomic weights. These varieties of an element's atoms are called **isotopes** of that element.

Isotopes

The number of neutrons in the nucleus of any atom can vary within certain limits. Changing the number of neutrons does not change the chemical properties of the atom, but it does alter the atomic weight and sometimes the stability of the nucleus. Hydrogen, which normally has one proton and one electron, provides a simple example. Three isotopes of hydrogen occur in nature, with atomic weights of 1, 2, and 3 (Figure 1-3). These isotopes are *hydrogen* (no neutrons), *deuterium* (one neutron),

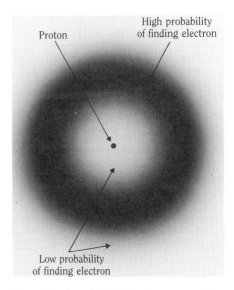

Proton

High probability
of finding electron

Low probability
of finding electron

Figure 1-2 A model of the hydrogen atom. The electron's negative charge is distributed like a cloud around the positively charged proton.

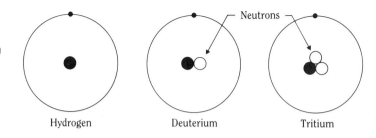

Neutrons

Hydrogen   Deuterium   Tritium

Figure 1-3 The isotopes of hydrogen all have one proton in their nucleus but differ in the number of neutrons. Deuterium is a stable isotope, whereas tritium is unstable and emits radioactivity when the nuclei decay.

and *tritium* (two neutrons). All three isotopes of hydrogen have identical chemical properties; they differ only in atomic weight and in the stability of the nucleus. Deuterium is a stable isotope of hydrogen; when it chemically bonds to oxygen, it behaves chemically like hydrogen and forms molecules of *heavy water* (D_2O). Tritium is an unstable isotope of hydrogen. Periodically, a tritium nucleus spontaneously "decays," or breaks apart; in the process, an electron is formed and emitted from the tritium nucleus. Emissions of subatomic particles and of the electromagnetic energy called x-rays and gamma rays are referred to as **radioactivity**; isotopes that emit such particles and energy are called **radioisotopes**. This nuclear decay converts a neutron into a proton, and the loss of the negatively charged electron is equivalent to the acquisition of a positive electrical charge in the nucleus. Thus, the nuclear decay converts tritium into an isotope of helium (atomic number 2).

Radioisotopes have a characteristic **half-life**, which is defined as the time it takes for half of the nuclei of any given sample of atoms of a particular isotope to decay. Tritium has a half-life of 12.3 years; in that length of time a random half of all the tritium atoms in a sample will have decayed. In the next 12.3 years, half of the remaining tritium nuclei will decay, and so on. Although the half-life of any radioactive isotope can be determined with great accuracy, it is impossible to predict which atoms will decay and which ones will remain stable during any given period of time (although all the atoms will decay eventually). This situation is analogous to that of traffic fatalities in the United States: We can predict with great certainty that at least 50,000 persons will die in automobile crashes in the United States during the next twelve months, but we cannot predict which individuals will be in those crashes. The half-lives of some useful radioactive elements are listed in Table 1-2.

Because there are three isotopes of hydrogen, the average weight of hydrogen listed in the periodic table is not exactly 1. Rather, this atomic weight represents the average of the weights of the three hydrogen isotopes found in nature. More than 99 percent of all hydrogen atoms have one proton and no neutrons and, therefore, an atomic weight of 1. Because the amounts of the other isotopes are small, they contribute very little to the

Table 1-2. Half-lives and Radiation Emitted by Some Radioisotopes

Isotope	Half-life	Radiation Emitted*		
		Alpha	Beta	Gamma
Radium-226	1620 years	√		√
Sulfur-35	87.1 days	√	√	
Radon-217	0.001 second	√		
Uranium-238	4.5 billion years	√		√
Uranium-235	710 million years	√		√
Plutonium-239	24,300 years	√		
Cesium-137	30 years		√	√
Iodine-131	8 days		√	√
Gold-198	2.7 days		√	√
Phosphorus-32	14.3 days		√	
Strontium-90	28 years		√	
Cobalt-60	5.2 years		√	√
Carbon-14	5730 years		√	
Tritium (^{3}H)	12.3 years		√	

*The decay of the nucleus of a radioisotope releases radiation. Alpha radiation consists of helium nuclei (a particle composed of two protons and two neutrons); beta radiation consists of electrons; gamma rays and x-rays consist of photons and are forms of electromagnetic radiation.

average weight, which is 1.008. Other elements also have several isotopes that vary in their natural abundance. Carbon, for example, has six naturally occurring isotopes. The most prevalent isotope, ^{12}C (six protons and six neutrons), accounts for 98.6 percent of all carbon atoms in nature. ^{14}C, a radioactive isotope, is rare in nature but is manufactured in large amounts for research and medical purposes and is a by-product of atomic bomb explosions.

Because many important and instructive experiments in genetics, biochemistry, and medicine use either radioactive isotopes or heavy isotopes, it is important to understand the differences between the two. Radioactive isotopes are often called *tracer elements* because their emission of subatomic particles enables experimenters to locate the radioactive molecules in particular cells or tissues and to follow the sequence of chemical reactions. The degree of radioactivity of any substance is easily measured by special detectors that "count" the particles the radioactive substance emits. *Heavy isotopes,* on the other hand, are incorporated into molecules to make them more dense than molecules synthesized from ordinary atoms (that is, the common isotopes of those elements). The term "heavy isotope" denotes a heavier-than-normal isotope that is not radioactive. These heavier molecules are easily separated by centrifugation

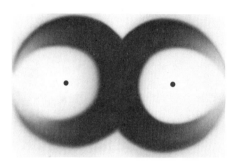

Figure 1-4 Electron probability distribution (cloud) around two hydrogen atoms in hydrogen gas. A covalent bond is formed because the negative charge is attracted to both protons in the hydrogen nuclei.

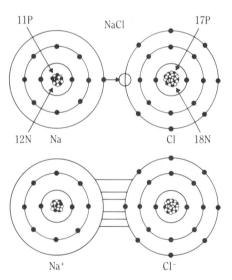

Figure 1-5 Sodium and chlorine atoms are chemically joined by ionic bonds. Because the electron is more strongly attracted to the chlorine atom, the chlorine atom becomes negatively charged while the sodium atom becomes positively charged.

from the millions of lighter molecules in cells and provide another tool for monitoring cellular chemical reactions.

Isotopes became generally available for biological and medical research only after World War II, when atomic reactors were developed in which various isotopes could be manufactured in large quantities and at low cost. Today thousands of radioactive and heavy isotopes are available for research and for other applications. Numerous medical diagnostic techniques and treatments, especially for cancers, use radioactive isotopes, and radioisotopes have been an invaluable tool for investigating the formation of molecules by chemical reactions in cells. More will be said about the experimental use of isotopes in later chapters.

Chemical Bonds

As mentioned earlier, the formation of chemical bonds between atoms is the result of interactions between the atoms' outermost orbital electrons. According to the laws of nature, electrons must occupy specific *orbital shells* around the atomic nucleus. Indeed, electrons can be thought of as positioned around the nucleus in much the same way that airplanes are stacked at different altitudes in the vicinity of airports so as to avoid crashes. An atom is most stable when each orbital shell is filled with a certain number of electrons. The first (innermost) shell becomes filled with only two electrons; the next two orbital shells are filled when they contain eight electrons each.

The number of an atom's electrons and, therefore, the number of its electron shells is determined by its atomic number. For example, hydrogen's atomic number is 1; therefore, an atom of hydrogen has a single electron in its one (unfilled) electron shell. Helium, with an atomic number of 2, has one (filled) electron shell. Carbon, whose atomic number is 6, has two electrons in its inner shell and four in its (unfilled) second shell. Heavier atoms have more electrons and thus more shells.

If an atom's outer electron shell is filled, that atom is normally chemically inactive because it has little tendency to share electrons with other atoms. For example, helium's single (and therefore its "outer") shell normally contains the full complement of two electrons, so it has little tendency to react with other elements. Hydrogen, on the other hand, with its unfilled outer shell, will react explosively with oxygen—which is why helium is used instead of hydrogen to fill balloons, dirigibles, and other lighter-than-air craft.

COVALENT AND IONIC BONDS

A covalent bond results from the equal sharing of two electrons by two atoms. Thus, in the formation of a molecule of hydrogen gas, which is composed of two hydrogen atoms, the two electrons are shared equally by both atoms (Figure 1-4). In such a molecule it is not possible to determine which electron belongs to which hydrogen nucleus. The sharing of the negatively charged electrons chemically joins the hydrogen atoms together. The energy required to separate the two atoms determines the strength of covalent and other kinds of bonds.

An **ionic bond** is also the result of sharing of electrons, but in this case the sharing is unequal; as a result, the atoms are held together by electrical forces and the ionic bond is not as strong as the covalent bond. A good example is table salt (NaCl). Crystalline salt is composed of sodium (Na) and chlorine (Cl) atoms joined together by ionic bonds to form sodium chloride (Figure 1-5). Sodium atoms have a single electron in their outer shell, which is an unstable condition. Chlorine atoms have seven electrons in their outer shell and need only one more to become full. When atoms of sodium and chlorine unite, the shared electron is mainly associated with the chlorine atom. This leaves the sodium atom positively charged (that is, it becomes a positive *ion*) and the chlorine atom negatively charged (a negative ion), and it is actually the attraction of opposite charges that holds the atoms together. Chemical bonds between most atoms are usually a combination of both covalent and ionic bonds.

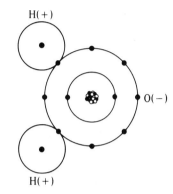

Figure 1-6 A molecule of water is formed when two hydrogen atoms are chemically joined to an oxygen atom. Because of the positive charge associated with the hydrogens and the negative charge associated with oxygen, water is a polar molecule.

HYDROGEN BONDS

Another kind of chemical bond that is extremely important in living cells is the **hydrogen bond**. Although hydrogen bonds are much weaker than either covalent or ionic bonds, they are responsible for the remarkable stability of the genetic information (discussed in Chapter 3) as well as for the structural conformation of proteins. To understand the hydrogen bond, it is easiest to consider the example of ordinary water (Figure 1-6), in which two hydrogen atoms are covalently bonded to an oxygen atom. The two hydrogen electrons fill the outer (second) shell of electrons around the oxygen atom. Because the electrons are more closely associated with the oxygen atom, the water molecule is said to be **polar**: that is, a positive electrical charge exists around the hydrogen atoms and a negative charge exists around the oxygen atom. As a result of this polarity, the hydrogens of one water molecule are attracted to the oxygens of other water molecules, as indicated by the dotted lines in Figure 1-7. Hydrogen bonds are the chemical bonds formed by this attraction.

Energy is required to separate water molecules, which is accomplished by breaking the hydrogen bonds between them; this is why ice turns to liquid on heating. Further heating causes the water molecules to dissociate, producing water vapor (steam) above the boiling point of water. Water is also a good solvent for electrically charged molecules, such as salt; the sodium and chloride ions can be surrounded by water molecules

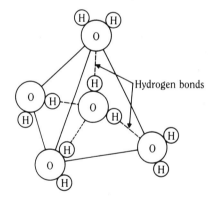

Figure 1-7 Hydrogen bonds hold water molecules together in particular orientations so that ice is formed. Heating ice destroys the crystalline structure and produces liquid water. Further heating allows water molecules to escape from the liquid in what is referred to as steam.

Figure 1-8 (*opposite page*) The twenty different amino acids used to construct proteins. The different amino acids can be linked together in any possible combination to produce essentially an unlimited variety of proteins.

and their positive or negative charges neutralized. Since most biological molecules are electrically charged, the solvent property of water largely determines how these molecules interact and function in living cells.

Because hydrogen bonds between molecules may be quite numerous, their overall effect can be great. The principle of using hydrogen bonds to hold molecules together is the same as that employed by the Lilliputians in tying up Gulliver: they used thousands of tiny threads that together were as strong as a few thick ropes. The three-dimensional shapes of most biological molecules are due to hydrogen bonds. When foods are cooked, it is primarily the disruption of hydrogen bonds in the protein molecules that changes the texture and palatability. Thus, most people eat eggs whose protein has been heated enough to make it white and solid (or yellow); few like their eggs uncooked.

Molecules and Macromolecules

Molecules are formed by combining relatively few atoms; **macromolecules** (from the Greek *macros,* meaning "large") are considerably larger and range from hundreds to millions of atoms all chemically bonded to one another. One general class of biologically important macromolecules is **proteins**, which consist of long chains of smaller molecules called **amino acids**. There are twenty different amino acids (Figure 1-8) that can be chemically linked together in any conceivable combination and length to form large protein molecules. Proteins have essential structural roles in the architecture of cells. However, the most significant function of proteins is to catalyze (speed up the rate of) chemical reactions. Proteins that catalyze chemical reactions are called **enzymes**, and every cell contains thousands of different enzymes. Enzymes can increase the rate at which a chemical reaction occurs by as much as a million-billion times. Without enzymes, biochemical reactions would proceed much too slowly to sustain life at temperatures the cell could tolerate.

Enzymes are long chains of 50 to 1000 amino acids joined together by covalent bonds. The complex mechanisms that join one amino acid to another are discussed in Chapter 5. Small chains of amino acids—six to twenty of them—also function as **hormones** in plants and animals. Hormones serve as chemical messengers; as they travel from one part of the organism to another and interact with the cells of various tissues, they regulate the way genes function in cells and also change the functions carried out by cells. Another class of macromolecules is the *nucleic acids,* which contain the genetic information in cells (Figure 1-9). Two types of biologically important nucleic acids are **DNA (deoxyribonucleic acid)** and **RNA (ribonucleic acid)**. DNA is the carrier of the genetic information in all cells in every kind of living organism and in some viruses. The RNA molecules

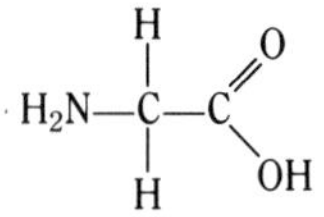

Glycine (gly)

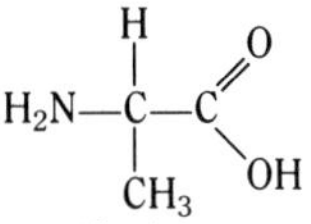

$$H_2N-\overset{\overset{\displaystyle H}{|}}{\underset{\underset{\displaystyle H}{|}}{C}}-\overset{\overset{\displaystyle O}{\parallel}}{C}-OH$$

Glycine
(gly)

Alanine
(ala)

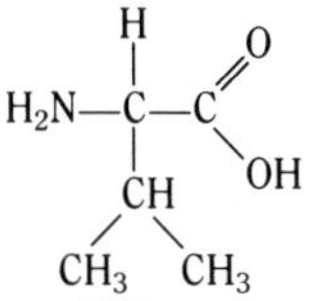

Valine
(val)

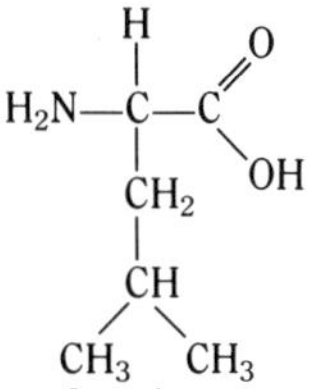

Leucine
(leu)

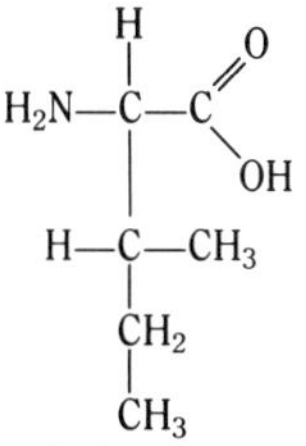

Isoleucine
(ile)

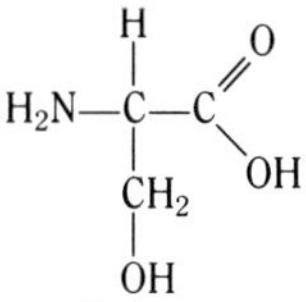

Serine
(ser)

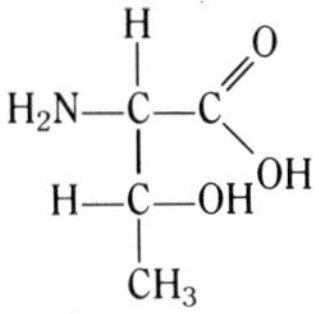

Threonine
(thr)

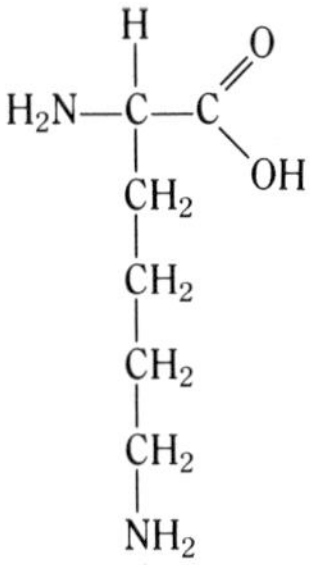

Lysine
(lys)

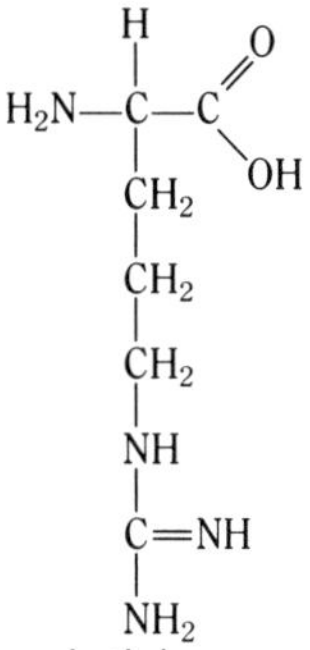

Arginine
(arg)

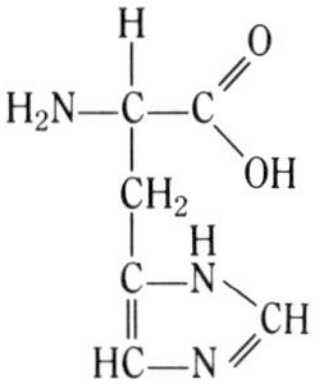

Histidine
(his)

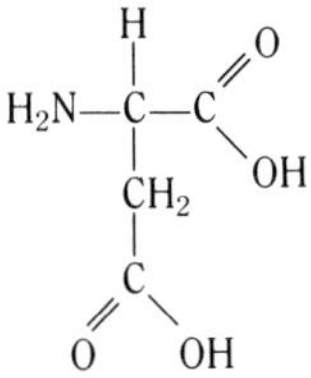

Aspartic acid
(asp)

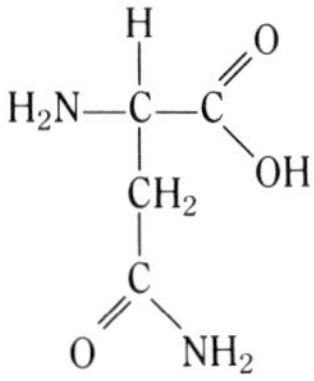

Asparagine
(asn)

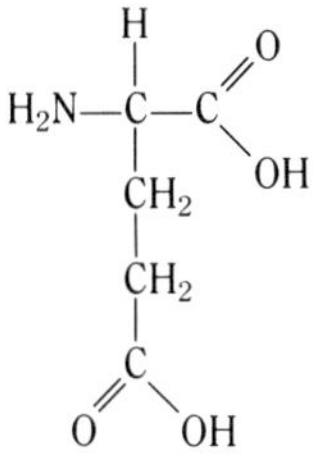

Glutamic acid
(glu)

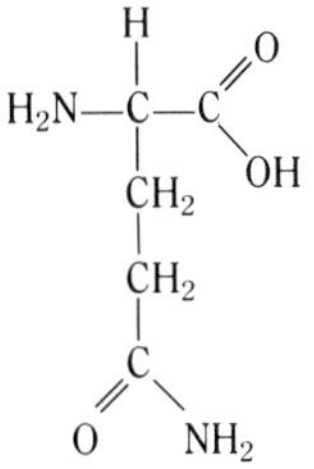

Glutamine
(gln)

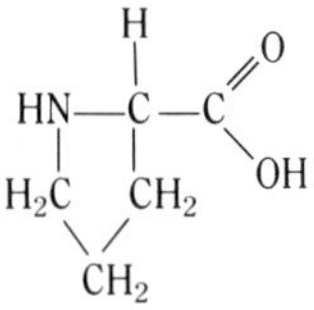

Proline
(pro)

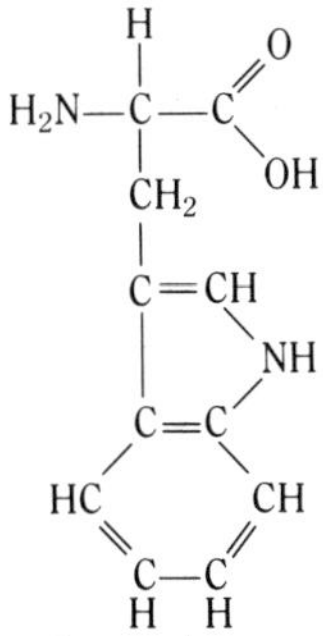

Tryptophan
(trp)

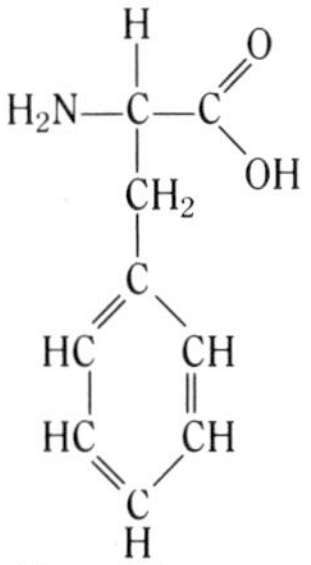

Phenylalanine
(phe)

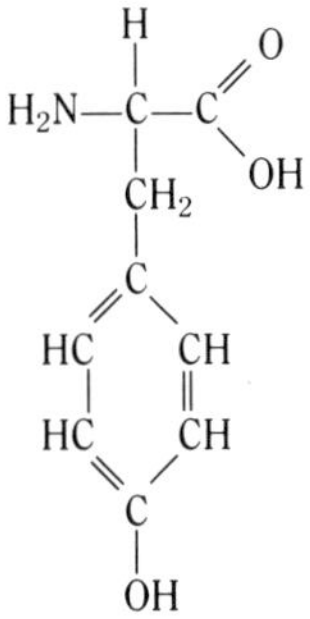

Tyrosine
(tyr)

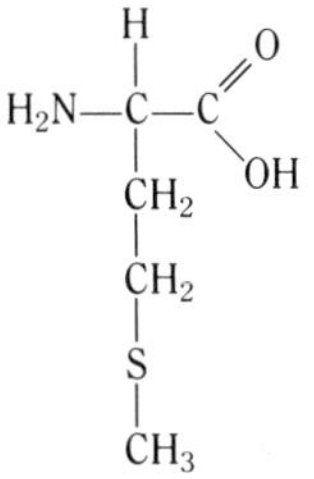

Methionine
(met)

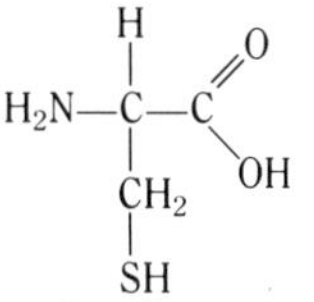

Cysteine
(cys)

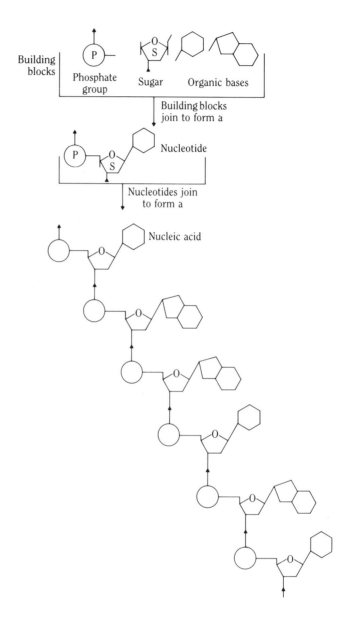

Figure 1-9 Macromolecules such as the nucleic acids DNA and RNA are constructed by the assembly of smaller molecules. DNA carries the genetic information in all cells and RNA is an essential intermediate in the expression of genetic information.

in cells perform a variety of functions that will be discussed in later chapters. The nucleic acids contain the information that directs the synthesis of enzymes and other proteins. The addition of each atom to a molecule is catalyzed by an enzyme that generally is specific for that chemical reaction and no other. Since many thousands of different chemical reactions occur in a cell every second, thousands of different enzymes are required— and the information for their synthesis is carried in DNA. Much of what follows in the next few chapters elaborates on how nucleic acids and proteins are synthesized and interact with one another to produce the enormous diversity of cells and organisms in nature. The other two classes of macromolecules found in cells are lipids and carbohydrates. The lipids are "fatty" molecules that are used as concentrated energy stores and for constructing cell membranes. The carbohydrates consist of starches and sugars, which are used by cells for energy, and of cellulose, which is found mostly in the cell walls of plants.

The chemical structures shown in Figures 1-8 and 1-9 are not meant to be memorized. However, recognizing their general patterns and similarities can be quite helpful. These molecular structures will become more familiar and useful after you have read a few more chapters. For now, think of them as notes on a piece of sheetmusic. After you read sheetmusic for a while, you stop seeing individual notes and begin to notice patterns of notes. Eventually, you just glance at the notes and simply hear the music. The nucleic acids and the proteins are the notes that make up the music of cells. And the variety of biological tunes that they can play in cells seems to be limitless.

Cells

Finally, we are ready to consider cells: the basic structural and functional units of all living organisms. In the simplest life forms, organism and cell are identical. Bacteria and blue-green algae (also called cyanobacteria) are single cells and belong to the general class of organisms called **procaryotes** (Figure 1-10). Yeasts, molds, and protozoa are also single-celled organisms, but their cellular structures and functions are more complex than those of bacteria and they belong to the other general class of organisms, the **eucaryotes** (Figure 1-11). All plants and animals are multicellular eucaryotes: their individual cells have the same basic properties that are found in yeasts and protozoa. Procaryotes have no nucleus and reproduce themselves by simply dividing in two. Eucaryotes have a nucleus and many other specialized cellular structures. All eucaryotes reproduce by a complex process called meiosis, which is described in a later chapter.

Biologists estimate that as many as 80 percent of all cells on earth are procaryotes, and multicellular plants and animals probably would not survive without the biochemical assistance of these microorganisms. We generally are not aware of the enormous number of bacteria around and in us, however, because we do not see them. The human body contains more bacterial than human cells, yet they add only a tiny fraction to total body weight. The skin is covered with numerous different kinds of bacteria that are adapted to various environments. In moist underarm areas, there are more than 2 million bacteria

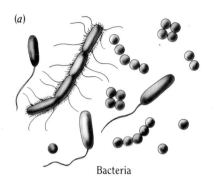

Bacteria

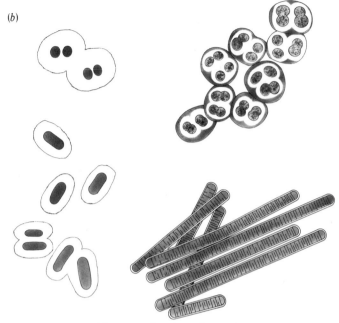

Blue-green algae (cyanobacteria)

Figure 1-10 Procaryotes—bacteria and cyanobacteria (blue-green algae). (*a*) Bacteria are single cells that come in a variety of shapes and can form long filaments in which individual cells remain attached to one another. (*b*) Blue-green algae are also single cells; they contain pigments and can usually carry out photosynthesis (produce chemical energy from light).

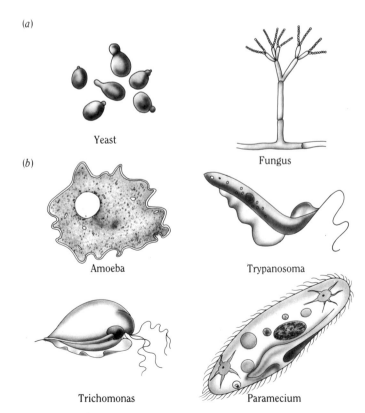

Yeast

Fungus

Amoeba

Trypanosoma

Trichomonas

Paramecium

Figure 1-11 Simple eucaryotes—fungi and protozoa. (*a*) Yeasts are single cells that reproduce by budding off new cells. Molds such as *Penicillium* reproduce by producing spores. (*b*) Protozoa such as amoebas are unicellular animals. All multicellular plants and animals are also eucaryotes.

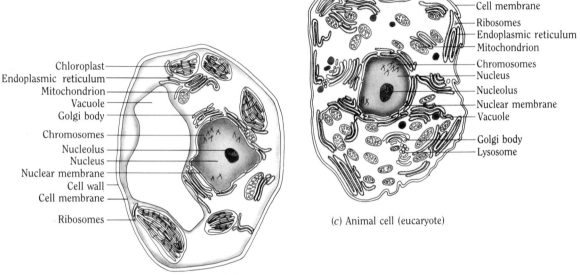

Cell membrane — Chromosome (DNA)

Cell wall — Ribosomes

(a) Bacterium (procaryote)

Chloroplast
Endoplasmic reticulum
Mitochondrion
Vacuole
Golgi body
Chromosomes
Nucleolus
Nucleus
Nuclear membrane
Cell wall
Cell membrane
Ribosomes

(b) Plant cell (eucaryote)

Cell membrane
Ribosomes
Endoplasmic reticulum
Mitochondrion
Chromosomes
Nucleus
Nucleolus
Nuclear membrane
Vacuole
Golgi body
Lysosome

(c) Animal cell (eucaryote)

Figure 1-12 The structures and organization of procaryotic and eucaryotic cells. In procaryotes all cellular components are in the cytoplasm. Eucaryotic cells have a defined nucleus and cytoplasm that are physically separated by the nuclear membrane. Other cellular organelles are also encased by membranes. The functions of the structures listed are described in Table 1-3.

per square centimeter of skin. The presence of normal skin bacteria often prevents invasion by harmful bacteria. Internally, bacteria often contribute to normal body functions, such as digestion. In fact, human stool consists mainly of billions of bacteria.

Cells are sometimes described as bags of molecules, but in reality they are highly organized structures whose components function and interact in precise ways. The cellular components of procaryotes such as bacteria coexist together in the cyto-plasm, the fluidlike interior that is encased by a cell membrane and cell wall (Figure 1-12a). In bacteria all components are free to move about in the cytoplasm, although larger structures such as DNA tend to be compartmentalized in certain sections of the cell.

Eucaryotic cells contain numerous specialized cellular structures called *organelles* (Figure 1-12b and c). In all eucaryotic cells the hereditary information is contained in the cell's **nucleus**, which is isolated from the cytoplasm by a nuclear membrane. The functions of the organelles and other cell structures shown in Figure 1-12 are described in Table 1-3.

17

Table 1-3. Eucaryotic Cellular Structures

Structure	Functions
Nucleus	Location of chromosomes containing DNA and associated molecules. Site of some cellular RNA and DNA synthesis.
Nuclear membrane	A porous double membrane that completely encloses the nucleus. The nuclear envelope allows passage of metabolites in and out of the nucleus but constrains DNA and nuclear proteins.
Nucleolus	An area of the nucleus that is the site for ribosomal RNA synthesis (ribosomal RNA is a structural component of ribosomes).
Ribosomes	Small particles, present in the cytoplasm, that are the sites of protein synthesis. If the protein being synthesized is destined for secretion, the ribosomes are associated with the endoplasmic reticulum.
Endoplasmic reticulum	Membranous sacs that transport proteins and other macromolecules in the cytoplasm of eucaryotic cells.
Cell membrane	A single membrane that completely encloses the contents of the cell and determines which molecules enter and leave the cell.
Mitochondria	Organelles within the cytoplasm where chemical energy (ATP) is generated.
Chloroplasts	Organelles found in the cytoplasm of plant cells that convert sunlight into chemical energy.
Golgi body	A cluster of parallel membranous sacs involved in the transport of proteins and other large molecules within the cytoplasm.
Cell wall	Outer layer of plant cells that gives them their shape. Animal cells do not have cell walls.
Vacuoles	Fluid-filled sacs in plant and animal cells. In plant cells vacuoles are generally quite large.
Lysosomes	Organelles found mainly in animal cells that are responsible for the degradation and disposal of worn-out or defective cellular components.

By stretching our imagination, we can compare a eucaryotic cell to a room in a house. The functions of a house are compartmentalized into its rooms and into specific objects in rooms, such as a stove, refrigerator, piano, or bathtub. These objects may be equated with the major organelles of cells—nucleolus, mitochondria, chloroplasts, Golgi bodies, and so on. Just as smaller objects, such as furniture, lamps, and rugs, may be moved about the house, so proteins and other macromolecules may be transported to different sites and organelles in the cell. Finally, small objects in a house and small molecules in a cell may move freely about and even enter and leave. The important thing about the cell is that cellular structures and functions are highly ordered by the laws of chemistry and physics—each molecule carries out its appropriate role in the designated site.

The Forces of Nature

Only four physical forces are known to exist in the universe; these four forces account for *all* interactions of matter, from the smallest subatomic particle to the largest galaxy. The four fundamental forces of nature, their relative strengths and the distances over which they act, are listed in Table 1-4. The *strong nuclear force* is responsible for the stability of matter; all atomic nuclei are held together by this force. Note, however, that the distance over which this force can be exerted is so small that it is only meaningful within the nuclei of atoms. The *weak nuclear force* is effective over even smaller distances, and compared to the strength of the strong nuclear force it seems almost insignificant; nevertheless, it, too, is essential in maintaining the stability of matter.

Both the *electromagnetic* and *gravitational forces* are extraordinarily weak compared to the strong nuclear force. However, they are effective over immense distances. Electromagnetic energy includes (going from lower to higher energy levels) infrared (heat) radiation, visible light, ultraviolet (UV) radiation, x-rays, and gamma rays. (Cosmic rays, which have the highest energy levels of all, are not electromagnetic radiation but actually atomic nuclei and other subatomic particles.) The particles of electromagnetic radiation are called *photons;* they contain energy but have no rest mass. Many forms of electromagnetic radiation can change the genetic information in organisms because photons can penetrate cells, interact with the electrons in molecules, and disrupt chemical bonds. Photons in UV light have sufficient energy to penetrate skin cells; photons in x-rays can pass completely through the body and expose a film behind it as they exit. Because bones are more dense than other tissues, they absorb more photons, causing an image of the skeleton—its shadow, so to speak—to be produced on the film.

Except in the presence of highly condensed matter (as in the interiors of stars), the gravitational force is weak compared with the other three fundamental forces—so weak, in fact, that it can be ignored for most purposes at the cellular level. Astronauts, for example, function at zero gravity. The gravitational force is primarily responsible for the formation and motions of giant objects in the universe, such as planets, stars, and galaxies. It also, of course, has played an important role in the evolution of multicellular life forms, and it places limits on the growth patterns and behaviors of these forms.

At the beginning of this century, scientists speculated that

Table 1-4. The Four Fundamental Forces of Nature

Force	Relative strength	Range of action (cm)
Strong nuclear	1	10^{-13}
Weak nuclear	10^{-5}	10^{-15}
Electromagnetic	10^{-2}	infinite
Gravity	10^{-39}	infinite

Do the Structures and Functions of Molecules Create "Life"?

Box 1-3

Of keen interest to scientists, philosophers, theologians, and indeed all curious people is the question formally posed by Louis Pasteur in 1864: "Can matter organize itself?" That is, can all the properties of living cells be explained solely in terms of the atoms and molecules they contain and of how those atoms and elements are organized into cells? If so, are all your thoughts, emotions, actions, and behaviors as well as your biology ultimately explicable in terms of the way matter is organized in your body's cells and tissues and the ways in which those cells interact with the environment? There are arguments for both sides of this question.

In the view of chemists Richard Dickerson and Irving Geis of the California Institute of Technology, "life" is

a behavior pattern of organized chemical systems of the proper degree and kind of complexity. An individual molecule of such a system can no more be alive than an individual bolt, wheel, or aileron in a jet airliner can fly. To say that DNA or any other molecule has the potential capacity for life in the right setting is as true, and nearly as misleading, as to say that the wings, fuel tank, and landing lights of a jet have the potential capacity for flight in the right setting. Moreover, even given the total collection of components in either of the systems we have been comparing, it is equally essential that these components be organized in precisely the right manner in order that the system be able to operate properly.

Other scientists, including the eminent biologist Ernst Mayr, do not believe that this reductionist viewpoint—the idea that life can be explained by the properties and organization of molecules—can ever scientifically account for all of the properties of organisms, especially human thoughts, beliefs, and behaviors. Complex biological systems are invariably more than the sum of their individual components. Science is unable to make any accurate predic-

(a)

(*a*) Dried tardigrade.
(*b*) Live tardigrade (Courtesy of Robert Schuster, University of California, Davis.)

tions about the nature of the whole organism, even with a complete understanding of all of the biological components, just as an electronics expert cannot predict what music or sounds will emerge from a stereo system that has been constructed.

Is there any evidence for the reductionist position that the properties of life are determined solely by the way molecules are arranged in cells and the way those cells are organized into tissues? Some experiments seem to suggest that life does derive solely from the structure and function of molecules within cells. At a temperature of absolute zero (−273°C), the chemical activity of atoms and molecules stops: all matter is frozen into one structural state, and molecules no longer function, since there is no energy available for chemical reactions. Yet many organisms can survive temperatures close to absolute zero and be revived. Thus, the property we call life is not lost by temporarily stopping all cellular functions and reactions. Bacteria, seeds, and protozoa resume their normal biological activities and growth once they are thawed out. (The temperatures in these experiments are much lower than those used to freeze sperm and even animal embryos, which can also later be thawed and live.)

A complex microscopic animal called a tardigrade (water bear), can be completely desiccated (dried out until almost all its water molecules have been removed) for months or years and later restored to an active living state. Tardigrades are not single cells but complex multicellular animals equipped with a nervous system and behaviors, yet they survive this desiccation.

These different kinds of experiments are interpreted by scientists and others as supporting the idea—the reductionist view—that the properties of life derive from the way molecules are organized and function in cells. According to the reductionist view, even complex human behaviors may eventually be explained by understanding how molecules, cells and tissues are organized and function in a person. However, there are many people who discount the reductionist view and hold that life is a mystery that will never be explained by science.

Readings: Boyce Rensberger, "Life in Limbo." *Science 80,* November 1980. J.H. Crowe and A. F. Cooper, Jr., "Cryptobiosis," *Scientific American,* December 1971.

(b)

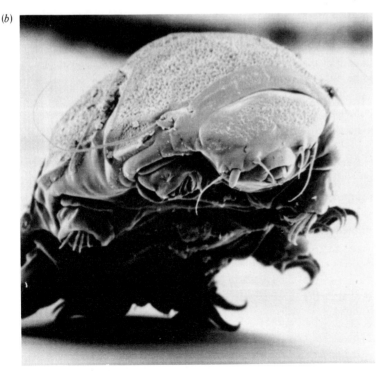

there might be some other force or energy in living organisms that is different from the physical forces found in inanimate (nonliving) matter. Today, all scientific evidence supports the conclusion that the forces of nature are identical in the cells of people, in the cells of plants, in rocks, and in distant stars. No physical forces other than these four are known to exist, nor are others expected to be found. The chemical and biological processes in the simplest bacteria as well as in human beings are explained by the same laws of chemistry and physics (see Box 1-3). As far as science has been able to determine, we are no more and no less than a complex aggregation of molecules obeying chemical and physical laws. To some people this is a satisfying idea; to others it is unappealing, since it may conflict with their belief in the existence of a human soul or of a divine creator.

Key Terms

element a substance that cannot be broken down further by ordinary chemical or physical means.

atoms the smallest units of any substance.

protons positively charged particles in the nuclei of atoms.

neutrons particles in the nuclei of atoms that are identical to protons but do not carry an electrical charge.

electrons negatively charged particles that orbit the nuclei of atoms.

ion a positively or negatively charged atom.

isotopes atoms with identical chemical properties but with different weights because of differences in the number of neutrons in the nucleus.

radioactivity the release of energy from an atom's nucleus in the form of particles or radiation.

radioisotope an isotope whose nucleus is unstable and decays spontaneously, emitting energy.

half-life the time it takes for one-half of the nuclei of a radioactive isotope to decay.

covalent bond a chemical bond created by the equal sharing of electrons by two atoms.

ionic bond a bond caused by the electrical attraction between two atoms. Weaker than a covalent bond.

hydrogen bond the sharing of a hydrogen atom between two other atoms. Weaker than a covalent or ionic bond.

DNA deoxyribonucleic acid, the macromolecule carrying the hereditary information in cells.

RNA a nucleic acid composed of a chain of ribonucleotides.

protein a macromolecule consisting of one or more chains of amino acids that perform catalytic or structural functions in cells.

amino acids the twenty different molecules that are found in proteins.

enzymes proteins that catalyze chemical reactions.

hormones chemicals that change the function of genes in cells and tissues.

cells the fundamental units of life. All cells are capable of reproducing themselves.

procaryotes simple single-celled organisms such as bacteria.

eucaryotes complex unicellular or multicellular organisms.

cytoplasm the part of the cell that lies outside the nucleus.

Additional Reading

Jacob, F. *The Possible and the Actual.* New York: Pantheon Books, 1982.

Platt, J. R. "Strong Inference." *Science*, October 16, 1964.

Thomas, L. "Hubris in Science." *Science*, June 30, 1978.

Weinberg, S. *The First Three Minutes: A Modern View of the Origin of the Universe.* New York: Bantam Books, 1979.

Weisskopf, V. F. "The Significance of Science," *Science*, April 14, 1972.

Review Questions

1. How can the difference between chemically identical isotopes be measured?

2. What are the three kinds of chemical bonds that hold atoms together?

3. What elements are present in significant quantities in all human cells?

4. How were the heavier elements in the periodic table created?

5. What cellular structures distinguish procaryotic from eucaryotic cells?

6. What is meant by the term "big bang"?

7. What are the three kinds of radiation emitted by radioisotopes?

8. Which of the three atomic particles governs all chemical interactions?

Answers to questions for all chapters are on page 439.

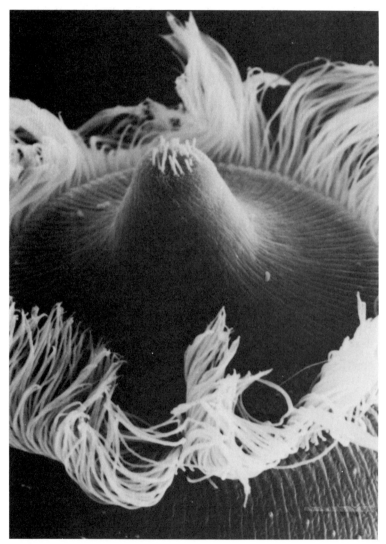

THE MOUTH OF A PROTOZOAN—A MICROSCOPIC, SINGLE-CELLED EUCARYOTE. (2,285 ×)

2

CELLS

The Smallest Living Organisms

Nature, in her blind search for life, has filled every possible cranny of the earth with some sort of fantastic creature.
JOSEPH WOOD KRUTCH

2
CELLS:
THE SMALLEST LIVING ORGANISMS

What impresses people most when they first observe and study nature is the enormous diversity of living organisms. What, if anything, do all these millions of diverse kinds of organisms— from vanishingly small microorganisms to awesomely large whales and redwood trees—have in common? Are there any fundamental unifying principles that are shared by all organisms? Are the chemical structures and biological functions of the molecules in all organisms similar? There are, in fact, common principles; indeed, the most extraordinary fact about living organisms is not that they are so different but that they are so similar.

One of the basic unifying principles of biology is that all living organisms, from the smallest and simplest to the largest and most complex, are composed of *cells*, which, as we saw in Chapter 1, are the smallest units of matter that are able to grow and reproduce. Cells are highly organized units of molecules and macromolecules in which chemical reactions are carried out that together produce the unique property we define as life. Cells may grow and reproduce independently, as do bacteria or yeast; or they may grow cooperatively, as in the tissues of plants and animals.

Another unifying principle in biology is that the overall chemical composition of all cells, no matter what their function or type, is basically the same. The proteins and other macromolecules found in the simplest bacteria are structurally and functionally quite similar to the macromolecules found in a human cell. Of particular significance is the fact that the chemical structure of DNA is the same, as is the genetic code (see Chapter 6). Proteins in all cells are synthesized from the same twenty different amino acids. Even the biochemical mechanisms by which proteins are synthesized are quite alike in all types of cells.

The third unifying concept in biology is that all organisms are related to one another and have evolved from some common ancestor. As mentioned in Chapter 1, simple cells presumably arose on earth once there were sufficient kinds and amounts of the appropriate molecules. More complex cells and eventually multicellular organisms are presumed to have evolved from these simple cells by various mechanisms, including **natural selection**, which is defined as the differential reproduction of genetically different individuals as a result of their fitness in particular environments. This important biological concept was proposed and documented by Charles Darwin (see Box 3-1 for a discussion of the history of Darwin's ideas). Many other scientists before him had similar ideas, but it was Darwin who presented the theory of natural selection in 1859 in his book *On the Origin of Species by Means of Natural Selection*. There, he developed the idea in such detail that the theory of evolution quickly became generally accepted. Evolutionary studies since Darwin's time have lent strong support to the basic concept of natural selection and to the idea that all organisms evolved from simple cells that arose on earth billions of years ago (Box 2-1).

When Is Matter Alive?

Because all organisms are composed of cells, it is helpful to understand what it means to say that a cell or an organism is alive. Instinctively, each of us knows when a plant, animal, or person is alive—or dead. What properties or characteristics do we measure to determine whether an organism is, in fact, living? No single, all-encompassing definition of life is acceptable to everyone, but there are certain properties of living cells that distinguish them from other forms of nonliving matter. The following six properties are unique to living cells:

1. **Metabolism.** Metabolism is the sum of all the chemical processes in a living organism. Cells are able to extract energy from the environment to fuel the chemical reactions they need to grow and reproduce. Most bacteria utilize chemicals that are provided in their environment, whether it be soil, water, or a human stomach. Some forms of bacteria, as well as the blue-green algae (cyanobacteria), are able to carry out **photosynthesis**, the process of converting the energy in sunlight to chemical energy that can be used for growth. The ultimate source of energy for growth in all higher plants and animals is sunlight, which can be used directly by all plants and indirectly by animals, who obtain energy from the environment by eating plants or other animals.

2. **Growth.** As a result of the utilization of energy and synthesis of new molecules, cells increase in size and weight.

3. **Reproduction.** Cells and microorganisms reproduce, giving rise to two identical copies of themselves. Growing cells eventually reach a size at which they divide, giving rise to two progeny cells that also grow and reproduce.

4. **Mutation.** During the process of growth and reproduction, cells occasionally undergo a **mutation**, which is a permanent change in the genetic information in the DNA. Although each cell's DNA has a small probability of undergoing a mutation, among the billions of cells in a population the chance that one or a few cells may mutate is quite high. Mutations that occur in sperm and egg cells, giving rise to progeny with new characteristics, are the ultimate source of genetic differences and the diversity of organisms.

5. **Response.** Organisms and even individual cells respond and react to their environment. Stimuli change the chemical reactions in a cell and change behaviors of organisms.

6. **Evolution.** Because of mutations and other biological mechanisms, the genetic information in a population of organ-

How Did Cells Arise?

BOX 2-1

Fossils of primitive bacteria and algae have been preserved in sedimentary rocks. Electron micrographs of these fossils show structures that are similar to ones in modern microorganisms. Some of the rocks in which such bacterial and algal fossils are found are more than 3 billion years old. Thus, it seems likely that the first primitive cells arose shortly after the earth was formed, or at least within a billion years or so. For most of the earth's history, life forms consisted of single cells; all of the complex plants and animals, including those that have become extinct, arose within the most recent half billion years, according to the fossil records.

How did the first simple cells arise? Most scientists believe that life began with *protocells*—clusters of complex proteinoid molecules similar in many ways to primitive bacteria and algae but not yet "alive." Under laboratory conditions thought to resemble the conditions on the primitive earth, some simple proteinoid microspheres have been formed that have many of the properties of present-day bacteria (see figure). They absorb dyes in the same manner as bacteria and often divide like bacteria. They can carry out a limited number of enzymelike chemical activities, and they have a well-defined structure, including membranes that regulate the entry and exit of small molecules and of electrically charged atoms. Although these proteinoid microspheres contain no genetic information and cannot regulate their growth, and thus are not alive, they do have a significant number of the important properties found in living cells today. It is enticing to speculate that such proteinoid microspheres are models of the protocells that evolved into the first true cells.

The origin of cells on earth is believed to have occurred in a series of fortuitous events that may be unique to our planet. There is no evidence so far that similar chance events have occurred elsewhere; the universe contains enormous numbers of galaxies and stars, but astronomers have yet to find any signs that these other stars possess planets on which the physical and chemical conditions would be hospitable to the evolution and survival of life forms as we know them. Failure to find evidence of life elsewhere in the universe, however, is not proof that life does not exist on some far-off planet.

Readings: Cheryl Simon, "Living Fossils." *Science News*, April 24, 1982. Sidney W. Fox, "New Missing Links." *The Sciences*, January 1980.

(a)

(b)

Proteinoid microspheres (*a*) can be created in the laboratory and have some of the properties of living cells, including the ability to form buds. These buds (*b*) can separate from the original microspheres to form new ones. (Photo courtesy of Institute for Molecular and Cellular Evolution, University of Miami, and Steven Brooke Studios.)

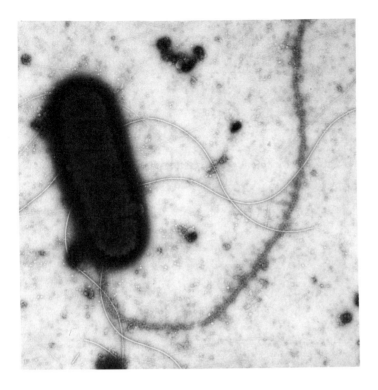

Figure 2-1 Electron micrograph of an *E. coli* bacterium showing its sex pilus (long, dark appendage at end of cell) and flagella (light fibers) used for movement. (Courtesy of Barry Eisenstein.)

isms changes over time. Some of these genetic changes in turn cause structural and functional changes in certain members of the population that make those members better able to survive and reproduce in their particular environments. This process of biological change and diversification over time is called **evolution.**

These six characteristics apply to all living cells; inanimate matter may carry out some but not all of these processes. Because viruses—particles that infect, grow, and reproduce in living cells—fail to exhibit all of these properties, they fall somewhere in between living and nonliving matter. Outside a living cell, viruses are not able to grow and reproduce. However, once they infect cells, they are able to employ the cellular machinery of their hosts in such a way as to enable them to reproduce themselves in great numbers—thereby qualifying them as being alive.

Genetic Information Governs Biochemical Processes

In many respects, some fundamental advances in genetics have emerged from studying bacteria because they are the simplest cells, and easily grown and manipulated in the laboratory. One species in particular, *Escherichia coli* (Figure 2-1), a microorganism occurring naturally in the human gut, has been the subject of extensive research. Because *E. coli* reproduces every half hour or so under laboratory conditions, and because the

Figure 2-2 Electron micrograph of human chromosome number 2 that has been stained to show the characteristic bands. (Courtesy of Carl B. Mankinen, Texas Department of Mental Health and Mental Retardation.)

chemical environment in which *E. coli* cells are grown can be manipulated experimentally, this organism is used for all sorts of genetic and biochemical experiments. And since there are male and female forms of *E. coli,* matings between genetically different bacteria can be performed to determine the location of genes in the bacterial DNA. (Bacterial mating is discussed in Chapter 4.) An *E. coli* bacterium contains 4,000–5,000 genes in a single DNA molecule. The precise location and functions of about one-quarter of the genes in *E. coli* are known, making it the best understood organism in nature from a genetic standpoint.

In some respects, bacteria are biochemically more efficient than human cells. When supplied with water, a few minerals, and the simple sugar glucose, bacteria grow and reproduce rapidly. Using only the carbon, oxygen, and hydrogen atoms in glucose and various minerals supplying a few essential salts such as sodium, potassium, and magnesium, bacteria can synthesize the many thousands of biological molecules and macromolecules they need. Among the biological molecules synthesized by enzymes in *E. coli* are vitamins, amino acids, fats, proteins, RNA, and DNA. Human cells, on the other hand, are unable to synthesize certain vitamins and certain amino acids, which must be obtained from food.

How are bacteria able to direct and coordinate the synthesis of all the different biological molecules they need in order to grow and reproduce, using only the sugar glucose and the elements supplied by a few minerals? The information that directs the chemical reactions in *E. coli* (and in all other organisms) is determined by its **genome**, the total amount of genetic information contained in an organism. In *E. coli* and other bacteria, the genome is a single molecule of DNA. In more complex organisms, the genome is defined by one complete set of **chromosomes**. Strictly speaking, the molecule of DNA in bacteria is not equivalent to the chromosomes in plant and animal cells, since eucaryotic chromosomes contain not only DNA but also proteins and other molecules that are associated with the DNA. Moreover, eucaryotic chromosomes are in the nuclei of cells whereas bacteria have no nuclei. The genetic information encoded in chromosomes is actually contained in the DNA molecule that is threaded from one end of the chromosome to the other (see Chapter 4). Much smaller DNA molecules are also present in the mitochondria and chloroplasts of plant and animal cells, but these are not called chromosomes since they reside outside of the nucleus and are not inherited in the same way as the nuclear chromosomes. Chromosomes in the nuclei of cells have characteristic lengths, shapes, and banding patterns (light and dark areas) that can be seen when the chromosomes are stained with dyes (Figure 2-2).

The genotypes of millions of the same species of bacteria

growing in the same environment are usually identical. However, in a population of hundreds of millions of bacteria, the genotype of some cells will be different from those in the majority of cells because of mutations. If the environment should change appreciably, some of the mutant bacteria may be able to adapt and reproduce, whereas the majority of the other bacteria may be eliminated. The observable characteristics of a bacterium, a plant, or an animal that result from the interaction of the organism's genotype (the particular set of genes present in the DNA) with the environment are referred to as the organism's **phenotype**. Thus, bacteria may have the phenotype of appearing round, rodlike, or filamentous under the microscope. When grown on certain kinds of agar medium, **bacterial colonies** may have a characteristic color (red, blue, or yellow) or may grow to a particular size. The phenotypes of plants and animals are of greater variety and are much more complex. The phenotype of a plant may be described by the size and shape of its leaves; the color, shape, and smell of its flowers; the color, shape, and size of its seeds; and so on. The phenotypes of animals include not only color, size, and shape but also such characteristics as temperament, personality, and intelligence.

The environment in which any organism develops and lives plays a crucial role in determining its phenotype. For example, bacteria having identical genotypes will produce colonies of different colors depending on the kind of chemicals added to the growth medium. Plants that are genetically identical will grow to different heights or yield different amounts of seeds depending on the environmental conditions of soil, water, and light. Similarly, a person's weight, height, strength, and intelligence are all strongly affected by the environments he or she experiences both before and after birth.

To appreciate how the expression of organisms' genotypes determines their phenotypes in different environments, it is necessary to understand the concept of **gene expression**. In terms of function, a **gene** is a section of DNA that codes for (directs the synthesis of) a specific protein, which is often an enzyme that is required to carry out a particular cellular chemical reaction. The environment that a bacterial cell experiences determines which of its genes are expressed, which enzymes are synthesized, and thus ultimately the phenotype of the cell. Gene expression, then, refers to the flow of information from genes in DNA into RNA and ultimately into proteins. The mechanisms of gene expression are discussed in Chapters 5 and 6.

Cells Adapt to Environments

Bacteria are among the most adaptable of all organisms. *E. coli,* which was originally isolated from hay, grows not only in laboratory test tubes but also in the human gut (where it aids in

Table 2-1. Differences in the Amounts of RNA, DNA, and Protein in *E. coli* growing with Different Generation Times

Generation time (minutes)	DNA	RNA	Protein
	(milligrams per gram of bacteria)		
25	30	310	670
50	35	220	740
100	37	180	780
300	40	120	830

digestion) and in various other tissues (where it can cause disease). The chemical reactions carried out by *E. coli* bacteria in these different environments may also be quite different. How do bacteria manage to adapt their enzyme-mediated biochemical reactions to their growth needs in different environments?

Bacteria regulate the synthesis of important macromolecules such as DNA, RNA, and protein according to the maximum rate of growth attainable in a given environment. The **generation time**, or time interval during which the number of bacteria doubles, determines the overall amounts of macromolecules the bacteria synthesize. Table 2-1 shows the amounts of RNA, DNA, and protein in bacteria growing at three different generation times. The chemical environment determines the generation time, which, in turn, dictates the overall macromolecular composition of the bacterial cells. Of course, in different environments bacteria will express different genes, synthesize different enzymes, and exhibit different phenotypes. Whereas the overall *amounts* of DNA, RNA, and proteins are determined by the generation time, the *kinds* of enzymes synthesized depend on the specific nutrients in the medium.

Bacterial adaptation to the chemical environment is accomplished by regulating the expression of genes so as to produce only those enzymes that are required for optimal growth in a particular environment. If, in addition to glucose, bacteria are supplied with all of the twenty different amino acids, they are able to grow much more rapidly because they do not need to produce the nearly 100 enzymes required to synthesize these amino acids, and so the amino acid–synthesizing genes remain unexpressed. The energy saved by the bacteria is redirected into increased synthesis of other molecules that enable the bacteria to grow twice as rapidly—that is, in half the former generation time (Figure 2-3).

The alteration of the size and macromolecular composition of bacteria in response to a nutritional change is a simple example of the way in which a cell's phenotype is determined by the interaction of its genotype with the environment. In the case of

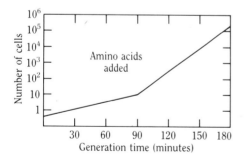

Figure 2-3 Effect of medium on bacterial growth. Growth of bacteria in a simple medium consisting of glucose and various minerals can be increased by adding nutrients such as the twenty different amino acids. Faster growth means the generation time is reduced, since the bacteria will divide more often.

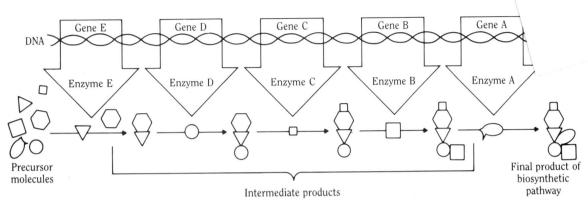

Figure 2-4 A biosynthetic pathway. Most molecules in cells are synthesized by a series of chemical reactions carried out by enzymes. The genes that direct the synthesis of the particular enzymes in a biosynthetic pathway may be located adjacent to one another in the DNA as shown here, or they may be separated. At each enzymatic step in the pathway the molecule becomes larger until the final product is synthesized. The end product may be an amino acid, a vitamin, a nucleotide, or another molecule essential for growth.

bacteria growing in an amino acid–supplemented medium, a number of genes are "switched off," but they can be switched back on again if the amino acids in the medium are used up or removed. Specific genes in animal cells are usually switched on or off for good, since specialized cells generally perform the same functions throughout the animal's life. Heart cells and nerve cells are not required to change their functions—indeed, for them to do so might lead to the animal's death. And heart cells do not synthesize the digestive enzymes that are synthesized in stomach cells even though both kinds of cells have identical genotypes.

Biochemical Pathways

Most cells, including bacteria and the cells in eucaryotic organisms, use the simple sugar glucose as the source of the carbon atoms they need for the synthesis of new organic molecules. Consequently, cells must be able to break down (degrade) the glucose in order to obtain carbon atoms for new syntheses. A series of cellular enzymes act sequentially in what is called a **degradative pathway** to break down glucose and other small molecules. The synthesis of new molecules in a cell occurs via a **biosynthetic pathway** (Figure 2-4). In such a pathway, atoms or groups of atoms are added stepwise in a series of enzyme-catalyzed chemical reactions to synthesize an essential molecule. In bacteria, biosynthetic pathways are used to synthesize each of the twenty amino acids, all of the vitamins, and the various components of RNA and DNA, to mention the most common products. The genes that govern the synthesis of the enzymes in each pathway are often—but not always—grouped together in the DNA. The genes in a biochemical pathway are usually expressed as a group; that is, either all of the genes in the biosynthetic pathway are expressed or else none are. The regulation of the expression of genes for biosynthetic and degradative pathways is discussed in Chapter 9.

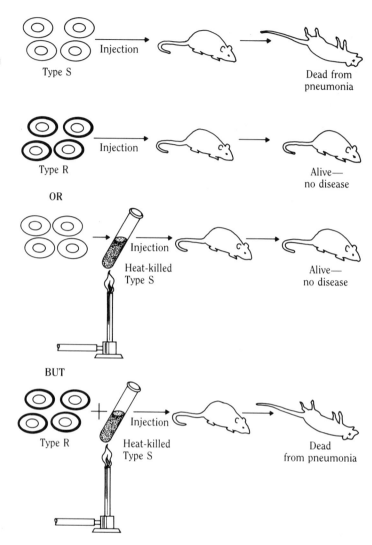

Figure 2-5 Griffith's experiments in which mice were injected with type S or type R pneumococci. Type S causes pneumonia and death of mice. Type R is harmless, as are heat-killed type S. However, a mixture of type R and heat-killed type S causes pneumonia and death of mice.

DNA Carries the Genetic Information

We have said that the genetic information of cells resides in their DNA molecules. How was this crucial fact, nowadays taken for granted, discovered? As with several other fundamental concepts of genetics, including Gregor Mendel's discoveries of the patterns of inheritance and Charles Darwin's idea of natural selection, DNA was not readily accepted as the carrier of the genetic information until many years after its initial discovery.

GRIFFITH'S DISCOVERY

One of the key observations pointing to DNA as the genetic material was made in England in 1928 by a microbiologist, Frederick Griffith. He had noticed that two major strains of the bacterium *Diplococcus pneumoniae* could be isolated from patients with symptoms of pneumonia. He called these strains type R pneumococci (*rough* colony phenotype) and type S pneumococci (*smooth* colony phenotype).

When Griffith inoculated mice with type S bacteria, they contracted pneumonia and died. When he injected mice with type R bacteria, however, they did not get the disease. Knowing that bacteria are killed by heating (which is why surgical instruments are sterilized by flaming or boiling), Griffith next heated the type S bacteria before injecting them into the mice. This time no disease was produced, proving that the pneumonia was indeed *caused* by live, but not dead, type S bacteria (Figure 2-5).

Then Griffith made a remarkable discovery. When he took type R bacteria that were unable to cause disease and mixed them with heat-killed type S bacteria that were also unable to cause disease and then injected the mixture into mice, the mice contracted pneumonia and died. Moreover, Griffith was able to isolate millions of live type S bacteria from the dead mice. The puzzling question was: How did the live type S bacteria arise in the mice? Griffith was unable to answer this question.

AVERY'S RESEARCH

Oswald T. Avery and his collaborators at Rockefeller University pursued this question for many years. Eventually they became convinced that pieces of DNA from the heat-killed type S bacteria could somehow enter the type R bacteria in the infected mice and genetically transform type R into type S pneumococci (Figure 2-6). They assumed that this genetic change was heritable—that the bacteria would continue to grow and reproduce as type S pneumococci and thus be able to cause disease. This process of changing both the genotype and phenotype of bacteria by transferring DNA from one type of cell into another is called **transformation**. As we shall see in Chapter 10, bacterial transformation using pieces of DNA is one of the key tools in genetic engineering, which permits genes from virtually any organism to be inserted in microorganisms.

In a paper published in 1944, Avery stated that "among microorganisms the most striking example of heritable and specific alterations in cell structure and function that can be experimentally induced and are reproducible in well defined and adequately controlled conditions is the transformation of specific types of pneumococcus." In later reports, Avery and his colleagues continued to claim that the transformation of bacteria was due to DNA and that DNA carried the genetic information. Yet most biologists at the time did not accept their view.

In the 1940s proteins were favored as the molecular carriers of genetic information. After all, proteins were known to be extremely complex molecules and to vary tremendously in size, structure, and activity. And were not the traits of organisms also

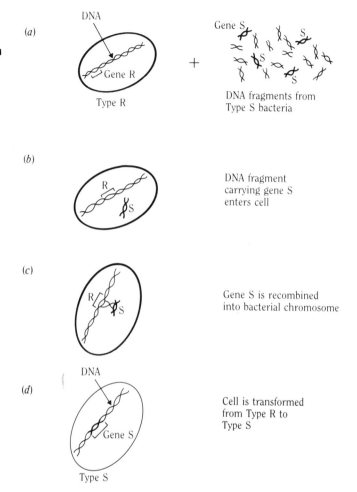

(a) DNA

Type R

+

Gene S

DNA fragments from
Type S bacteria

(b)

DNA fragment
carrying gene S
enters cell

(c)

Gene S is recombined
into bacterial chromosome

(d) DNA

Type S

Cell is transformed
from Type R to
Type S

Figure 2-6 Type R bacteria are transformed to type S. (*a*) Fragments of DNA from type S bacteria are added to type R bacteria. (*b*) Some type R cells receive DNA carrying gene S and (*c*) recombine that fragment of DNA into their chromosome. The expression of gene S (the fragment of DNA containing gene R is lost) changes type R to type S bacteria. This process is called bacterial transformation.

complex and enormously variable? The complexity of proteins and the complexity of organisms was too obvious a connection for most scientists to ignore.

Moreover, DNA appeared to be a chemically uninteresting molecule. At that time in history, DNA was thought to consist of a monotonous, repeating structure consisting of only four different bases. Thus, it was not possible in the 1940s for scientists to envisage how such a molecule could contain the enormity of genetic information needed by cells. As a final argument against DNA being the genetic material, many scientists held that the experiments done in Avery's laboratory were technically flawed. Even using the best chemical techniques available at the time, the preparations of bacterial DNA used to transform one type of pneumococcus into the other were contaminated with about 1 percent of protein. Many critics claimed that this small amount of protein, not the DNA, was actually responsible for bacterial transformation.

Scientific Discovery and the Prepared Mind

BOX 2-2

Nobel laureate and biochemist Albert Szent-Györgyi is often quoted as saying, "The creative scientist sees what everyone else has seen, but thinks what no one else has thought." A classic example of this kind of scientist is Alexander Fleming, best known for his discovery of the antibiotic penicillin.

Fleming noticed in 1929 that some Petri plates used to culture bacteria were contaminated with molds. He also noticed that bacterial growth was inhibited in the moldy areas. Now, this observation had undoubtedly been made before in many laboratories, but there is no evidence that any scientist before Fleming had ascribed any significance to the phenomenon or grasped its implications. Fleming, however, reasoned or knew intuitively that the molds growing on the plate were producing some chemical substance that could inhibit bacterial growth, and he also realized the potential value of isolating such a substance.

Fleming proceeded to isolate the inhibitory compound, which he named after *Penicillium,* the species of mold that produced it. For a number of reasons, the world was not prepared for this miracle drug. One reason was that sulfanilamides (sulfa drugs) were in widespread medical use at the time and were effective in treating many bacterial infections. Also, penicillin was chemically unstable and difficult to purify. As a result, penicillin did not come into widespread use until World War II, when it was used to treat soldiers for gonorrhea and other bacterial diseases. Today, penicillin and similar chemically synthesized derivatives are among the most effective drugs for treating many kinds of bacterial infections.

What is less well known about Alexander Fleming is that his mind had been prepared for this discovery by another important observation he had made years earlier. In 1922 Fleming reported on "a remarkable bacteriolytic element found in tissues and secretions." While routinely spreading on nutrient agar plates bacteria-containing nasal secretions from a patient, he noticed that something in the secretions seemed to "dissolve" the bacteria and prevent their growth. He followed up on this chance observation and found that there was a bacterial inhibitory substance in many human secretions, including tears and saliva, that would destroy almost any kind of bacteria. He called this substance, which proved to be an enzyme, *lysozyme*—that is, an enzyme that disintegrates bacteria and other kinds of cells.

Lysozyme is produced by organisms ranging from bacteriophages to human cells. This protein has contributed much to our understanding of the genetic code (Chapter 6) and to protein evolution (Chapter 18). It was Fleming's discovery of lysozyme some seven years before his discovery of penicillin that alerted him to be on the lookout for substances that inhibit bacterial growth. Thus, where other scientists threw out moldy cultures, he seized the opportunity to study them and went on to make a major contribution to science and medicine.

Reading: Gunther S. Stent, "Prematurity and Uniqueness in Scientific Discovery." *Scientific American,* December 1972.

However, Avery and his collaborators were correct in their conclusions, and history now accords the work of Griffith, Avery, and others the recognition they did not receive at the time. DNA does indeed contain the genetic information, as was proved with research involving viruses.

THE HERSHEY-CHASE EXPERIMENT: FINAL PROOF

We now turn to the experiments that are generally cited as proving that DNA does, in fact, carry the genetic information. To understand these experiments it is necessary to know what **bacteriophages** (usually called *phages,* from the Greek *phagein,* "to eat") are and how they grow and reproduce in cells that they infect. Phages are viruses—particles consisting of a DNA mole-

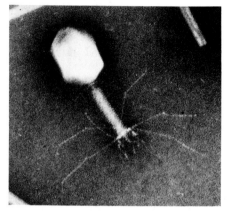

Figure 2-7 Electron micrograph of a T4 phage. (Courtesy of Robley Williams.)

cule (some have an RNA molecule instead of DNA) that is packaged inside a head or coat made up of proteins (Figure 2-7). Phages are viruses that grow and reproduce only inside living bacteria; although most phages survive outside cells, they are inert (chemically inactive) there. The phage DNA molecule is extremely long compared with the size of the head; the DNA must first be condensed and packaged in the protein head when the phages are synthesized in bacteria. The coat proteins of viruses determine what kinds of plants or animals a virus can infect, since it is the viral coat proteins that attach to specific receptor proteins on the surfaces of cells. That is why tobacco mosaic virus attacks and infects only tobacco leaves and why *Herpes simplex* virus attaches to and infects only certain human cells.

Phages are too small to be seen except in an electron microscope, but they are easily detected and titered (counted) because of their ability to cause **lysis** (disintegration) of bacteria. To titer a phage suspension, researchers mix a sample with billions of uninfected bacteria and spread the mixture onto solid medium (Petri plates containing agar and nutrients). After an incubation period of 12–24 hours, the surface of the Petri plate is covered with a dense lawn of bacteria. Wherever a phage particle was present initially, a bacterium becomes infected. New phages are synthesized in the original infected bacterium, and when it lyses (bursts open), hundreds of phages are released, which then infect and lyse other bacteria nearby. Eventually this process produces a hole in the bacterial lawn called a *plaque* that is visible to the naked eye. **Plaque assay**, or counting the number of plaques on the Petri plate and multiplying by the dilution factor, permits researchers to calculate the number of phages in the original suspension (Figure 2-8*a*).

Bacteria are counted in a similar way except that a much smaller number of bacteria are spread on the Petri plate instead of the billions that are added to the phage sample. Wherever a single bacterium lands on the surface of the nutrient agar, it begins to grow. Within 12 to 24 hours millions of bacteria have grown in that spot, forming a visible bacterial colony (Figure 2-8*b*). By counting the number of bacterial colonies arising from a series of dilutions, the number of bacteria in the original solution can be calculated.

One phage extensively used in genetic studies is the T-phage that infects *E. coli* bacteria. A classic experiment using T-phages was performed in 1952 by Alfred Hershey and Martha Chase at the Cold Spring Harbor Laboratories in New York. It was known at that time that T-phages carried the genetic information for directing the growth and reproduction of new phages upon infecting *E. coli* bacteria. Because chemical analyses showed T-phages to be composed almost exclusively of DNA and protein,

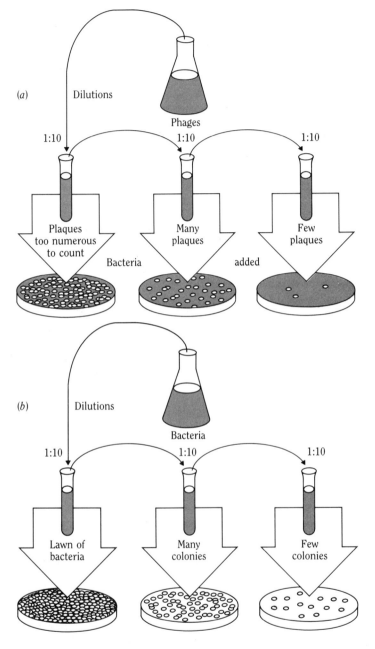

Figure 2-8 Indirect methods are used to titer (count) the number of phages, viruses, or cells in a solution. (a) Dilutions of phages are made so that successive tubes have progressively fewer particles. A sample from each tube is added to billions of bacteria and the mixture is spread onto a soiid medium in a Petri plate. After incubation, the bacteria completely cover the surface of the plate except in spots where a bacterium was infected by a phage. A single phage particle will grow and multiply in that area so that small clear spots called plaques appear in the bacterial lawn. Since each plaque is initiated by a single phage, counting the number of plaques on the plate and multiplying by the dilution factor can reveal the number of phage particles present in the original solution. (b) Bacteria are titered in a similar manner except that the number of bacterial colonies are counted. Each visible colony on the Petri plate is initiated by a single bacterium.

the important question became: Which part of the phage carries the genetic information—the DNA or the protein molecules?

Proteins in the tail fibers of T-phages recognize specific proteins on the surface of the bacteria, allowing the phages to attach to the bacterial cell wall and inject their DNA. From this point on, the infection is irreversible, and the bacteria are destined to lyse. However, for about 15 minutes after infection, no

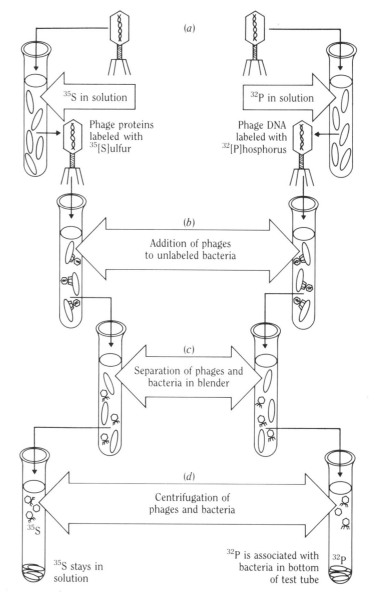

Figure 2-9 The Hershey-Chase experiment, which showed that phage DNA carries the genetic information. (*a*) Phage proteins are labeled with radioactive sulfur (^{35}S); phage DNA is labeled with radioactive phosphorus (^{32}P). (*b*) Bacteria are infected with these radioactive phages, and (*c*) the phage and bacteria are physically separated by mixing in a blender. (*d*) The solution of phage and bacteria is centrifuged so that the heavier bacteria form a pellet at the bottom of the tube. Most of the ^{32}P is found associated with the bacteria, proving that the phage DNA entered the bacteria and must carry the genetic information necessary for synthesis of new phages.

Labels within figure:
(*a*) ^{35}S in solution — ^{32}P in solution
Phage proteins labeled with ^{35}[S]ulfur — Phage DNA labeled with ^{32}[P]hosphorus
(*b*) Addition of phages to unlabeled bacteria
(*c*) Separation of phages and bacteria in blender
(*d*) Centrifugation of phages and bacteria
^{35}S stays in solution — ^{32}P is associated with bacteria in bottom of test tube

phage particles can be detected inside the infected bacteria, even if the cells are broken open and the contents analyzed for phages. New phage particles begin to appear in the bacteria after 15 minutes; by 30–40 minutes most of the bacteria have lysed, releasing several hundred new phages per infected cell.

The steps in Hershey and Chase's experiment to determine whether it is DNA or protein that allows phages to exactly reproduce themselves are outlined in Figure 2-9. First, T-phages were grown in bacteria in the presence of either a radioactive isotope of sulfur (^{35}S) or a radioactive isotope of phosphorus (^{32}P).

Sulfur atoms but not phosphorus atoms are present in phage proteins, whereas phosphorus atoms but not sulfur atoms are present in phage DNA. Phages were then prepared that had incorporated either the ^{32}P or ^{35}S atoms and consequently were radioactively labeled in either their proteins or DNA. The radioactively labeled phages were then added to nonradioactive bacteria growing in a liquid medium in two different tubes. After a few minutes the process of phage infection was complete and the phage-bacteria mixtures were agitated in a kitchen blender. The agitation detached the phages from the bacteria but did not harm the phage-infected or uninfected cells. Next, the radioactive phage-bacteria mixtures were centrifuged (spun at high velocity) to sediment all the bacteria to the bottom of the tubes. The phages, being much smaller and lighter, would remain in solution unless they were irreversibly attached to the bacteria. Following this separation, Hershey and Chase measured the amount of radioactivity in the solution and in the bacteria at the bottom of the tube to determine where the DNA and protein molecules of the phages had ended up.

Hershey described the results of the experiment as follows:

1. Most of the phage DNA remains with the bacterial cells. 2. Most of the phage protein is found in the supernatant fluid. 3. Most of the initially infected bacteria remain competent to produce phage. 4. If the mechanical stirring is omitted, both protein and DNA sediment with the bacteria. 5. The phage protein removed from the cells by stirring consists of more-or-less intact, empty phage coats, which may be therefore thought of as passive vehicles for the transport of DNA from cell to cell, and which, having performed that task, play no further role in phage growth.

This experiment thus showed that it is mainly phage DNA that enters bacterial cells and plays a role in the synthesis of new phages. Hershey and Chase concluded that phage DNA molecules contain the genetic information for directing the synthesis of new phages when the DNA is injected into bacteria.

Surprisingly, the actual data obtained in the Hershey-Chase phage experiment were less conclusive than the data obtained in the bacterial transformation experiments. In the phage experiment, about 20 percent of the radioactive sulfur—and consequently of the phage protein—actually remained associated with the bacteria, so that the contaminating protein could still have contained the genetic information, as had been argued by the critics of the transformation experiments. By 1952, however, the scientific community was more ready to embrace DNA as the genetic material. And with the announcement of the structure

Table 2-2. Number of Chromosomes and Amount of DNA in Various Organisms

Organism	Amount of DNA (number of base pairs)	Number of DNA molecules or number of chromosomes (haploid)
Tumor virus SV40	5.1×10^3	1
Phage T4	1.8×10^5	1
E. coli	4.0×10^6	1
Yeast (baker's)	1.35×10^7	17 or 18
Drosophila (fruit fly)	1.65×10^8	4
Human	2.9×10^9	23
South American lungfish	1.0×10^{11}	19

of DNA by Watson and Crick the following year (see Chapter 3), DNA became universally accepted as the carrier of the genetic information.

Genetic Information in Plant and Animal Cells

Phages and bacteria carry their genetic information in a single molecule of DNA or single chromosome. However, many plant cells and nearly all animal cells contain two copies of each chromosome (indicated by 2N). Thus, plants and animals are said to be composed of **diploid** cells. However, the *gametes*—pollen and eggs in plants or sperm and eggs in animals—are **haploid** cells; that is, each cell contains only a single copy of each chromosome (indicated by 1N). When the gametes produced by males and females unite following a sexual mating, the diploid chromosome number (2N) is restored: Two copies of each chromosome are present in a fertilized egg and in all cells derived from it. The related chromosome pairs in diploid organisms are referred to as **homologous** chromosomes, and the homologous pairs are identical with respect to the position of genes on the visible chromosomal structure, although the information in the genes they carry may be either identical or different.

The diploid number of chromosomes in every human cell is 46 except for sperm and egg cells, which are haploid and contain 23 different chromosomes. The increasing biological complexity of various organisms from phage and bacteria to flies and humans is usually matched by an increase in the total amount of genetic information contained in DNA, although this is not always the case (Table 2-2). Chromosomes come in a great variety of sizes, so that the number of chromosomes does not always reflect the amount of genetic information carried in different organisms. Flying insects such as the mosquito (N = 3) and the common fruit fly (N = 4) rank among animals with the fewest

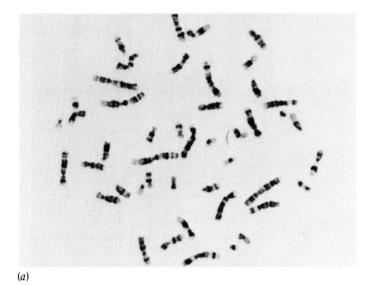

(a)

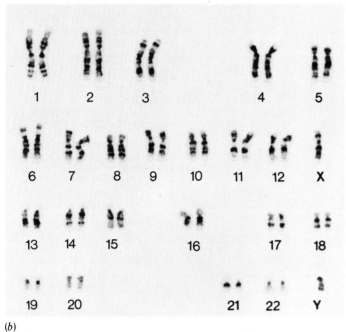

(b)

Figure 2-10 Karyotyping of chromosomes. Human blood cells that are arrested during mitosis are stained and examined under a microscope. (a) When the chromosomes are located, they are photographed and printed. Each chromosome is cut out, paired with its homologous chromosome, and given its numerical assignment. (b) The ordered display of all the chromosomes is called the karyotype, which, in this case, is the karyotype of a human male. (Courtesy of Patricia Jacobs, University of Hawaii.)

chromosomes. Turkeys (N = 41) and donkeys (N = 31) are among animals with the greatest number of chromosomes. The chromosome complement of humans (N = 23) ranks us in the middle of the scale, somewhere between mosquitoes and turkeys.

The **karyotype** of an organism is the visual display and arrangement of all of the chromosomes from a single cell. The normal karyotype of a human cell with its 46 chromosomes arranged in homologous pairs is shown in Figure 2-10.

Nondividing Cells: The Cause of Aging?

BOX 2-3

Many kinds of animal cells are now grown in tissue cultures—nutrient-rich liquid mediums contained in Petri plates or in flasks. Unlike bacteria, animal cells grow best when they can attach to a surface; thus, a layer one cell thick forms on the bottom of a Petri plate as animal cells grow and divide. Most animal cells stop dividing after forming a continuous layer of tissue, at which point all cells are in physical contact with one another. The cessation of cell division on contact with other cells is a characteristic property of normal cells. (One distinguishing characteristic of cancer cells is that they lose the ability to stop growing on contact and instead pile up in tissue culture plates, yielding dense masses of cells.)

In order to continue to grow normal animal cells in vitro (outside of living organisms), researchers put nondividing cells from a dormant culture into fresh nutrient medium. These few cells again attach to the bottom of the Petri plate and begin growing.

This sequential transfer of animal cells from a nondividing culture to new mediums allows certain strains of animal cells to be grown almost indefinitely.

About twenty years ago two scientists discovered that cells from a human embryo can be grown and transferred in this manner but that cell division ceases after about fifty doublings, even if the cells are placed in fresh mediums and are not physically touching. More surprising was the finding that the same type of cells taken from adult tissues are capable of fewer doublings and that the older the person from whom the cells are taken, the fewer the number of cell divisions in vitro. Similarly, cells from other animals that have average life spans longer or shorter than the average human life span have been shown to have a range of cell doublings that correlates with these animals' maximum life spans. Some examples are listed in the accompanying table.

It appears, then, that body cells might be genetically programmed to undergo a finite number of cell divisions and then die. If this is the case, then the process of aging must be largely under genetic control. The idea that the rate of aging and the maximum life span of any animal species is controlled by a kind of "genetic clock" has some evidence to support it. By comparing different strains of mice, Roy Walford, a scientist at the UCLA School of Medicine, has been able to identify a group of mouse genes that affect the aging process. In humans, genes on chromosome 6 correspond to the same group of genes in mice. Further studies may determine how significant the genetic contribution to aging is—and perhaps whether the human aging process can be changed.

Readings: Leonard Hayflick, "The Cell Biology of Human Aging." *Scientific American*, January 1980. Richard Conniff, "Living Longer." *Next Magazine*, May/June 1981.

Species	Number of cell doublings	Estimated maximum life span (years)
Galapagos tortoise	90–125	175
Human	40–60	110
Chicken	15–35	30
Mouse	14–28	3.5

Many plants are diploid, but some—cultivated crop plants in particular—have more than two copies of each different chromosome and are referred to as **polyploid** organisms. Plants are better able to tolerate extra chromosomes in their cells than are animals, which generally cannot survive if they are polyploid. Plant breeders have developed commercially valuable varieties of fruits and vegetables such as apples, potatoes, strawberries, watermelons, cherries, and tomatoes by increasing the chromosome numbers in the plant cells through genetic manipulations and breeding. Some created varieties of wheat are hexaploid (six copies of each chromosome), and each cell contains 42 chromosomes ($N = 7$). Potatoes are tetraploid (four copies), as are many varieties of rice. Cultivated strawberries are octaploid (eight copies), which is partly why commercially grown strawberries are so much larger than strawberries produced by diploid plants growing naturally in the wild.

Eucaryotic Cell Division

Because of the large number of different chromosomes in eucaryotic cells and the need to duplicate and separate them with precision for transmission to daughter cells, a special process called **mitosis** (pronounced my-tō-sis) occurs when a eucaryotic cell divides. During mitosis the chromosomes are duplicated and a complete set of chromosomes is segregated into each of the two daughter cells when the parent cell divides. Chapter 11 presents a more detailed discussion of mitosis and an analogous process called **meiosis** (pronounced my-ō-sis), which is responsible for the segregation of a haploid chromosome set into the gametes as a prelude to the sexual mating that produces a new diploid individual.

In all cells—procaryotes and eucaryotes—the duplicaton of DNA and subsequent cell division must be coordinated so that each daughter cell receives a complete set of genetic instructions. The growth and reproduction of eucaryotic cells occur in four distinct phases that together are called the cell cycle (Figure 2-11). DNA synthesis (S) occurs only during part of the cell cycle and is later followed by the mitosis (M) phase, which includes cell division. The other two phases of the cell cycle, G_1 and G_2, are needed not only to synthesize the molecules and cellular structures necessary for reproduction but also to carry out the cell's normal functions. As Figure 2-11 indicates, the M phase, in which the chromosomes are segregated and the parent cell divides, is the briefest part of the cell cycle. The complete cell cycle, or generation time, for animal cells varies from about 19 to 24 hours.

All animal cells (except cancer cells) regulate their growth and reproduction (see Box 2-3); whenever growing cells come into contact with other cells, further cell division is inhibited by some as yet unidentified form of cell–cell communication. If cell division were not regulated in plants and animals, the size of various tissues and even the eventual size of organisms could not be maintained.

Within each cell are DNA molecules (one per chromosome) that together carry all of the organism's genetic information.

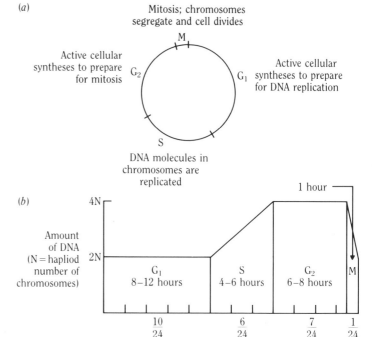

Figure 2-11 The cell cycle of eucaryotes. (a) The growth and division of cells occurs in four phases. The amount of DNA synthesis in cells undergoing division is shown in (b).

The growth, division, and biological activities of every cell in the bacterium, plant, or animal are determined by the expression (or lack of expression) of certain genes on those DNA molecules. Thus, before we can understand how genes function, it is necessary to understand the structure and functions of DNA.

Key Terms

natural selection the differential reproduction of genetically different individuals as a result of their fitness in particular environments.

metabolism the sum of all the chemical reactions in a cell or organism.

photosynthesis the cellular process of converting light energy to chemical energy.

mutation any permanent change in the genetic information of a cell.

evolution the process of change and diversification of organisms over time.

viruses particles containing DNA or RNA that infect living cells.

genome the total amount of genetic information contained in an organism.

chromosomes structures in the nuclei of eucaryotic cells that carry the genetic information in the form of DNA molecules.

genotype the particular set of genes present in the DNA of organisms.

phenotype the observable characteristics (traits) of an organism that result from the interaction of its genotype with the environment.

gene expression the flow of information from genes in DNA into RNA and ultimately into proteins.

gene a sequence of bases in DNA that codes for a functional cellular product, usually a protein.

generation time   the time required for growing cells to double their mass or to divide into two cells.

degradative pathway   a series of enzymes that act sequentially to break down small molecules.

biosynthetic pathway   a series of enzymes that act sequentially to synthesize new molecules.

transformation   the transfer of DNA from one bacterial strain into a genetically different strain, thereby transforming the genotype and phenotype of the recipient.

bacteriophages   infectious particles containing DNA or RNA that are able to grow and reproduce in bacteria. Phages generally destroy the bacteria they infect.

lysis   the bursting open of a cell due to disruption of the membrane that surrounds it.

plaque assay   a technique for counting viruses.

bacterial colony   a visible spot of bacterial growth consisting of millions of cells.

diploid cells   cells that contain two copies of each different chromosome.

haploid cells   cells in which each different chromosome is present in single copy. Sperm and eggs are haploid cells of animals.

karyotype   visual arrangement of all of the chromosomes from a single cell so that they can be identified and counted.

polyploid cells   cells that contain more than two copies of each different chromosome.

mitosis   process of chromosome segregation and cell division.

meiosis   process by which the haploid set of chromosomes winds up in gametes (sex cells).

cell cycle   the four stages of growth and reproduction of eucaryotic cells.

***

_____ Additional Reading _____

Block, I. "The Worlds Within You." *Science Digest,* September/October 1980.

Crick, F. *Of Molecules and Men.* Seattle: University of Washington Press, 1966.

Fox, S. W. "New Missing Links." *The Sciences,* January 1980.

Lee, J. "Timekeepers of the Solar System." *Science 80,* May/June 1980.

Lovelock, J. E. *Gaia: A New Look at Life on Earth.* New York: Oxford University Press, 1979.

Margulis, L. *Early Life.* Boston: Science Books International, 1982.

_____ Review Questions _____

1. What are the three unifying principles in biology?
2. What class of macromolecules is most affected by the rate of growth of procaryotic cells?
3. What was proved by the transformation experiments involving pneumonia-causing bacteria?
4. In what ways are phages different from bacteria?
5. What radioisotopes were used in the Hershey-Chase experiment and why were they chosen?
6. How many chromosomes are present in human brain cells? in skin cells? in heart cells? in sperm?
7. Which phase of the eucaryotic cell cycle is the shortest?

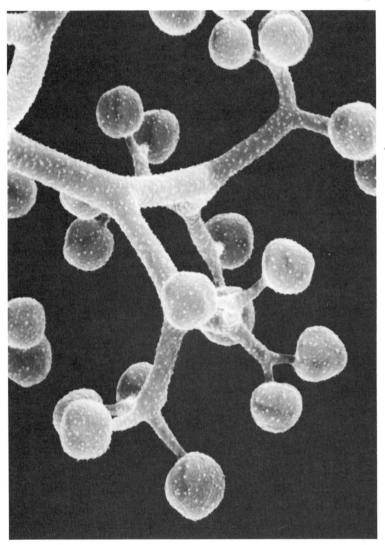

THE SPORES OF A MOLD CLUSTER (CALLED *SPORANGIOLA*) ON BRANCHES. (4,080 ×)

3

DNA

*The Carrier
of Genetic
Information*

*I have often thought how much more
interesting science would be if those who
created it told how it really happened,
rather than reported it logically and
impersonally, as they so often do in
scientific papers.*

GEORGE W. BEADLE

Figure 3-1   James Watson (*left*) and Francis Crick (*right*) in 1962 at the time they were awarded the Nobel Prize. (United Press International.)

A one-page article titled "Molecular Structure of Nucleic Acids" that appeared in the British journal *Nature* in April 1953 produced a revolution in biology, the consequences of which are still being explored. Francis Crick, an English physicist, and James Watson, an American biologist, began their paper with this modest introduction: "We wish to suggest a structure for the salt of deoxyribose nucleic acid (D.N.A.). This structure has novel features which are of considerable biological interest." The biological interest was indeed considerable, and the scientific implications were revolutionary, because, for the first time, the structure and functions of the genetic information in all living organisms became comprehensible according to the laws of chemistry and physics.

The structure of the DNA molecule proposed by Watson and Crick has survived the test of time in all of its essential details. Implicit in their proposed DNA structure was the chemical basis for the three unexplained functions of the genetic material: duplication, recombination, and mutation. Watson and Crick concluded their now famous article with this coy understatement: "It has not escaped our notice that the specific pairing we have postulated immediately suggests a possible copying mechanism for the genetic material."

With the structure of the DNA molecule before them, scientists could readily understand how genetic information could be encoded in DNA and how the hereditary information could be accurately duplicated. Within a few years of the discovery of DNA's structure, the chemistry of that structure and of gene functions also became understood, and biologists achieved a basic understanding of how genetic information is encoded, deciphered, and used to direct the synthesis of other molecules in all living cells. The importance of the Watson-Crick model of DNA to genetics and to all of the biological sciences can hardly be overstated.

The Structure of DNA

The combined insights of Watson and Crick (Figure 3-1) enabled them to assemble an array of observations made by a number of scientists into a model that was an accurate representation of the DNA molecule. They were aware of two crucial facts concerning the four **bases** in DNA—adenine (A), thymine (T), guanine (G), and cytosine (C)—which are the chemical entities in DNA whose sequence encodes the genetic information. First, Watson and Crick knew that the order of the four bases in DNA is random; that is, any one of the four bases is free to follow any other base in the chain of DNA. Second, from the work of the biochemist Erwin Chargaff, they knew that the number of adenine bases in DNA always equals the number of thymine bases and that the number of guanine bases always equals the number .

Table 3-1. Base Composition of Various Species of Bacteria

Bacterium	A + T (%)	G + C (%)
Micrococcus pyogenes	70	30
Streptococcus pneumoniae	62	38
Hemophilus influenzae	60	40
Salmonella typhimurium	50	50
Escherichia coli	48	52
Shigella dysenteriae	46	54
Serratia marcescens	42	58
Mycobacterium tuberculosis	32	68
Streptococcus griseus	26	74

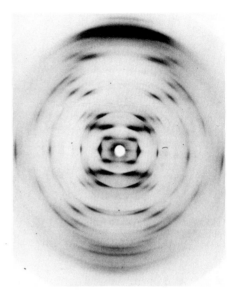

Figure 3–2. A photograph of a DNA diffraction pattern obtained by shining a beam of x-rays on a partially crystalized sample of DNA. The structure of the DNA can be deduced from the regular pattern and positions of the dark spots on the film. (Courtesy of R. Langridge, University of California San Francisco Medical School.)

of cytosine bases. That A = T and G = C is known as the **equivalence rule.** This rule played a crucial role in the Watson-Crick model, since it suggested to them that A must always pair with T and G must always pair with C in DNA—a fact that placed restrictions on the possible structure.

The relative amounts of the four bases in the DNA of bacteria (as well as in other organisms) may vary greatly, as shown in Table 3-1, but regardless of the relative amounts the equivalence rule always applies. The base composition among all vertebrates (fishes, birds, cats, humans, and so on) has a narrower range of variation: about 40–44 percent, which is much less than the variation observed among bacterial species. The base composition of human DNA is 40 percent G–C base pairs and 60 percent A–T base pairs.

By 1950 the idea that molecules could assume a *helical* configuration was well known. (If you want to observe a helix, examine a coil spring or the threads on a screw.) Linus Pauling, a chemist at the California Institute of Technology, had demonstrated in the late 1940s that protein molecules, made of long chains of amino acids, are twisted into helical structures. Several years later, two English scientists, Maurice Wilkins and Rosalind Franklin, obtained x-ray photographs of purified crystals of DNA. If the pattern of spots on these x-ray films (Figure 3-2) could be interpreted, the physical structure of the DNA molecule could be deciphered. Watson and Crick examined the x-ray photographs obtained by Wilkins and Franklin and drew the following conclusions:

1. DNA is composed of two helical chains of molecules twisted together to form a double helix.

2. The width of the double helix is about 20 angstroms (1 angstrom = 10^{-10} meter, which is in the range of the radius of an atom).

3. There are ten base pairs per turn of the helix and the

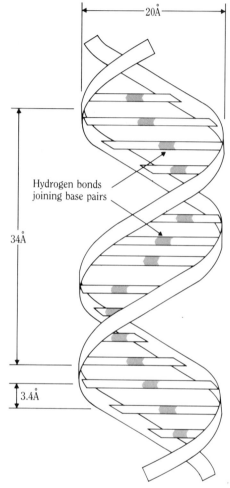

Figure 3-3 The Watson and Crick model for DNA. From the x-ray photographs they were able to deduce that the molecule is a double-helix with the dimensions indicated. The complementary base pairs are separated by 3.4Å (3.4 nm) along the chains, and one complete helical turn occurs every Å (3.4 nm).

20Å

Hydrogen bonds joining base pairs

34Å

3.4Å

helix makes one complete turn every 34 angstroms along the length of the DNA chain (Figure 3-3).

4. In the two chains of the helix, adenine must always appear opposite thymine and guanine opposite cytosine—that is, adenine and thymine are *complementary*, as are guanine and cytosine.

5. The stability of the two long helical chains is accounted for by the hydrogen bonds that connect the complementary bases along the entire length of the double helix.

It is the particular order of the bases in the two DNA chains that determines the genetic information in the DNA molecule; how this information is encoded and decoded is discussed in Chapter 6. However, some other important structural features of the DNA molecule should be mentioned at this point. The backbone of each helical chain consists of repeating units of sugar molecules (deoxyribose) and phosphates (PO_4^- groups). Also, each chain of the double helix has a defined chemical orientation; at one end of each chain, called the 3-prime (3') end, is a deoxyribose sugar, and at the other end of the chain, the 5-prime (5') end, is a phosphate group. The two chains of the helix are chemically *antiparallel* parallel but in opposite directions) as shown in Figure 3-4.

Not only were Watson and Crick able to figure out the structure of DNA, but they also pointed out how the genetic information could be accurately duplicated generation after generation in all organisms.

Functions of DNA: Replication

Each single chain of the DNA double helix actually contains all of the genetic information, since the bases in one chain determine what the bases in the other chain must be. **Replication** of the genetic information is accomplished by synthesizing two new chains of DNA in which bases pair up with their complementary bases in the existing chains. In this way replication gives rise to two double-helical DNA molecules that are informationally identical because they contain the same sequence of bases as the original DNA molecule.

Examination of the Watson-Crick model (Figures 3-3 and 3-4) leads to the prediction that DNA molecules must replicate in a *semiconservative* manner; that is, after each chromosome (DNA) duplication, newly replicated DNA molecules should contain one preexisting chain and one newly synthesized chain. Several years after the Watson-Crick structure of DNA was announced, Matthew Meselson and Franklin Stahl at the California Institute of Technology devised an experiment to test whether replication is indeed semiconservative. In theory, DNA could replicate in a conservative manner; that is, after replication the two original chains could remain hydrogen-bonded

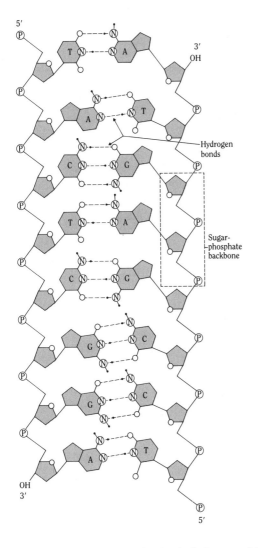

Figure 3-4 DNA chains joined by hydrogen bonds (dashed lines) between the complementary base pairs. Note that the sugar-phosphate backbones of the separate chains are in opposite (antiparallel) orientation.

together and the two newly synthesized chains would become partners. However, this does not occur, as the Meselson-Stahl experiment demonstrated.

When a solution of cesium chloride (CsCl)—a salt similar to sodium chloride but heavier—is centrifuged for a long time (24 hours or more), a density gradient is established in the tube, because the very high velocities (30,000 rpm) cause more of the salt molecules to be pulled to the bottom of the tube. If DNA is added to the CsCl solution before the centrifugation, it, too, will sink in the salt solution, eventually settling and forming a narrow band at the position in the gradient where its density precisely matches the density of the salt solution. (The principle is the same as that governing the level at which your body floats. In fresh water you usually sink below the water because the weight of water you displace is less than the weight of your body.

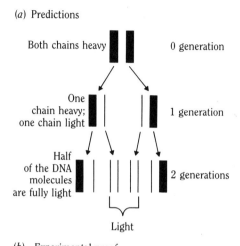

(a) Predictions

Both chains heavy — 0 generation

One chain heavy; one chain light — 1 generation

Half of the DNA molecules are fully light — 2 generations

Light

(b) Experimental proof

Generations

0   1.0   1.9   3.0   4.1

DNA bands in CsCl gradient

Heavy   Intermediate density   Light

Figure 3-5 The Meselson-Stahl experiment. This experiment confirmed the prediction of the Watson-Crick model that DNA replicates by a semiconservative mechanism. After each generation, one old chain of DNA is joined to a newly synthesized chain. When the bacterial DNA was analyzed by centrifugation the heavy, intermediate, and light DNA molecules could be visualized.

But if there is salt in the water, as there is in sea water, part of your body floats above water because salt is heavier than water. And the saltier the water, the higher you float, because the submerged part of your body displaces an amount of salt water equal to the weight of your entire body.)

The Meselson-Stahl experiment was designed to produce "heavy" DNA by growing *E. coli* bacteria in a medium containing a stable, heavy isotope of nitrogen (^{15}N). This is referred to as "labeling" the DNA. The more abundant isotope of nitrogen (^{14}N) is "light" because it contains one neutron less than ^{15}N. Thus, it follows that any molecules synthesized with ^{15}N atoms will be heavier than those synthesized with ^{14}N. After many generations of growth in a medium containing $^{15}NH_4Cl$ (heavy ammonium chloride), bacteria will all have DNA heavier than ordinary ^{14}N-containing DNA. The bacteria are then collected by centrifugation and the ^{15}N medium is removed. Growth of the bacteria is then continued in ^{14}N medium.

If DNA replicates in a semiconservative manner, after one generation of growth in the light medium all of the DNA molecules in the bacteria should be of hybrid, or intermediate, density because all should consist of one heavy (old) chain and one light (new) chain (Figure 3-5a). After two generations of growth, half of the DNA molecules should be of hybrid density and half should be fully light. The results of the Meselson-Stahl experiment confirmed these predictions (Figure 3-5b): Bacterial DNA does indeed replicate semiconservatively, as Watson and Crick suggested; because of complementary base-pairing, one preexisting chain is joined to the newly synthesized (complementary) chain. Since this experiment was performed, it has been shown that semiconservative DNA replication occurs in cells of plants and animals as well as bacteria. We now know that the Watson-Crick model for the structure of DNA and its replication is valid for all organisms.

Functions of DNA: Mutation

Another important property of DNA is its ability to change informational content. The word **mutation** means change, as described in Chapter 2; any change in the sequence of the bases in DNA constitutes a mutation because it changes the genetic information contained in the DNA molecule. Changing even a single base pair in the hundreds of thousands of bases in a DNA molecule can affect—sometimes fatally—the phenotype of a bacterium, plant, or animal.

An examination of the Watson-Crick model of DNA helps us see how, in the normal process of DNA replication, mutations may occur. Mutations arise whenever a mistake occurs in the replication mechanism. This mechanism normally ensures the pairing of A with T and G with C. However, no process, includ-

Table 3-2. Estimates of Mutation Rates for Some Human Genes

Phenotype and description	Mutations (average per million genes per generation)
Hemophilia A—severe bleeding due to lack of factor VIII in blood	44
Hemophilia B—mild bleeding due to lack of factor IX in blood	2–3
Achondroplasia—dwarfism	10
Retinoblastoma—cancer of the retina of one or both eyes	8
Duchenne muscular dystrophy—progressive degeneration of muscle tissue	67

ing DNA replication, is 100 percent accurate. Mistakes in the pairing of bases during DNA replication are extremely rare; in bacteria, for example, mistakes in base-pairing (mutation rates) are usually on the order of 1 in 10 million generations or less for any gene. Mutation rates in humans are more difficult to estimate because of the long human generation time and small human population size relative to bacteria. However, mutation rates have been measured for some human genes; they are listed in Table 3-2. Many environmental chemicals and ionizing radiation such as x-rays or radioactivity can increase the likelihood that DNA will be damaged and undergo a change in one or more base pairs. How mutations arise and how they are corrected are discussed in Chapter 7.

WHAT CAUSES MUTATIONS?

Do mutations occur randomly and spontaneously during DNA replication, or are they caused by specific interactions between an organism and something in its environment? Scientists have debated for centuries whether or not traits acquired by an individual can be passed on to its progeny, a phenomenon referred to as **Lamarckianism** (see Box 3-1). To some people it seems obvious that their learned talents and acquired tastes will be passed on to their sons and daughters—that learned traits can be inherited by succeeding generations. Many nineteenth-century scientists observed that the children of musicians tended to be musicians, the sons of athletes were usually athletic, and the children of wealthy parents generally became wealthy themselves. Thus, some scientists argued that acquired traits could be inherited. How various people have answered this question at different times has had serious consequences in the management of human diseases. Some of their answers have been used to support erroneous notions about race and intelli-

How New Species and Organisms Arise

BOX 3-1

Why do certain species of plants and animals become extinct? How could complex animals have evolved from simpler ones? How did the enormous variety of life forms on earth arise? These questions have intrigued philosophers and scientists ever since they first began to systematically study nature.

In the early 1800s, a renowned French zoologist, Jean Baptiste Lamarck, put forth the idea that animals become better adapted to their environment with each generation, eventually giving rise to new types and new species. Lamarck's basic idea is summed up in what he called the law of inheritance of acquired characteristics: "All that has been acquired or altered in the organization of individuals during their life is preserved and transmitted to new individuals who proceed from those who have undergone these changes."

Lamarck used this idea to explain in ingenious ways the origin of animal characteristics. For example, he suggested that webbed feet in aquatic birds arose because they spread their toes in order to swim faster to catch fish. Gradually the skin between the toes stretched to form a web, and this useful acquired character was passed on to progeny. Snakes sup-

posedly arose from animals that were accustomed to crawling under bushes and along the ground in order to hide themselves from predators. Eventually, their legs became useless, and subsequent generations were born without legs. Moles gradually lost their eyes because eyes were of no use in their underground environment.

The classic Lamarckian example of an acquired trait is the giraffe's long neck. Lamarck reasoned that because giraffes lived in arid regions where there was often little grass to graze, they were forced to eat the leaves of trees. This resulted in their having to stretch their necks to reach higher and higher leaves. The higher the animals had to stretch to obtain leaves, the longer their necks became, and elongated necks were then passed on to progeny, until finally the giraffes' necks became long enough to reach leaves even in the tallest trees. While these examples may seem foolish today, Lamarck was actually a very astute scientist and must be credited with one of the first rational attempts to account for the evolution of organisms.

It was Charles Darwin, born in 1809, the same year Lamarck's

book was published, who eventually conceived of the idea of natural selection and "survival of the fittest." Neither Lamarck nor Darwin knew anything about the mechanisms of inheritance. Yet from years of careful observations of the characteristics of plants and animals, Darwin was able to deduce the basic mechanism of evolution. Experiments have shown that Darwin was right and Lamarck was wrong. The fundamental ideas of natural selection cannot be expressed more clearly than in Darwin's own words:

As many more individuals of each species are born than possibly can survive, and, as a consequence, there is a frequently recurring struggle for existence, it follows that any being, if it varies however slightly in any manner profitable to itself, under the complex and sometimes varying conditions of life, will have a better chance of surviving, and thus be naturally selected. From the standpoint of inheritance, any selected variety will tend to propagate its new and modified form.

Reading: R. Lewin, "Lamarck Will not Lie Down." *Science* 213, (1981).

gence and to justify prejudices and existing social structures (see Chapter 19 for further discussion of these issues). However, if the environment does not specifically direct where mutations occur in DNA, then acquired traits *cannot* be inherited.

That mutations do occur spontaneously and randomly in the DNA of bacteria was demonstrated in 1943 in an experiment carried out by Max Delbrück and Salvador Luria. Their experi-

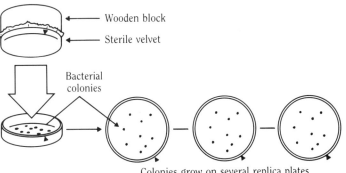

Figure 3-6 Replica plating. A velvet pad is used to transfer bacteria from colonies on the original Petri plate to several replicas. If the orientation of the wooden block is not changed, bacterial colonies will grow up on the replica plates in the same position as on the original plate.

Colonies grow on several replica plates

ment involved a rather sophisticated statistical analysis of the number of bacteria that became resistant to infection by T-phages. However, their experiment did not convince many people at the time that exposing bacteria to phages has no bearing whatsoever on mutations to phage resistance. A few years later, however, simpler experiments showed unequivocally that mutations do arise by chance in bacteria.

As discussed in Chapter 2, *E. coli* can be infected and lysed by virulent T-phages. If a sufficiently large number of bacteria are spread onto an agar medium in a Petri plate that contains millions of T-phages, almost all of the bacteria will be killed. However, a few isolated bacterial colonies will grow up on the surface of the medium. All these bacteria—and all of their descendants—are resistant to infection by T-phages, suggesting that a heritable change, a mutation, has occurred in the DNA of the bacteria. Did the phage-resistant bacteria arise because they were exposed to the T-phages, or did the mutations that fostered resistance arise by chance in the DNA *before* the bacteria were exposed to the T-phages? If mutant organisms arise by chance and are better adapted in a particular environment as a consequence of their altered phenotype, it cannot be argued that the environment caused the mutation or the change (see Box 3-2).

In 1952 Esther and Joshua Lederberg developed a simple replica plating technique (outlined in Figure 3-6) that proved bacterial mutants arise by chance. In this technique, several hundred bacteria are spread on the surface of agar medium in a Petri plate and are incubated overnight. By the next day each bacterium has grown into a visible colony containing more than a million bacteria. Samples of the bacteria in each colony are then transferred via a velvet-covered wooden block to several Petri plates that contain millions of T-phages. (These plates are called *replica plates* because they are used to *replicate* the bacterial colonies.) Some bacteria from each colony stick to the threads of the velvet and thus are transferred to each of the plates. Because of the millions of T-phages on the replica plates,

Determining Cause and Effect in Scientific Experiments

BOX 3-2

How many times have you heard statements such as "I dropped the dish because I was frightened by that loud noise"; "The baby's constant crying made me furious"; "My car crashed because the pavement was wet"; "I got sick because of the infection." All such statements suggest that certain observable *effects*—the broken dish, anger, car crash, or sickness—are *caused* by something in the environment. Cause-effect relationships also suggest a temporal connection: one event (the cause) occurs before the other (the effect). If you think about the above statements, you will realize that the effects are not quite so simply related to the apparent causes. For example, wet pavements don't cause accidents but may be a contributing factor, along with carelessness, slick tires, drunkenness, poor eyesight, and so forth.

Scientists have the difficult task of sorting out cause-and-effect relationships in their experiments. They often find it difficult to ascertain the actual cause of the phenomena they measure or observe, and the logic that leads them to attribute causality to this or that factor may sometimes be faulty.

An often-told tale makes this point more clearly. Once upon a time, a scientist wanted to study the physiology of the jumping behavior of fleas. He first devoted a great deal of time and effort to training a particularly adept flea

to jump through a tiny hoop whenever he gave the verbal command, "Jump." After weeks of training, the flea was able to jump through the hoop every time the scientist said, "Jump." Once he had successfully trained the flea, he pulled one leg from the flea and recorded the number of successful jumps the flea made whenever he gave the command. Even with only five legs the flea scored 10 for 10. The scientist then removed a second leg and again commanded the flea to jump. Now the score fell to only 9 successful jumps in 10 attempts. The scientist removed a third leg, and the flea's success rate dropped to 7 out of 10. Even with only one leg, the flea managed to jump through the hoop 3 times out of 10. Finally, the last leg was removed from the flea. The scientist repeatedly commanded the flea to jump, but to no avail. The flea did not move. No matter how many times the scientist shouted, "Jump," the flea did not respond.

The scientist wrote up his conclusions as follows: "Removing the legs one by one from a flea that had been trained to jump when given a verbal command caused the flea to become progressively hard of hearing. After removal of all the legs, the flea became completely deaf."

As another example of how scientists must be cautious in concluding that one thing causes

another, consider the disorder of high blood pressure, which affects millions of people. Various causes of high blood pressure have been proposed: stress, obesity, cigarette smoking, even genetic factors. But is it not possible that it is the high blood pressure in the first place that causes people to overeat, to smoke, or to feel stressed, rather than the other way around?

Designing and carrying out experiments that give unambiguous and truthful answers is not easy. As the story of the flea suggests, results of experiments can be misinterpreted. Unfortunately, this has sometimes been true in the application of genetics to human problems. It is important in genetic experiments to always be critical and to carefully consider all possible cause-effect relationships, particularly before deciding that genes are the cause of disease, aggression, laziness, or mental retardation. The controversy over the genetic basis of intelligence and the horrifying consequences of the idea of *eugenics* (the selective breeding of people with desirable traits) that culminated in the Nazi genocide programs in World War II are but two examples of how genetic ideas affect individuals and society. Genetic experiments were not always interpreted and applied correctly in the past, and, as we will see later in the book, certain abuses of genetics have not yet disappeared (discussed in Chapter 19).

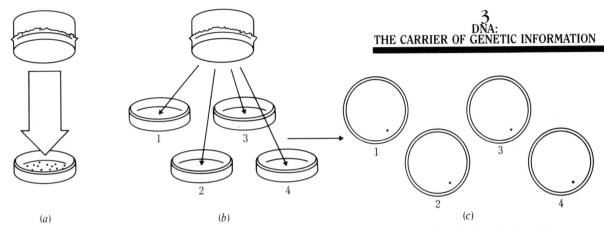

(a)   (b)   (c)

Figure 3-7 The Leberbergs' replica plating experiment. (a) A velvet-covered block is pressed against plate containing many bacterial colonies. (b) The block is then pressed against each of several replica plates, all of which contain millions of T-phages. This transfers some of the bacteria from each of the colonies on the original plate to each of the replica plates. (c) After incubation, a single bacterial colony has grown in the same position on each replica plate. This experiment proves that bacterial resistance to infection by T-phages occurs by chance and is not caused by exposure to phages.

almost all of the bacteria are killed. However, as shown in Figure 3-7, one bacterial colony does appear on the replica plates and invariably grows in the same position on each replica plate; this position corresponds to that of a bacterial colony present on the original plate that was not exposed to the T-phages. The only conclusion that can be drawn is that phage-resistant bacteria already existed in the original colony, were picked up by the velvet, and were transferred to the replica plates, where they subsequently grew to form new colonies. The mutation to phage resistance must have arisen in that bacterial colony during growth of cells on the original Petri plate *before* the bacteria were exposed to the T-phages.

It might still be argued that the mutation occurs by chance on the replica plates *after* the bacteria are exposed to the phages. But then one must ask why the phage-resistant colonies always appear in the *identical position,* corresponding to one particular colony on the original plate. If the experiment is repeated with a number of different original plates, phage-resistant colonies always appear in a spot on the replica plates that corresponds to a particular colony (or colonies) on the original plate, and that spot will vary for each set of experiments. The only reasonable conclusion is that as the bacteria grow on the original plates, a mutation occasionally arises in the DNA that makes that bacterium and all of its descendants in the colony resistant to infection. When these resistant bacteria are subsequently exposed to the T-phages on the replica plates, they are the only ones able to grow and reproduce.

The replica plating experiment can be performed with any microorganism, including eucaryotic cells such as yeast. The conclusion that mutations arise randomly and by chance in DNA molecules has been found to hold true for any trait that can be tested in this fashion, such as resistance to phages, resistance to antibiotics, or occurrence of nutritional deficiencies. We now know that mutations occur by chance in the DNA of all organisms.

CONSEQUENCES OF RANDOM MUTATIONS

Why is this experiment so important? Partly because it provides an explanation for some modern ecological disasters, such as the emergence of DDT-resistant mosquitoes, flies, and other insects. It also explains why organisms eventually become resistant to any pesticide. Mutant organisms continuously arise by chance in natural populations; in the right environment, the mutant types will flourish and the original nonmutant organisms will die out. The proof that mutant organisms arise spontaneously in populations also provides the mechanism needed to explain the principle of natural selection proposed by Darwin. It is a tribute to Darwin's genius that he was able to convincingly document his idea of natural selection even though he knew nothing about chromosomes and genes and had no knowledge of DNA or of mutations. Today, given the Watson-Crick model of DNA, it is easy to demonstrate how genetic errors give rise to organisms that may have a reproductive advantage in a particular environment.

The widely accepted theory of mutations and natural selection has replaced the less tenable ideas of the French scientist Jean Baptiste Lamarck (Box 3-1). However, the idea that acquired traits and not simply chance mutations may be inherited by future generations still attracts the attention of some scientists. In 1981 Edward Steele, an immunologist, provoked a scientific controversy by claiming that a particular kind of acquired immunity in mice (the immune system is discussed in Chapter 16) could be inherited by progeny mice in a Lamarckian fashion. Steele's results could not be repeated by other scientists, but the last word on the subject of Lamarckian inheritance has not yet been heard. Recent findings of "jumping genes" and other kinds of movable genetic elements (discussed in Chapter 4) still keep alive the possibility that some molecular mechanism may someday be discovered whereby an acquired trait could become encoded in a piece of RNA or DNA that might subsequently be incorporated into the chromosomes of sperm or eggs. Such a finding would not negate the fact that most evolutionary changes result from random mutations and natural selection; it would simply demonstrate another biological mechanism by which organisms can diversify and evolve.

Functions of DNA Recombination

If organisms are to survive in a rapidly changing environment, genetic diversity must exist among individual members of the population; in this way, there always will be individuals able to survive and reproduce under conditions that may be lethal for the majority. Mutation is the ultimate source of all genetic diversity, but the amount of genetic variability in a population can be increased enormously if new combinations of genetic information can be constructed from the already existing DNA of different individuals. The process of **recombination** accomplishes precisely this. DNA molecules are reconstructed by recombination so that new arrangements of genes and thereby new genetic information are passed from parents to progeny.

Genes located at the same position—a **genetic locus**—in a homologous chromosome pair are called alleles. For example, we mentioned earlier that bacteria can be either sensitive or resistant to infection by T-phages. A single gene in the bacterial DNA determines this sensitivity or resistance. Sensitivity is determined by the normal, or **wild-type** allele; phage resistance results if the DNA contains a mutant allele. In genetic notation, wild-type alleles, which are the ones commonly found in nature, are indicated by a (+) sign, whereas mutant alleles are indicated by the absence of this sign. Many genetic loci in plants and animals have more than two allelic states, which presumably arose by successive mutations. Multiple alleles persist in a population of individuals because they are all functional to a greater or lesser degree (the genetics of populations is discussed in Chapter 17).

A hypothetical bacterial example will help explain how recombination between DNA molecules increases genetic diversity and the likelihood of selecting a new bacterium. Imagine that bacteria are growing in a liquid medium to which are added two antibiotics, penicillin and streptomycin, both of which are able to kill the bacteria. Now, suppose that in the culture a mutant bacterium arises spontaneously that is resistant to penicillin; it will still be killed by the streptomycin. Conversely, if a mutant arises that is resistant to streptomycin, it will still be killed by the penicillin. What is the probability that mutations giving rise to resistance to both antibiotics will occur by chance in the same bacterium? The probability of two different mutations arising by chance in the same DNA molecule is the product of the separate probabilities of each event (the probability of a coin coming up heads three times in succession is $1/2 \times 1/2 \times 1/2$, or 1/8). If the mutation rate to penicillin resistance is one in a million and the same is true for resistance to streptomycin, then the product is a million times a million, or a thousand billion (written as $10^{-6} = 10^{-6} \times 10^{-12}$). Thus, it is highly unlikely that both mutations would occur in the same DNA molecule even in a population of billions of bacteria, an amount commonly grown in test tubes.

Now imagine that as the penicillin-resistant mutants arise, they are killed by the streptomycin and their DNA is released

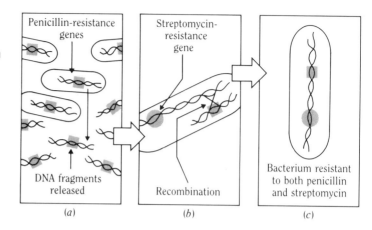

Figure 3-8 Recombination between DNA molecules can unite independently arising mutations in the same chromosome. (*a*) Mutations give rise to bacteria resistant to killing by penicillin. These cells are destroyed by streptomycin and fragments of DNA are released. (*b*) Other mutations give rise to streptomycin-resistant bacteria. If these mutants take up the penicillin-resistant DNA fragments, recombination will produce (*c*) a bacterium resistant to both antibiotics.

into the liquid culture. Recall from the *Pneumococcus* transformation experiment (Chapter 2) that bacteria can take up DNA from genetically different bacteria and recombine pieces of it into their own DNA. It is possible then, that the DNA carrying genes for penicillin resistance will be taken up by another mutant bacterium that is resistant to streptomycin (Figure 3-8). A single bacterium now contains genes conferring resistance to both antibiotics; it will survive, as will all of its progeny. Within a few hours, the culture will consist almost entirely of doubly mutant bacteria that are able to reproduce because they are resistant to both penicillin and streptomycin; the others are destroyed. The process of recombination thus helps produce new types of organisms that would be very unlikely to occur by mutation alone. Recombination is important in bacteria but even more important in diploid organisms, where the recombination that occurs during meiosis contributes significantly to genetic diversity.

DNA TRANSFER AMONG BACTERIA

Three mechanisms are used by bacteria to transfer pieces of DNA from one bacterium to another, possibly changing its genotype and phenotype (Figure 3-9). (Not all bacterial species use all three mechanisms, although *E. coli* does.) **Transformation**, as was mentioned in Chapter 2, involves the incorporation of a fragment of DNA into a bacterium, usually followed by recombination of the DNA fragment into the recipient chromosome. **Transduction** accomplishes the same thing, but in this process a piece of DNA is transferred from one bacterium to another by means of a phage. In transduction, pieces of bacterial DNA are packaged by chance into phages that are growing in an infected bacterium. After the infected bacterium lyses, these transducing phages may infect other bacteria and transfer bacterial genes, which become integrated into the chromosome by recombina-

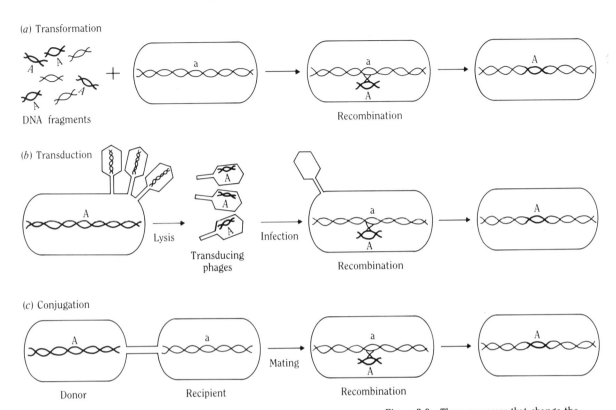

(a) Transformation

DNA fragments

Recombination

(b) Transduction

Lysis

Transducing phages

Infection

Recombination

(c) Conjugation

Donor

Recipient

Mating

Recombination

Figure 3-9 Three processes that change the genotypes and phenotypes of bacteria. (a) In transformation, fragments of DNA are incorporated into the chromosome of recipient cells. (b) In transduction, fragments of DNA are transferred from one bacterium to another by phages. (c) In conjugation, male and female bacteria physically unite. DNA from the male donor is transferred to the female recipient.

tion. Transducing phages do not kill the cell they infect, since they have accidentally packaged bacterial DNA instead of the phage DNA and thus carry few if any phage genes. **Conjugation** is the third process by which DNA can be transferred from one bacterium to another. Some strains of bacteria can attach to and mate with bacteria of the same or different species and thereby transfer DNA from cell to cell. (Sexual conjugation between bacteria, which is discussed in Chapter 4, should not be confused with the more complex sexual matings that occur in plants and animals.)

THE MECHANISM OF RECOMBINATION

Genetic recombination involves a physical exchange between two DNA molecules and can be detected by both physical and genetic techniques. The Watson-Crick model does not really provide any insight into the mechanism by which DNA molecules interact and exchange pieces, but a number of other models have been proposed over the years to explain how genetic recombination occurs in cells.

In 1961 M. Meselson and J. Weigle performed an experiment, using *lambda* (λ) phages, showing that recombination results from the actual physical breaking apart of DNA mole-

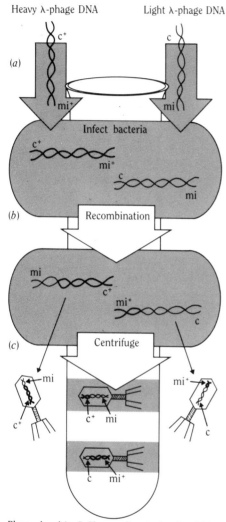

Phages band in CsCl according to density of DNA

Figure 3-10 Recombination as a result of the break-rejoin mechanism. (*a*) The DNA carried in wild-type λ-phages ($c^+\ mi^+$) is heavy; the DNA carried in mutant phages ($c\ mi$) is light. (*b*) When bacteria are infected, the light and heavy DNAs break and rejoin (recombine) to produce recombinant molecules and recombinant genotypes. (*c*) The recombinant phages can be separated because of their density differences and the genotypes identified by the phenotypes of the plaques formed by recombinant phages.

cules and the reassembly of pieces of DNA carrying different genetic information into recombinant DNA molecules of mixed parentage. This general mechanism for genetic recombination is referred to as the **break-rejoin model**. It is surprising that DNA molecules can be fragmented and put back together accurately so that no genetic information is lost. Yet this is what happens during genetic recombination in all cells—from recombination in the simplest bacterium to the formation of sperm and eggs in people.

The experiment showing that recombination occurs by a break-rejoin mechanism, like the Meselson-Stahl experiment, utilizes the density labeling of DNA to physically distinguish λ-phage DNA molecules grown in different cells. In this case, λ-phage DNA is labeled with two heavy isotopes, ^{13}C and ^{15}N, to make the DNA much heavier than the normal light DNA, which contains ^{12}C and ^{14}N (Figure 3-10). The density labeling of molecules is a valuable biological technique because it allows the physical separation and identification of particular molecules of DNA.

As Figure 3-10 shows, λ-phages carrying wild-type alleles of two different genes called *c* and *mi* are grown in the presence of the heavy isotopes so that all of the DNA in these wild-type phages is heavy. Mutant λ-phages are prepared in the same way except that light isotopes are used and the DNA of the mutant phages is of normal density. These phages carry mutations in two different genes: the *c* gene, which causes the phages to form clear plaques on the Petri plates (the wild-type plaques are cloudy), and the *mi* gene, which causes the phage plaques to be minute (smaller in size than wild-type). Both wild-type phages (carrying heavy DNA) and mutant phages (carrying light DNA) are then used to infect bacteria growing in light medium. Enough phages of both types are added to the bacterial culture that every bacterium becomes infected by at least one heavy and one light DNA molecule. As new λ-phages are synthesized inside the bacteria, all of the newly synthesized phage DNA will consist of light DNA, since the bacteria are growing in light medium. However, if recombination has occurred between the original infecting DNA molecules by a break-rejoin mechanism, then some λ-phages should contain pieces of light and heavy DNA and should have undergone recombination between genes *c* and *mi*. The recombinant phages should also show new phenotypes that can be seen by the kinds of plaques that are formed.

After the infected bacteria lyse, λ-phages of different densities are released, and those carrying heavy DNA are separated from the lighter ones by centrifugation in CsCl. Like DNAs of different densities, phages of different densities form separate bands in the CsCl density gradient. The phenotypes and genotypes of the λ-phages in each band can be determined by collect-

Table 3-3. Genotypes and Phenotypes of Original and Recombinant Lambda
Phages

Genotype	Phenotype of plaque
$c^+ mi^+$ (wild-type)	Cloudy plaque—normal size
$c\ mi$ (mutant)	Clear plaque—small size
$c\ mi^+$ (recombinant)	Clear plaque—normal size
$c^+ mi$ (recombinant)	Cloudy plaque—small size

ing the bands of separated phages and spreading samples of the phages on Petri plates to observe the phenotypes of the plaques produced (Table 3-3). By means of this experiment, heavy and light DNA molecules can be correlated with the presence of either mutant or wild-type alleles in the phage DNA. Since the *mi* gene is located at the extreme end of the λ-DNA, some recombinant DNA molecules that are formed from the original parental phage DNAs will be almost completely heavy and yet show a recombinant phenotype ($c^+ mi$).

The results of the λ-phage experiment agree with the predictions of a break-rejoin model of recombination and are not consistent with other models of recombination. During recombination, DNA molecules are broken apart and segments from genotypically different DNA molecules are rejoined to give new arrangements of alleles. These recombinant DNA molecules are subsequently replicated, giving rise to other DNA molecules, which may undergo additional recombination. Recombination occurs by breakage and rejoining of DNA in all organisms. Recombination of DNA molecules during meiosis in plants and animals increases the genetic diversity of these organisms as it does in phages and bacteria.

The molecular and chemical details of the three genetic processes of DNA described in this chapter—replication, mutation, and recombination—are extremely complex. Dozens of enzymes are involved in each process, and the mechanisms of action of all of these enzymes are still not completely understood, although some enzyme reactions have been studied extensively.

The Watson-Crick model for the structure of DNA provided the framework for understanding these three principal functions of DNA. As DNA came to be investigated and understood in more and more detail, it was discovered that extremely small pieces of DNA move from cell to cell and from organism to organism (see Box 3-3). Thirty years after the discovery of the structure of DNA, scientists are beginning to understand how small pieces of DNA move in and out of chromosomes, causing organisms to change their genotypes and phenotypes. Scientists and genetic

Viruses Are Small but Viroids Are Smaller

BOX 3-3

Until quite recently, viruses were believed to be the smallest agents that carry genetic information and that are able to reproduce inside cells. However, a new class of infectious agents called *viroids* have been discovered that cause serious diseases in plants, although it has yet to be demonstrated that they can infect animals.

In the 1920s, researchers described spindle-tuber potato disease, which causes potato plants to produce cracked, gnarled potatoes. A virus infection was thought to be responsible, but numerous efforts over the years failed to detect or isolate any virus in connection with the disease. In the 1960s scientists working at the U.S. Agricultural Research Center in Beltsville, Maryland were able to show that the infectious particle in diseased potatoes was much smaller than any pre-

viously isolated virus. They also showed that the agent causing the disease in potatoes was transmissible to many other species of plants. It was finally shown that the infectious agent is a tiny single strand of RNA—a viroid. By the 1970s a group of German scientists determined that the single RNA strand of the potato-spindle viroid is so small and simple that it is scarcely large enough to code for even a single protein. The viroid apparently does not code for any proteins; it simply uses existing cellular enzymes to replicate itself, destroying cells and whole plants in the process.

There are a few animal diseases that appear to be caused by

viruslike agents that scientists have so far been unable to isolate. *Scrapie* is a disease that kills sheep and goats; *kuru* is a rare disease of natives of New Guinea. Both diseases affect cells of the brain and nervous system and can be transmitted to other animals by injecting them with brain tissue from diseased individuals. Some animal diseases, then, may be caused by tiny molecules of RNA. How did these tiny molecules originate, and what was their original cellular function? As time goes on, you will undoubtedly hear more about this intriguing scientific detective story, as the plot thickens and the evidence accumulates.

Reading: Garrett Epps, "Viroids Among Us." *Science 81*, September 1981.

---

engineers are beginning to use these pieces of DNA to create useful new organisms and products. From a basic understanding of DNA has emerged a brand new biotechnology industry, and society is just beginning to appreciate the many possible applications (and the potential risks) that derive from being able to move small pieces of DNA from one organism to another.

Key Terms

bases   the chemical entities in DNA whose sequence encodes the genetic information: adenine (A), thymine (T), cytosine (C), and guanine (G).

equivalence rule   the rule that the number of adenine bases in all DNA molecules equals the thymine bases and the number of cytosine bases equals the guanine bases.

replication   the duplication of DNA molecules.

mutation   any change in one or more base pairs in DNA; a heritable change in the genetic information.

Lamarckianism   the inheritance of acquired traits; an idea named after the French scientist Jean Baptiste Lamarck.

replica plating   a technique in which bacteria are transferred to a series of Petri plates from an original plate by means of a velvet pad.

recombination   the breaking and rejoining of genetically different DNA molecules; the appearance of traits in progeny that were not observed in parents.

genetic locus   the specific location of a gene on a chromosome.

alleles   alternative functional states of the same gene occurring on homologous chromosomes.

wild-type alleles   alleles normally found in nature, as opposed to mutant alleles.

transformation   transfer of DNA from one bacterial strain into a genetically different strain.

transduction   transfer of pieces of DNA from one bacterium to another by means of a phage.

conjugation   mating between a male and female bacterium in which DNA is transferred from the male and incorporated into the DNA of the female by recombination.

break-rejoin model   model explaining how genetic recombination results from the physical exchange of segments of DNA.

_____ Additional Reading _____

Crick, F. H. C. "The Structure of the Hereditary Material." *Scientific American*, October 1954.

Epps, G. "Viroids Among Us." *Science 81*, September 1981.

Portugal, F. H. and J. S. Cohen. *A Century of DNA: A History of the Discovery of the Structure and Function of the Genetic Substance.* Cambridge, Mass.: MIT Press, 1979.

Van't Hof, J., et al. "Chromosome DNA Replication in Higher Plants," *BioScience*, June 1979.

Watson, J. D. *The Double Helix: A Personal Account of the Discovery of the Structure of DNA.* New York: Atheneum, 1968.

Watson, J. D. and F. H. C. Crick. "Molecular Structure of Nucleic Acids." *Nature*, April 25, 1953.

_____ Review Questions _____

1. What are the three classes of molecules used to construct DNA?

2. How many different bases are present in DNA molecules? How many different sugars?

3. What are the three basic functions performed by DNA?

4. What kind of molecule is involved in bacterial conjugation? in bacterial transformation?

5. What kind of molecule carries the genetic information in bacteria? in viruses? in viroids?

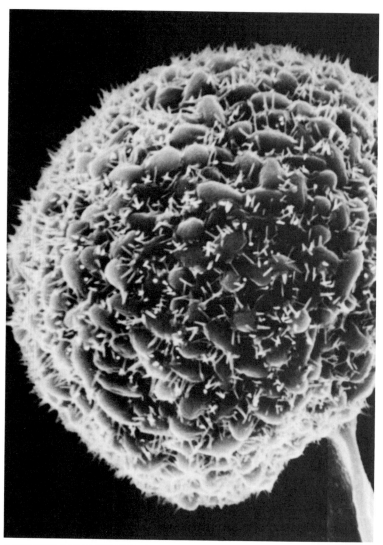

A *SPORANGIUM*, A SPORE-CONTAINING ORGAN IN MOLDS. (2,900 ×)

4

GENETIC INFORMATION

*How Genes
Are
Exchanged*

*One of the most difficult issues in science
is to decide when a particular phenomenon
is worth investigating.*
HERMAN BONDI

The biochemical functions in every cell in the human body as well as in other organisms are governed by information contained in DNA molecules. Not only does DNA contain all of the cell's genetic information, but this information is shared by the exchange of DNA or pieces of DNA with other cells. Among bacteria, DNA can be transferred to cells of the same species or other species. Because DNA molecules or pieces of DNA move from bacterium to bacterium, recombination—the exchange of pieces of DNA between chromosomes—generates new gene arrangements, thereby providing cells with new genotypes and new phenotypes. Even plant and animal cells are able to rearrange their genetic information as well as transmit it via sexual matings.

The arrangement of genes in DNA molecules can be changed by several genetic mechanisms. If by chance some new arrangement of genes proves to be advantageous to the cell or organism, that particular organization of genes will be maintained and may result in the survival and reproduction of both the revised DNA and of the new organisms. Chance events in nature cause the formation of new gene combinations—some of which, in turn, create organisms better adapted to their environments.

The transfer of DNA molecules from cell to cell and the incorporation of fragments of DNA from one organism into the chromosomes of another creates a biological paradox. On the one hand, DNA molecules must conserve and reproduce their genetic information with utmost fidelity; otherwise, the mutant organisms that arise may not survive and reproduce. Therefore, the mechanism of DNA duplication must be exceptionally accurate; any alterations or mistakes must be kept to a minimum. On the other hand, genetic diversity is essential to the survival of individual organisms and ultimately to populations of organisms, so the genetic information in DNA molecules must be susceptible to change. The need for genetic information to be preserved and the necessity for it to change constitutes a biological paradox that is resolved by the unique structure and properties of DNA molecules.

Environments can change rapidly. Bacteria growing in soil may have to survive without water in times of severe drought, while a sudden rainfall may inundate them with water, stimulating growth. Bacteria that live in an animal's stomach experience changing environments each time the animal eats. Bacteria have evolved genetic mechanisms that allow them to cope with these "feast or famine" environmental conditions. Macroscopic environments may also change in a more or less permanent fashion. For example, vast areas of the earth that were once covered with water are now barren desert. Over the billions of years of geological history, the temperatures of large regions of the earth's surface have fluctuated from cold that produced ice ages to warmth sufficient to sustain tropical climates. Changing environments have caused many millions of species of plants and animals to become extinct, as is demonstrated by the fossil record (see Chapter 18).

The enormous variety of organisms that are alive today and the even greater numbers of species that have become extinct suggest that DNA must continually change if organisms are to adapt and survive. New genes and combinations of genes are necessary if organisms are to flourish and new species are to evolve. Therein lies the paradox. How can a DNA molecule strictly maintain the accuracy of its information and yet still change often enough to provide new genetic information that will allow organisms to adapt to changing environments? This chapter discusses how the structure of DNA permits the solution of this paradox by means of ingenious genetic mechanisms.

Bacterial Conjugation
as a Means of Genetic Exchange

Nobody had thought of bacteria as sexual organisms until 1946 when Joshua Lederberg and Edward Tatum showed that genetically different strains of *E. coli* could mate with each other in the process called **conjugation**. It was fortunate for Lederberg and Tatum that they happened to choose strains of *E. coli* that are able to exchange DNA molecules upon cell–cell contact, since not all strains of *E. coli* and not all species of bacteria are able to conjugate with one another and thereby exchange pieces of DNA.

In the original experiment carried out by Lederberg and Tatum, the two bacterial strains each had two different nutritional requirements that resulted from mutations in different genes (Figure 4-1). When the strains were grown separately, the mediums had to be supplemented with the nutrients methionine and biotin (for strain A) or leucine and threonine (for strain B). When the two strains were mixed together, the medium had to be supplemented with all four nutrients in addition to the required minerals and glucose. When both strains were grown in the mixed culture, some of the bacteria exchanged genetic information, producing a few wild-type bacteria having no nutritional requirements. These relatively rare recombinant bacteria could be detected by spreading samples onto Petri plates that did not contain any of the four nutritional supplements. Under this condition neither of the original mutant bacterial strains could grow, but wild-type recombinants did grow, eventually producing visible colonies. By recombining appropriate pieces of DNA, these bacteria no longer carried the original mutations but had become wild-type for all genes essential for growth on the minimal medium.

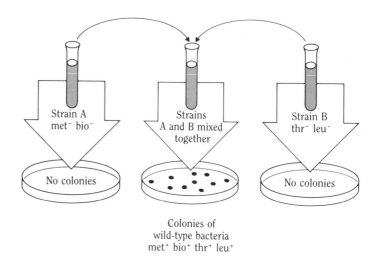

Figure 4-1 Recombination in bacteria. Strain A carries two mutations in genes that are necessary for synthesis of the amino acid methionine and the vitamin biotin. Strain B carries two mutations in genes necessary for the synthesis of the amino acids threonine and leucine. Wild-type bacteria are produced when the two mutant strains are grown together.

Where did the wild-type bacteria come from? One strain of *E. coli* (the male) occasionally transfers some of its DNA to the other bacterial strain (the female) through conjugation (Figure 4-2*a*). Once the DNA from the male enters the female bacterium, recombination may occur between the two DNA molecules, eventually giving rise to a bacterium that contains all wild-type genes (Figure 4-2*b*). Thus, in bacteria, conjugation and recombination provide a mechanism for generating genetic diversity. In a population of hundreds of millions of *E. coli* cells, some of which may be mutant, many new gene combinations can be produced by conjugation and recombination. (Of course, many other kinds of recombination also occur that are not detected by this particular type of experiment.)

Plasmids Determine the Sex of Bacteria

Soon after the initial discovery that bacteria can conjugate and recombine their DNA molecules, a variety of related genetic and biochemical experiments led to the discovery that what distinguishes male from female bacteria is the presence in the male of a small, circular DNA molecule called the F factor (fertility factor), or **F plasmid**. Any bacterium containing this F plasmid has the male phenotype and can transfer DNA; any bacterium without the F plasmid has the female phenotype and can only accept DNA. The F plasmid contributes nothing to the growth and survival of the individual bacterium that contains it; male and female bacteria grow equally well. It is the F plasmid, however, that enables the male cell to attach to and conjugate with the female cell and then transfer all or part of the male DNA, which may include chromosomal genes as well as the genes contained in the F plasmid.

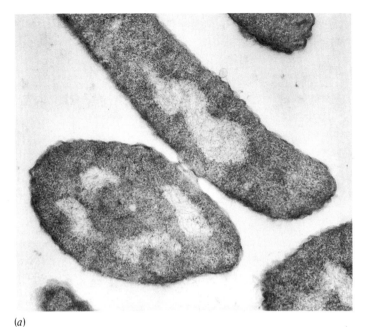

(a)

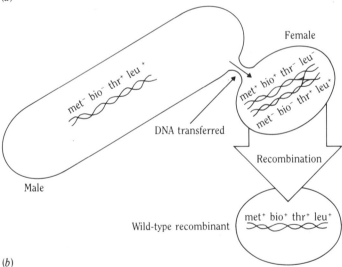

Female

met⁻ bio⁺ thr⁻ leu⁻

met⁺ bio⁺ thr⁻ leu⁻

met⁻ bio⁻ thr⁺ leu⁺

DNA transferred

met⁻ bio⁻ thr⁺ leu⁺

Male

Recombination

Wild-type recombinant

met⁺ bio⁺ thr⁺ leu⁺

(b)

Figure 4-2 How conjugation produces recombinant bacteria. (*a*) Male and female bacteria physically join and DNA is transferred from the male to the female. (*b*) Following transfer, the DNA of the male may be recombined into the DNA of the female, producing a wild-type DNA molecule. The male also survives, since a newly replicated DNA molecule is actually transferred to the female cell. (Photo courtesy of David P. Allison, Oak Ridge National Laboratory.)

The F plasmid DNA contains about 100 genes; the functions of all but a dozen or so are still unknown. The F plasmid genes that *have* been identified are involved in plasmid replication and transfer, including synthesis of the organ of attachment of the male to the female bacterium—a specialized appendage called a *conjugation tube,* or sex **pilus** (plural *pili*). Other genes carried on the F plasmid are responsible for the physical transfer of the donor's DNA into the female bacterium. Many other types of plasmids in bacteria carry genes that perform many different functions, some of which cause human diseases (see Box 4-1).

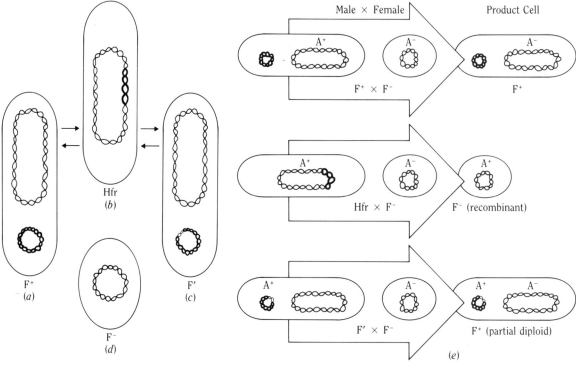

Figure 4-3 The F plasmid determines three types of male cells and one female. (a) An independent F plasmid determines a male cell. During conjugation only the F plasmid DNA is transferred to the female. (b) Occasionally, the plasmid DNA is recombined into the chromosome, where it is replicated along with other chromosomal genes. This type of male is called Hfr. During conjugation, part or all of the male's chromosome is transferred and recombination between any genes of the male and female is possible. (c) The F′ plasmid is occasionally removed from the chromosome of an Hfr male and carries with it a segment of chromosomal DNA. During conjugation, the chromosomal genes are transferred with the F plasmid. (d) Female cells (F⁻) lack the F plasmid. (e) Female cells may mate with any of the three male types to produce the product cell as indicated.

GENETICALLY DIFFERENT MALE
BACTERIA CREATED BY THE F PLASMID

It is possible to distinguish three genetically different kinds of male *E. coli* (Figure 4-3). One type of male, called F^+, transfers only the F plasmid to all the female (F^-) bacteria it conjugates with, and in the process, the females are converted into males. When DNA is transferred from a male to a female bacterium, the male cells do not lose the F plasmid or chromosomal DNA, because what is actually transferred is a newly replicated strand of the DNA. Because no chromosomal genes are transferred from the male to the female during conjugation between F^+ and F^- cells, no recombination for chromosomal genes is observed.

About one in every million F plasmids becomes integrated into the male's chromosomal DNA, producing another type of male called **Hfr** (high frequency of recombination). When Hfr bacteria conjugate with female bacteria, part or all of the male chromosomal DNA is transferred to the female along with the F plasmid, which still initiates the conjugation and transfer of DNA. The designation "Hfr" comes from the fact that recombination occurs frequently between the donated chromosomal DNA fragment and the corresponding segment in the female's chromosome.

Finally, the F plasmid can become dissociated from the chromosome in an Hfr bacterium by the same recombination

Bacterial Plasmids and Human Diseases

BOX 4-1

Plasmid DNA provides a powerful mechanism for changing the genetic information of bacteria. It benefits bacteria because in certain environments such changes may promote the survival and evolution of the bacteria. But what's good for bacteria may not be good for people. Travelers to foreign countries, especially in tropical climates that provide warm, moist environments in which bacteria thrive, frequently suffer from diarrhea, which can cause not only temporary discomfort but also severe illness. Severe diarrhea results in depletion of body fluids and loss of essential salts. If the fluids and salts are not replaced, the person may die.

Recently, it has been shown that many diarrhea victims are carrying in their digestive systems strains of *E. coli* that contain plasmids, some of whose genes direct the synthesis of the toxins that cause diarrhea. People born and raised in areas where the toxin-producing bacterial strains

occur probably develop an immunity to the toxins as a result of exposure to the bacterial strains. But visitors to these areas are sensitive to the toxins and thus suffer diarrhea.

Interestingly, the plasmid-produced toxin is related to the more dangerous toxin produced by the cholera-causing bacterium *Vibrio cholerae*. In cholera-causing bacteria, the toxin gene is not located on a plasmid but is carried in the chromosome. This suggests that in the distant past the *E. coli* plasmid containing the diarrhea-toxin gene might have been transferred to *Vibrio* bacteria. If that occurred, the toxin gene could have been recombined into the chromosomal DNA and evolved into the gene that produces the cholera toxin.

Occasionally, *E. coli* enter the human bloodstream, causing the destruction of blood cells and thereby a disease known as *hemolytic anemia*. These *E. coli*

strains have been found to possess another plasmid that carries a gene directing the production of *hemolysins,* proteins that cause the destruction of red blood cells. Still other plasmids discovered in *E. coli* and related bacteria have genes that direct the synthesis of a variety of appendages analogous to the sex pilus coded for by the F plasmid. These piluslike appendages help the bacteria attach to cells in the lining of animal stomachs and intestines. In this way bacteria are able to invade areas of the body that are not their customary habitats. Although this ability to populate new areas of the body may benefit the bacteria, it may also cause the animal to become quite sick. In short, plasmids provide one mechanism by which bacteria exchange genetic information and continuously adapt to new environments, sometimes at the expense of their hosts.

Reading: Richard P. Novick, "Plasmids." *Scientific American*, December 1980.

---

mechanism that permitted it to integrate in the first place. Again with low probability, some adjacent bacterial genes may remain attached to the F plasmid DNA when it exits from the bacterial chromosome. The male bacteria formed in this process are called *F prime* (F′) strains; they carry one or several chromosomal genes on the F prime plasmid, as shown in Figure 4-3c. When F prime male bacteria mate with females, the bacterial chromosomal genes carried on the F′ plasmid are transferred to the female bacteria each time the F plasmid itself is transferred.

The movement of the F plasmid into and out of the bacterial chromosome is a rare event, comparable in frequency to that of mutations. But it does guarantee the kind of genetic diversity that could be important for the survival of a population of bacterial cells in rapidly changing environments. (The importance of plasmids in bacterial evolution has not been firmly established, however.)

ISOLATING THE F PLASMID

If such a piece of DNA as a plasmid really exists, how can it be isolated? Because the F plasmid DNA in *E. coli* has the same base composition as chromosomal DNA, the two DNAs cannot easily be separated from each other. That is, the relative amounts of adenine-thymine pairs and guanine-cytosine pairs in the chromosome and plasmid DNAs are the same. Therefore, the chromosomal and plasmid DNAs of a particular bacterial strain have identical weights and cannot be separated by density centrifugation techniques such as were used in the Meselson-Stahl experiment (Chapter 3).

The F plasmid was eventually isolated from male (F$^+$) strains by transferring the plasmid from *E. coli* into another bacterial species, *Serratia marcescens*, whose base composition is different enough from *E. coli*'s that their DNAs can be physically separated from one another by density gradient centrifugation (see Table 3-1 for base compositions of bacteria). After the F plasmid was transferred from *E. coli* to *S. marcescens*, all DNA from the *S. marcescens* was centrifuged in cesium chloride solution, producing the DNA banding patterns shown in Figure 4-4. Now that the plasmid DNA molecules had been isolated, they could be physically separated from the other DNA and analyzed. The surprising finding of this analysis was that the F plasmid DNA is a single closed circular DNA molecule, something like a rubber band (Figure 4-5).

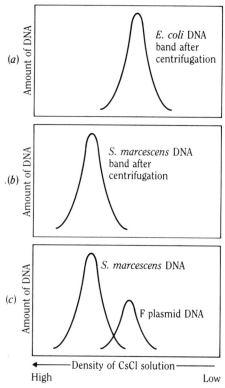

Figure 4-4 Isolation of F plasmids. (*a*) DNA from *E. coli* forms a band in a density gradient solution that corresponds to the base composition of its DNA. (*b*) DNA from *S. marcescens* has a different base composition and forms a band at a different position. (*c*) After mating F$^+$ strains of *E. coli* with *S. marcescens*, the F plasmid DNA is separated and can be isolated.

USING Hfr BACTERIA TO CONSTRUCT GENETIC MAPS

The mating of Hfr male bacteria with female bacteria can be used to locate the position of genes on *E. coli* DNA. This allows the construction of a **genetic map**, a diagram showing the location of genes in the bacterial chromosome (genetic maps can also be constructed for the chromosomes of other organisms, as we will see in Chapter 11). One way to conceptualize a genetic map is to think of genes as if they were towns located at different points along a highway (chromosome). On this linear genetic map (highway), each gene (town) has a name and the distance between genes (towns) varies.

When an Hfr bacterium conjugates with a female bacterium, the male's DNA is replicated and a strand of DNA is transferred, beginning at the site of F plasmid integration in the chromosome. Each male gene is transferred to the female at a particular time after the mating process begins, and the elapsed time can be measured because the mating can be interrupted at any stage by agitating the bacterial mixture in a kitchen blender

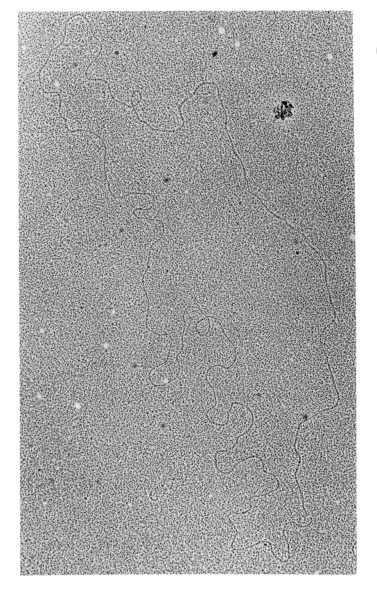

Figure 4-5 Electron micrograph of a circular DNA plasmid from a bacterium that causes tumors in plants. (Courtesy of C. I. Kado and R. Tait, University of California, Davis.)

(Figure 4-6). Only those genes that have been transferred to the female at the time the bacteria are physically separated can produce recombinant bacteria. Such recombinant strains are detected by spreading a sample of the mated bacteria onto mediums where only particular recombinant types are able to form colonies. As the time of entry of each gene is determined, its position can be located on the genetic map. It requires approximately 100 minutes for the entire male bacterial chromosome to be transferred to the female, so one can imagine traveling a highway 100 miles long and mapping the towns on it according to the time it takes to get from town to town.

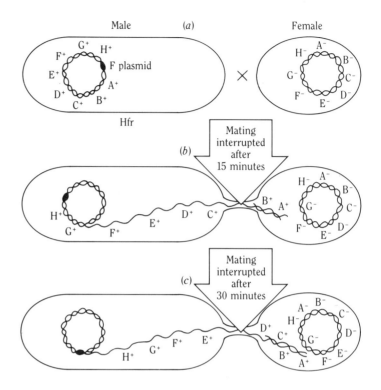

Figure 4-6 The interrupted mating technique for constructing the genetic map of *E. coli*. (*a*) Wild-type Hfr bacteria are mixed with female cells. (*b*) The Hfr chromosome is replicated and is transferred to the female cell. The mating can be interrupted at various times, and timing of wild-type gene transfer can be measured by detecting recombinant bacterial colonies on Petri plates. (*c*) The longer the mating continues, the more of the male chromosome is transferred and the greater the variety of recombinants that can be detected.

Bacteria are normally able to grow and divide every 30 minutes under optimal laboratory conditions, yet as we have just seen, they take at least 100 minutes to mate and transfer DNA. This observation led William Hayes, one of the discoverers of Hfr bacteria, to make the humorous observation that *E. coli* is the only organism in nature that engages in the sex act three times longer than its normal lifespan.

As more and more *E. coli* mutants became isolated and positioned on the genetic map using Hfr strains, researchers realized that every bacterial gene is linked to every other gene, and consequently the genetic map, and hence the chromosome, appears to have no ends (Figure 4-7). This discovery led some *E. coli* geneticists to suggest that chromosomes in bacteria are circular DNA molecules like the smaller plasmids. In fact, bacterial chromosomes were guessed to be circular before the circular structure of plasmid DNA was proven.

The fact that the bacterial chromosome, like the F$^+$ plasmid, is a physically closed loop of DNA was later confirmed by electron micrographs of DNA molecules (Figure 4-8). Chloroplasts (organelles in plant cells) and mitochrondria (organelles in both plant and animal cells) also contain physically closed circles of DNA. The enormously long DNA molecules in human chromosomes have not been proven to be physically circular, but they do appear to be continuous, unbroken threads of DNA,

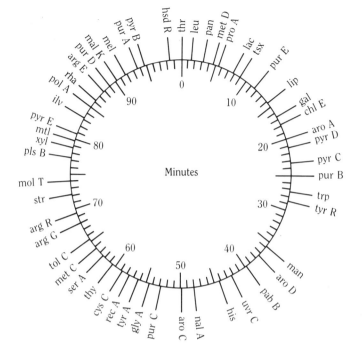

Figure 4-7 The genetic map of *E. coli*. Genes are indicated by three-letter abbreviations. If genes are part of a biochemical pathway, a capital letter is added (metD is a gene in the biosynthetic pathway for methionine). The *E. coli* genetic map is divided into units of one minute; the entire *E. coli* genetic map is set at 100 minutes, which is the time required for a male bacterium to transfer an entire chromosome to a female cell. Genetic maps for other organisms are assigned units based on the frequency of recombination between genes (see Chapter 11).

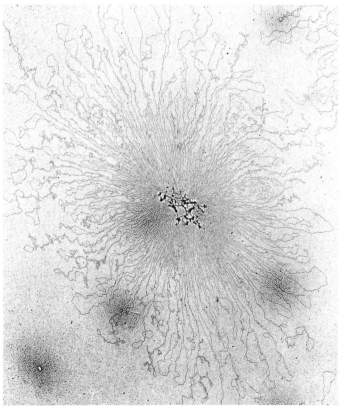

Figure 4-8 Electron micrograph of an *E. coli* DNA molecule. The molecule is an enormous circle of DNA that is somehow condensed and packaged into a bacterial cell. (Courtesy of Ruth Kavenoff, University of California, San Diego.)

Table 4-1. Size Comparison of DNA Molecules from Different Organisms

Organism	Number of base pairs ($\times$ 1,000)	Length in microns (.000001 meter)
Tumor virus	5.1	1.7
Phage ØX 174	5.4	1.8
Phage T4	116	55
Smallpox virus	190	63
E. coli	4,000	1,360
Yeast	13,500	4,600
Drosophila	165,000	56,000
Human	2,900,000	990,000

and each DNA molecule consists of many loops. If completely stretched out, the DNA from the smallest human chromosome would be about 1 centimeter long; each human cell contains forty-six DNA molecules of this length or greater. Yet in every human cell each DNA molecule is so tightly coiled that it is condensed down to a length of about .0001 centimeter. The amount and size of DNA increases as organisms become larger and more complex (Table 4-1).

DNA molecules are too small to be seen under the light microscope, but they can be detected indirectly by **autoradiography**, which is done as follows. If cells are grown *in vitro,* the DNA molecules can be radioactively labeled by adding radioisotopes to the medium that will be incorporated into newly synthesized DNA. For example, the DNA in Chinese hamster cells can be radioactively labeled with tritiated (^{3}H) thymine. After the radioactive thymine has been incorporated into the hamster cells, they are broken open and the DNA is gently extracted and placed onto the surface of a filter, which is then covered with a photographic film. The film is gradually exposed by the electrons that are emitted by the radioisotope in the thymine. After a few weeks, the exposure is sufficient for the image of the DNA to appear on the film when it is developed.

To determine how DNA is replicated in growing animal cells, radioactive thymine is added to the medium for a brief time. Cells are then broken open to release the DNA, and the replication of the DNA is monitored by autoradiography (Figure 4-9). As the figure shows, this segment of the hamster cell's DNA is replicating in both directions (bidirectional replication) from an origin of replication that is indicated by the arrow. Because only replicating regions of the DNA incorporate radioactive thymine, replicating regions of the DNA molecules expose more of the grains on the film than are exposed by the nonreplicating

Figure 4-9 Autoradiography of replicating DNA from animal cells. The dark spots represent regions of the DNA that were replicating when the radioactive thymine was added. (Courtesy of D. Prescott, University of Colorado, Boulder.)

regions of the DNA, which do not contain any radioactivity. In every DNA molecule in a hamster (or human) cell, there can be a hundred or so origins of DNA replication that become active during one stage of cell division.

In contrast to DNA replication in plant or animal cells, bacterial DNA replication begins at just one unique site called the **origin of replication**. Because plant and animal cell DNA is so much larger than bacterial DNA, its replication must begin at a number of different sites if all the DNA in the chromosomes is to be duplicated in the S phase of the cell cycle.

An intriguing feature of DNA replication that occurs in both bacteria and animal cells is that replication not only begins at specific sites but also proceeds in both directions from the origins: DNA replication is bidirectional. Details of the DNA replication mechanism, which are quite complex, are deferred until later in this chapter.

Mitochondrial DNA

All eucaryotes (fungi, plants, animals) have tiny organelles in the cytoplasm of their cells called **mitochondria**. These functionally specialized structures produce most of the cells' chemical energy in the form of **adenosine triphosphate (ATP)**. ATP molecules are as essential to the activities of the cell as gasoline is to the movement of an automobile: they provide the energy that sustains most of the other chemical reactions in the cell. Cells that require more than normal amounts of energy, such as heart and muscle cells, have correspondingly more mitochondria and produce more ATP.

A small DNA molecule in each mitochondrion provides some of the necessary genetic information that enables it to perform its functions, and genes in the chromosomal DNA in the cell nucleus supply the rest. A milestone in molecular biology was reached in 1981 with the announcement that the complete base sequence of human mitochondrial DNA had been determined. The human mitochondrial DNA was shown to be physically circular, like plasmid DNA. It contains 16,569 base pairs—enough coding information for about fifteen genes, most of whose functions and locations in the mitochondrial DNA have been identified.

Surprisingly, the mitochondrial DNA in human cells is among the smallest detected in any organism. The mitochondrial DNA in the fruit fly (Drosophila) is slightly larger—about 19,500 base pairs. Bread yeast (*Saccharomyces cerevisiae*) has the largest mitochondrial DNA—about 78,000 base pairs—even though it is a microorganism. There has been much speculation as to how mitochondria originated in cells and the evolutionary relationship of mitochondrial DNA to the DNA of microorganisms. Some scientists argue that mitochondria are closely related to primitive bacteria because mitochondria synthesize proteins that are quite similar to ones found in procaryotic cells. This argument has led to the suggestion that bacteria invaded primitive eucaryotic cells and evolved until they became essential organelles of such cells. Other scientists argue that mitochondria evolved independently in eucaryotic cells from fragments of eucaryotic chromosomal DNA.

Plasmids That Confer Resistance to Antibiotics

There is another kind of plasmid found in many different species of bacteria that, like the F plasmid, contributes to genetic diversity and permits dissemination of useful genetic information among bacteria much more rapidly and efficiently than can be accomplished by other mechanisms. Such **R plasmids** (resistance-transfer plasmids) confer to bacteria resistance to one or more different antibiotics. During the 1950s, Japanese microbiologists working with dysentery patients isolated bacterial strains of the genus *Shigella* that were resistant to most of the antibiotics then available. Moreover, these dysentery-causing *Shigella* bacteria were able to transfer their multiple antibiotic resistance to previously susceptible bacteria of other species that also cause diseases.

In many respects the R plasmids perform functions quite similar to those of the F plasmids. R plasmids also consist of

Use of Antibiotics: A Permanent or Temporary Benefit?

BOX 4-2

The discovery of antibiotics—drugs that destroy bacteria that cause infections and diseases in people and many other animals—has been an enormous boon to humankind. However, the widespread use of antibiotics has also caused the "natural selection" of bacteria that are resistant to many important antibiotics, including penicillin and tetracycline. Antibiotic resistance in bacteria is spread by R plasmids that are transferred among several different species of bacteria. R plasmids often confer multiple antibiotic resistance to pathogenic (disease-causing) bacteria, making effective treatment difficult. For example, *Shigella flexneri,* a dysentery-causing bacterium distantly related to *E. coli,* became increasingly drug-resistant as the use of antibiotics increased in Japan in the 1950s and 1960s (see graph). Similar patterns of increasing antibiotic resistance have been observed for other pathogenic strains of bacteria, including those that cause sexually transmitted diseases such as gonorrhea.

Over the past twenty years thousands of tons of antibiotics have been added routinely to animal feed in the United States because the producers of beef, pork, chicken, lamb, and other meats became convinced that antibiotics were needed to prevent diseases in their animals and to promote faster weight gain (animals continuously fed small doses of antibiotics gain more weight than animals whose feed does not contain antibiotics). However, we now realize that such widespread and indiscriminate use of antibiotics constitutes a serious health problem for animals and humans.

In 1977 the U.S. Food and Drug Administration proposed a ban on the routine addition of penicillin and tetracycline to animal feed. But farmers and pharmaceutical companies successfully lobbied against the FDA proposal. Addition of antibiotics to animal feed is still routine in the United States, although the practice has been banned in England.

Human bacterial diseases are much more difficult to treat nowadays. Antibiotics that were previously successful in destroying the bacteria and curing the diseases now may have no effect. Several days of microbiological tests may be required to find out which antibiotics a particular bacterium is sensitive to. But a few days may mean the difference between life and death.

Nosocomial infections (infections acquired in hospitals) are of particularly serious concern. About 3–4 percent of all patients admitted to hospitals in the United States today acquire a bacterial infection of one sort or another while being treated for some other ailment. Because antibiotics are used in large quantities in hospitals, the bacteria isolated in hospitals nowadays are usually resistant to many antibiotics in common use. Therefore, becoming infected by a strain of bacteria acquired in a hospital is particularly dangerous for patients. Despite the most stringent precautions, many hospital patients acquire staphylococcal and streptococcal infections that are difficult to treat and cure and that may complicate their recovery from other illnesses.

It is becoming increasingly evident that even the development of antibiotic "wonder" drugs may not always have wonderful consequences. Because plasmids provide bacteria with the genetic information they need to survive in antibiotic-containing environments, we need to reduce our wholesale, indiscriminate use of antibiotics so that when these drugs are really needed, they will be able to kill disease-causing bacteria.

Reading: Richard P. Novick, "Antibiotics: Wonder Drugs or Chicken Feed?" *The Sciences*, July/August 1979.

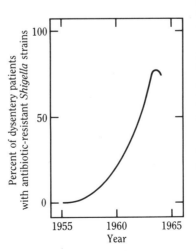

circular DNA molecules that replicate independently in the cytoplasm. They possess genes that direct the synthesis of the sex pili that enable the bacteria to conjugate with other bacteria and thereby transfer the plasmid DNA. In addition, some R plasmids direct the synthesis of "promiscuous pili" that permit mating with bacteria of many different species.

In environments containing antibiotics, bacteria containing R plasmids obviously have a distinct survival advantage over those that lack R plasmids. Since the 1950s, the often indiscriminate use of antibiotics has created a world-wide problem in the treatment of certain bacterial diseases in both domestic animals and in people because it has caused many pathogenic (disease-causing) bacteria to develop resistance to the commonly used and most effective antibiotics (see Box 4-2).

R plasmids were present in bacteria long before antibiotics were discovered, so it can be inferred that they have contributed in other ways to the survival and fitness of bacteria during evolutionary history. When Kalahari Bushmen in Africa—who do not intermingle with other tribes and who have never seen a doctor, much less taken an antibiotic—were tested for antibiotic resistance it was found that some of their intestinal bacteria contained R plasmids making them resistant to five different antibiotics. In nature, bacteria, fungi, and other microorganisms normally synthesize antibiotic substances. The microorganisms themselves are resistant to the chemicals they produce because they often carry an appropriate R plasmid, but other sensitive microorganisms are not so fortunate and are destroyed. In this way the antibiotic-resistant microorganisms secure their ecological niche. Thus, even microorganisms obey Darwin's rule of "survival of the fittest."

Jumping Genes

We have seen that bacteria exchange genetic information by conjugation, a mating process in which plasmid DNA as well as chromosomal DNA is transferred from cell to cell. Recently, another remarkable genetic mechanism has been discovered that permits groups of genes to move spontaneously from one DNA molecule to another. Initially it was discovered that a group of antibiotic-resistance genes could move by itself from one DNA molecule to another in the same bacterial cell. Because the segment of DNA appeared to "jump" around from DNA molecule to DNA molecule by some novel and unconventional genetic mechanism, these segments of DNA were referred to as "jumping genes" or **transposons** (Figure 4-10). The term transposon usually refers to a segment of DNA that carries antibiotic-resistance genes, but any group of genes that jumps from one DNA molecule to another without employing the usual mechanisms of recombination is called a transposon.

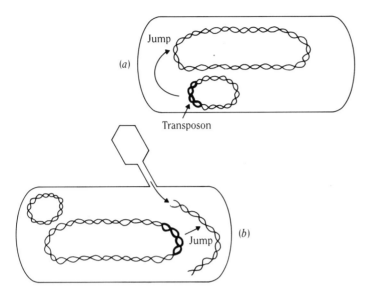

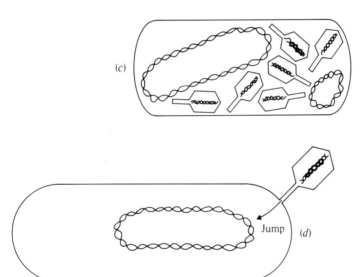

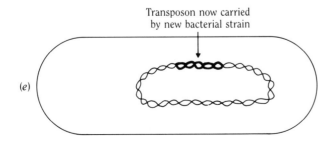

Figure 4-10 Jumping genes. (a) A group of antibiotic-resistance genes jumps from a plasmid into the chromosome. (b) A phage infects the cell and the group of genes (transposon) jumps into the phage DNA. (c) The virus reproduces inside the bacterium. (d) The bacterial cell bursts, releasing the phages, and a phage carrying the transposon infects another bacterium. (e) The transposon jumps into that cell's chromosome.

Table 4-2. Jumping of Antibiotic Resistance Genes from One DNA Molecule to Another

Transposon	Number of base pairs ($\times$ 1,000)	Number of base pairs in IS at ends	Antibiotic resistance carried
Tn 3	4.8	40	Penicillin
Tn 5	5.2	1400	Kanamycin
Tn 7	12.7	40	Streptomycin
Tn 9	2.5	700	Chloramphenicol
Tn 10	9.3	1400	Tetracyclines

Transposons move between these different kinds of bacteria:

Escherichia	*Pseudomonas*	*Shigella*
Haemophilus	*Rhizobium*	*Staphylococcus*
Klebsiella	*Salmonella*	*Streptococcus*
Proteus	*Serratia*	*Yersinia*

PLASMIDS, TRANSPOSONS, AND INSERTION SEQUENCE ELEMENTS
What enables segments of DNA to jump out of one chromosome and into another? It turns out that two short pieces of DNA with identical base sequences are located at the ends of every transposon. These **insertion sequences** (IS) facilitate the movement of the transposon from one chromosomal location to another because certain IS elements are common to DNA molecules in many different kinds of microorganisms (see Table 4-2). The IS elements contain as few as 40 and as many as 1,400 bases. Identical IS elements move about because they recognize one another and form hydrogen bonds between their complementary bases. To date, about a half dozen different IS elements have been identified in various species of bacteria.

The relationship between plasmids, transposons, and IS elements is shown in Figure 4-11. Transposons carrying antibiotic-resistance genes generally originate as part of plasmids. At the ends of each transposon are identical IS elements that are responsible for moving the transposon from the plasmid to some other DNA molecule. Sometimes only a single IS element jumps to another chromosome, but quite often the two IS elements move together, carrying along the genes located between them. Because identical IS elements are found in a wide variety of organisms, transposons can also move from one organism to another. The movement of IS elements and transposons contributes to the genetic diversity in microorganisms. However, the frequency of transposon movement is generally comparable to mutation rates, which are quite low for any particular gene.

Although transposons were first discovered and character-

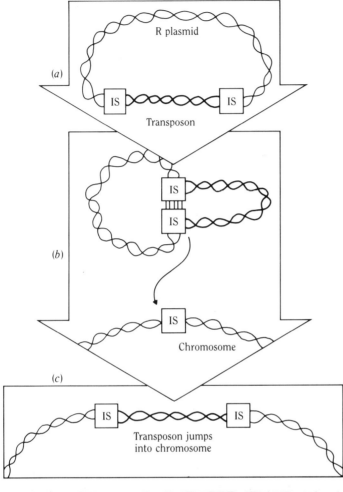

Figure 4-11 The role of IS elements in jumping genes. (*a*) The transposon carried in an R plasmid is flanked by IS elements. (*b*) The IS elements pair and recognize a similar IS element in another DNA molecule. (*c*) The transposon jumps from the plasmid to the chromosome.

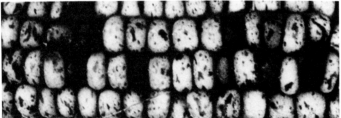

Figure 4-12 Variation in the color of corn kernels. Such variation is due to movable genetic elements that alter the expression of genes. (Courtesy of M. G. Neuffer, University of Missouri, and Crop Science Society of America.)

ized in bacteria, other kinds of jumping genes were discovered much earlier. In the 1940s a plant geneticist, Barbara McClintock, was able to show that the color differences in corn kernels had a genetic basis and resulted from what she called "movable genetic controlling elements" (Figure 4-12). Like many other genetic discoveries, her observations achieved little recognition and their importance was generally unappreciated until the discovery and characterization of similar genetic elements in bacteria provided the framework for understanding the movable genetic elements in corn.

The importance of plasmids, transposons, and IS elements in increasing genetic diversity in bacteria and in other kinds of organisms is just beginning to be appreciated and understood. These bits of genetic material move into and out of chromosomes and in so doing cause mutations and change the way genetic information is expressed. It may turn out that transposable elements contribute as much to genetic diversity and to the evolution of organisms as do mutations from other sources.

MOVABLE DNA ELEMENTS IN ANIMAL CELLS

Transposons and insertion sequences are not unique to bacteria. Movable DNA elements with similar properties have been found in fruit flies and in human cells. It seems more and more likely that movable DNA elements are necessary for genetic diversity and are important factors in the evolution of animals as well as of bacteria. Howard Temin, a Nobel laureate virologist, has suggested that RNA tumor viruses, which cause cancer in animals, actually arose from the animals' own DNA millions of years ago. If this is true, genes in chromosomal DNA and genes in viral RNA should be related and should have similar base sequences. And, in fact, experiments have confirmed just this—animal viruses do contain genes in their RNA molecules that are similar in base sequence to genes found in the DNA of normal cells.

When RNA tumor viruses infect animal cells, the genetic information carried in the RNA molecules is converted into a DNA copy by an enzyme called **reverse transcriptase** that is found only in these viruses. The DNA that is synthesized from the RNA can be incorporated into a chromosome, and in this way the genetic information carried by the virus becomes an integral part of animal cells' chromosomes (Figure 4-13). The viral genes can be detected at different locations in the cell's chromosomes, suggesting that these viruses behave in a manner analogous to transposons in bacteria. By moving from cell to cell, animal viruses can alter expression of genetic information and even transfer new genetic information from cell to cell or from animal to animal as a result of infection. When researchers have extracted the viral DNA molecules from the chromosomes of animal cells, they have found base sequences that are identical at both ends of the viral DNA and that may be analogous to the insertion sequences found in bacteria. This fact has led to speculation that certain types of animal viruses may be the counterpart of the jumping genes in bacteria. These exciting recent discoveries have given rise to a new concept of viruses: rather than merely being agents of disease, they may also increase genetic diversity and play an essential role in evolutionary processes.

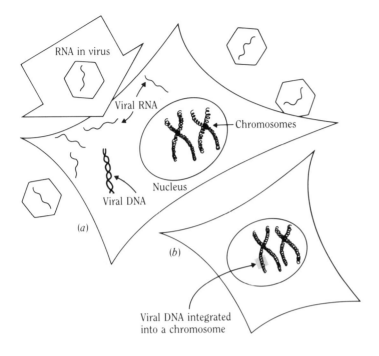

RNA in virus

Viral RNA

Chromosomes

Nucleus

Viral DNA

(a)

(b)

Viral DNA integrated
into a chromosome

Figure 4-13 Integration of RNA tumor virus genetic information into chromosomes of cells. (a) RNA viruses infect a cell and the single strands of RNA are injected. A DNA molecule corresponding to the information in the RNA is synthesized by the enzyme reverse transcriptase. (b) The viral DNA becomes integrated into a chromosome and subsequently is replicated with the chromosome.

How can the requirements for fidelity of genetic information and accuracy in DNA replication be reconciled with these new ideas of jumping genes and base changes in DNA? After all, bacteria beget bacteria, cats beget cats, and the recognizable appearance of organisms does not change appreciably from generation to generation, yet all of the genetic mechanisms we have discussed until now are designed to promote genetic change. This is precisely the paradoxical situation we mentioned at the beginning of the chapter. DNA molecules continually change their genetic information through mutations and gene rearrangements, yet at the same time they duplicate genetic information with extraordinary accuracy. The mechanisms of DNA replication provide a partial resolution of this paradox.

Replication of DNA

Constructing millions of identical copies of any object with many parts is a formidable task. Consider for a moment some of the problems encountered in the assembly-line production of automobiles. Thousands of different steps are involved, tests must be performed at various stages, and many parts must be manufactured in different locations and yet arrive at the right place at the right time in order to be assembled. Information is constantly fed back to control points in the assembly process so that the flow of parts and automobiles is properly synchronized. Sophisticated computers regulate the assembly line. Quality control inspections are performed at many steps along the line

Table 4-3. Functions of Some Bacterial DNA Replication Enzymes

Enzyme	Function
RNA primase	Synthesizes a short piece of RNA that is necessary to begin synthesis of each new DNA strand.
DNA gyrase	Twists circular DNA molecules into tight coils and untwists them.
DNA helicase	Unwinds strands of the double helix so that replication can proceed.
DNA replicase	Joins nucleotides together.
DNA polymerase I	Fills in gaps in newly replicated DNA strands and corrects errors in base pairing.
DNA ligase	Joins the sugar to the phosphate to make the sugar-phosphate backbone of each strand of DNA continuous.

to determine whether the number of defects is excessively high and whether corrections need to be made. Yet despite all these measures, the assembly-line process is still far from perfect: many of the products contain defects.

Accurate replication of a DNA molecule as long as that contained in a single human chromosome is a far more complex task than the construction of an automobile. The Watson-Crick model of DNA described in Chapter 3 makes the replication appear deceptively simple—merely a matter of inserting the correct complementary bases according to the preexisting single-strand templates. But even in a bacterium there are approximately 3 million base pairs and in a human egg or sperm about 3 *billion.* To appreciate the immensity of the task of synthesizing a DNA molecule in *E. coli,* realize that you would have to correctly line up 3 million base pairs in 30 minutes or less. And this doesn't take in account that elsewhere in the cell all the millions of bases and sugar molecules are being synthesized and must arrive at the right place at the right moment.

THE ROLE OF ENZYMES

The mechanism of DNA replication for some phages and for a few bacteria has been worked out in considerable detail, but how the DNA in the forty-six human chromosomes is synchronously replicated is still only partly known. About a hundred different enzymes are needed to synthesize the four bases (A, T, G, and C) and to join these bases to deoxyribose sugars and phosphates before they are inserted into a new strand of DNA. Other enzymes are needed to join the complementary bases together in the new DNA strand, to untwist the helix so that the replication can proceed, to repair gaps and breaks in the DNA, and to perform other functions essential for accurate replication. A few of the well-characterized enzymes involved in the process of DNA replication are listed in Table 4-3.

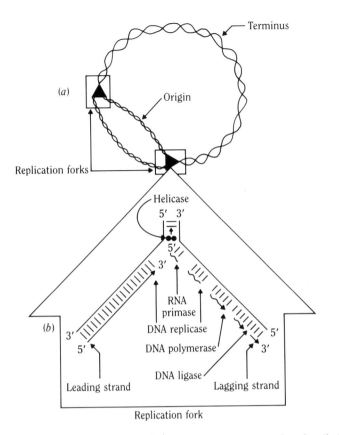

Figure 4-14 The mechanism of DNA replication in bacteria. (*a*) The DNA is a circular molecule. Replication begins at the origin and moves in both directions around the circle until the terminus is reached. Each replicating DNA molecule can have two or more replication forks. (*b*) Each new strand of DNA begins with a short piece of RNA synthesized by the enzyme primase. The piece of RNA is then lengthened by the enzyme replicase, which attaches DNA bases to the end of the RNA strand, thereby making a strand of DNA. Polymerase I removes all of the RNA bases and replaces them with correct complementary DNA bases. The enzyme ligase joins all the discontinuous segments together as the replication fork moves further along the DNA. Many other enzymes are needed for replication in addition to the four described here.

Inside bacterial cells, DNA molecules are twisted and coiled into a variety of shapes so that the long DNA molecule can be stuffed into the cell and still carry out its functions. The coiling and uncoiling of circular DNA molecules, the twisting and untwisting of DNA helices, and the condensation of DNA molecules into chromosomes are accomplished by a general class of enzymes called **topoisomerases**, so called because they catalyze changes in the form, shape, and structure of DNA molecules. DNA gyrase and helicase are just two of several different kinds of topoisomerases; gyrase twists and untwists coils in the DNA and helicase unwinds the DNA strands from each other during replication.

THE REPLICATION PROCESS

Bacterial chromosomal and plasmid DNA replication begins at the origin of replication, as already mentioned. Synthesis of new DNA strands proceeds in both directions along the DNA, eventually stopping at another site in the DNA molecule called the **terminus of replication**. Precisely how DNA replication begins and ends at specific sites is not well understood, although it is known that particular base sequences in the DNA molecule determine the origin and terminus of replication (Figure 4-14*a*).

The region of the DNA in which new synthesis occurs is called the **replication fork**; it moves continuously along the DNA molecule until replication is finished. Synthesis of new strands of DNA must always proceed in different directions because of the chemically reversed orientation of the sugar-phosphate backbones of the complementary strands of DNA (one strand is oriented in what is referred to as the 5′ to 3′ direction, whereas the opposite strand is oriented 3′ to 5′). Because of this chemical polarity, replication of one new DNA strand (the leading strand) can proceed continuously, whereas replication of the other strand (the lagging strand) is discontinuous. (This may seem overly complex, but nature has made it work.)

New strands of DNA in the replication fork are synthesized in the steps shown in Figure 4-14b. Each new strand of DNA must be initiated by the enzyme RNA primase, which synthesizes a short piece of RNA onto which DNA bases are subsequently added one by one by the enzyme DNA replicase. The leading strand requires only one RNA primer at the origin, whereas the discontinuous strand requires one RNA primer for each discontinuous segment.

DNA replicase inserts the complementary bases into the leading strand without leaving any gaps and into the lagging strand discontinuously. The short RNA primers must be removed from the lagging strand segments and replaced by strands of DNA; this is accomplished by the enzyme DNA polymerase I. The short stretches of DNA on the lagging strand are joined together by DNA ligase enzymes, and eventually both DNA strands become continuous from beginning to end.

DNA replication begins with one double-stranded helical molecule, which becomes two double-stranded helical molecules identical to each other and to the parent molecule from which they were replicated. If the replication process has copied the DNA accurately, the daughter molecules will contain genetic information identical to the parent molecules. However, an error in insertion of even a single base will produce a mutation and change the genetic information carried by the replicated DNA. To minimize replication errors, cells contain enzymes whose functions are to monitor the accuracy of the replication process and also to correct mistakes or damage that arise in DNA from other insults.

Thus, we see the resolution to our biological paradox. On the one hand, mechanisms have evolved to ensure the accuracy of the DNA replication process and to guarantee that the genetic information will be transmitted faithfully from generation to generation. On the other hand, cells have mechanisms that alter and disrupt the genetic information carried by DNA—recombination, transposons, insertion sequences, and integration of viral DNA into chromosomes all contribute to genetic changes

DNA From Space— A Science Fantasy?

BOX 4-3

How did life really begin on earth? Nobody knows, but one indisputable fact is that DNA molecules carry the genetic information in every organism on earth. Did today's DNA molecules evolve from simpler molecules that formed billions of years ago, or did they arise elsewhere in the universe? Was the earth "seeded" from space with cells or organisms? If it was, then perhaps there are other intelligent life forms elsewhere in the universe. And if they exist, perhaps we can communicate with them.

One logical way to communicate over interstellar distances would be to transmit information on one particular wavelength, much as we do here on earth for radio, television, and shortwave communication. According to many experts, the wavelength of choice to transmit signals into space or to receive them is at the wavelength of the excited hydrogen atom, because hydrogen is the most abundant element in the universe. We assume that any civilization as scientifically developed as our own would have discovered the spectrum of hydrogen light emitted by the stars. Many scientists believe that if we hope to detect meaningful signals—signals that would, of necessity, have traveled thousands or even millions of years—we should search the heavens with antennas tuned to the wavelength of the hydrogen atom. Even searching for a message from space at just one wavelength, however, would be an enormously expensive and time-consuming task.

A few years ago, some of the world's most renowned physicists, astronomers, engineers, and biologists gathered at an international meeting to consider alternative means of communication. Are there ways to communicate other than with electromagnetic signals traveling at the speed of light? What would be the most desirable properties for an information-containing signal that has to cover enormous distances of space and that may take millions of years to reach a receiver? Clearly what is needed is a signal that contains as much information as possible, in the smallest form possible, and in the stablest form possible so that its information content will not diminish over time or distance. Ideally, the information in the message should be able to reproduce itself if it is to be disseminated as widely as possible. And like the advertising messages that appear on radio or TV, interstellar messages should be repeated over and over to reach as many receivers as possible in the universe. Some kind of booster mechanism should also be involved—here on earth, signals that are transmitted over thousands of miles are electronically boosted so that they arrive at their destination with their informational content and signal strength intact.

Some scientists believe that DNA molecules have properties that make them ideally suited for transmitting information through eons of time and light-years of space. A molecule of DNA could wander forever through the void of space essentially unchanged except for an occasional mutation. If, by chance, a DNA molecule landed in a suitable environment, its message might begin to be reproduced in enormous numbers. Recently, Sidney W. Fox of the University of Miami proposed that the formation of amino acids and other molecules on the primitive earth might have provided just such an environment for the reproduction of nucleic acids such as DNA. Could a DNA molecule have arrived from space billions of years ago?

Suppose that a molecule of DNA containing a message from space did arrive on earth billions of years ago and became encapsulated within a protocell, began replicating, and eventually evolved into humans. Although this idea makes for a good science fiction story, it ultimately pushes the question of how life began further out into space.

Reading: Francis Crick. *Life Itself: Its Origins and Nature.* New York: Simon & Schuster, 1981.

and genetic diversity. It appears that individual organisms can afford to "experiment" genetically. Failure in one organism does not affect the continued success of others. However, if the genetic alteration should prove favorable for the survival and reproduction of one individual, the new genetic information will be faithfully replicated and passed on to progeny. The mechanisms that generate genetic diversity, coupled with a replication machinery of exceptional precision, have given the DNA molecule its unique and indispensable role as carrier of genetic information in all organisms. How DNA arose in cells and became the repository of genetic information is still a mystery, although some "far out" suggestions have been advanced (see Box 4-3).

————————————— Key Terms —————————————

conjugation   transfer of DNA from a male to a female bacterium during cell-cell contact.

F plasmids   small, circular DNA molecules in bacteria that can be transferred to other bacteria. Occasionally, an F plasmid facilitates transfer of bacterial chromosomal genes in addition to its own.

pilus   the tube through which DNA is transferred from one bacterium to another.

Hfr bacteria   male bacteria that transfer their chromosomes with high frequency to female bacteria.

genetic map   the locations of genes with respect to one another in chromosomes.

autoradiography   visualization of DNA by incorporation of radioactive chemicals that afterward expose a photographic film.

origin of replication   site or sites in DNA where synthesis of a new molecule is initiated.

mitochondria   organelles in the cytoplasm of eucaryotic cells whose special function is to produce chemical energy in the form of ATP.

ATP   adenosine triphosphate, the molecule in all cells that supplies the energy for most chemical reactions.

R plasmid   a small circular DNA molecule that carries genes whose products make bacteria resistant to antibiotics. R plasmids can be transferred to other bacteria by conjugation.

transposon   a group of genes, usually ones that confer antibiotic resistance, that move as a unit from one DNA molecule to another.

insertion sequence (IS)   small pieces of DNA at the ends of transposons that have identical base sequences. These sequences pair with identical sequences in other DNAs and are responsible for transposon movement.

reverse transcriptase   an enzyme present in certain RNA viruses that transcribes the information in the RNA molecule into DNA after the RNA has infected an animal cell.

topoisomerases   enzymes that change the form, shape, or structure of DNA molecules.

terminus of replication   site or sites in DNA where replication stops.

replication fork   the region in DNA molecules in which synthesis of new DNA is occurring. The replication fork moves continuously until replication is completed.

_____ Additional Reading _____

Check, W. "Plasmids: More than Genetic Debris." *Mosaic*, March/April 1982.

Cohen, S. N. and J. A. Shapiro. "Transposable Genetic Elements." *Scientific American*, February 1980.

Day, M. and N. Burton. "Plasmids." *Endeavor*, vol. 6 (no. 3), 1982.

Grivell, L. A. "Mitochondrial DNA." *Scientific American*, March 1983.

Lewin, R. "Can Genes Jump Between Eucaryotic Species?" *Science*, July 2, 1982.

Novick, R. P. "Plasmids." *Scientific American*, December 1980.

Zimmerman, D. "Genes in the Cytoplasm: Mitochondria and Chloroplasts." *Mosaic*, July/August 1981.

_____ Review Questions _____

1. What are the four mating types of *E. coli*?

2. What three mechanisms are used by bacteria to exchange genetic information?

3. By what genetic mechanism do bacteria adapt to new environments?

4. What size is a tumor virus in comparison to a human cell?

5. What molecules contain most of the chemical energy in cells?

6. What is the function of mitochondria?

7. What is the function of choloroplasts?

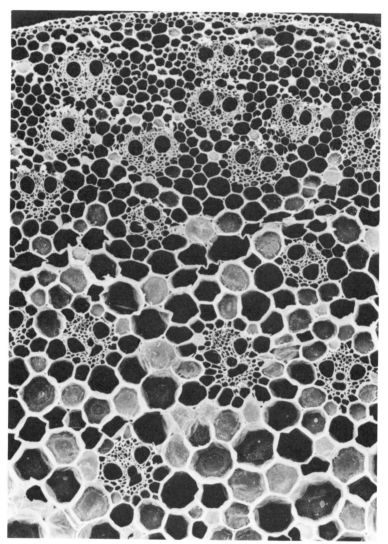

CELLS IN THE STEM OF A CORN PLANT. (87 ×)

5

GENETIC INFORMATION

*How Genes
Are
Expressed*

*There is a great difference between the
approach of the priest, the politician and
the scientist. The priest persuades humble
people to endure their hard lot, the
politician urges them to rebel against it,
and the scientist thinks of
a method that does away with
the hard lot altogether.*
MAX PERUTZ

DNA molecules contain all the information that determines the growth and reproduction of cells, in addition to directing their specialized functions within the organism. DNA is responsible, directly or indirectly, for the synthesis of all other molecules in cells. In particular, it supplies the information for the synthesis of the two other major classes of information-containing molecules, ribonucleic acids (RNA) and proteins. Together DNA, RNA, and proteins are the informational macromolecules of cells (Table 5-1).

To understand how the genetic information is extracted from DNA and utilized by the cell, it is necessary to understand how RNA and protein molecules are synthesized and what functions they perform. It is obvious that in order for genetic information to be used by cells, the information carried by particular genes must be expressed. Therefore, genes must produce other kinds of molecules that actually carry out the genetic instructions.

Gene expression occurs in two steps. The first step is the process of **transcription**, which involves the synthesis of many different RNA molecules from genes in DNA. The chemical composition of RNA is quite similar to that of DNA, but there are some important differences between RNA and DNA molecules. These differences are shown in Figure 5-1. Transcription is the process by which the genetic information contained in sequences of bases (genes) in DNA is transferred into a complementary sequence of bases in RNA. The information in some of the RNA molecules is further deciphered in the next step of gene expression, **translation**, which converts the genetic information that is now carried in the sequence of bases in RNA into a sequence of amino acids that determines a particular protein's structure and activity in the cell.

The Central Dogma

A few simple rules govern the exchange of information among DNA, RNA, and protein macromolecules in cells of all organisms. The information exchanges that can be carried out are shown in a simple diagram devised by Francis Crick. The rules laid down in this diagram hold true for every cell and organism, and together they are known as the **central dogma** of molecular biology:

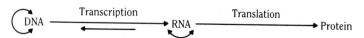

The long arrows in the diagram symbolize the fact that information always flows, in a two-step process, from DNA to RNA (transcription) and from RNA to protein (translation). The short arrow pointing from RNA to DNA indicates that in a few exceptional circumstances, information can flow backward from

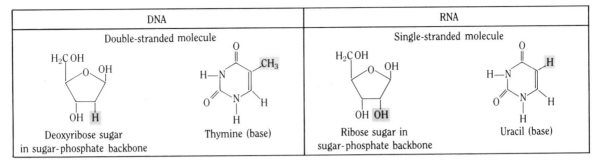

DNA	RNA
Double-stranded molecule	Single-stranded molecule
Deoxyribose sugar in sugar-phosphate backbone — Thymine (base)	Ribose sugar in sugar-phosphate backbone — Uracil (base)

Figure 5-1 Chemical differences between DNA and RNA.

Table 5-1. The Informational Macromolecules in Cells

Informational molecules	Properties
DNA A C G T T C C T G C A A G G	Double-stranded, helical nucleic acid. Information is contained in the sequence of bases A, T, G, C.
RNA A U U C G C U G G A G C	Single-stranded nucleic acid. May have some helical sections. Information is contained in the sequence of bases A, U, G, C.
Protein trp phe arg gly pro ala gly ser ala ser leu ala gly arg ala pro leu trp pro	Chain of amino acids. Information is contained in the sequence of the twenty different amino acids, which also determines the structure and function of the particular protein.

an RNA molecule into DNA. This occurs only with certain RNA viruses, particularly those that are able to cause tumors in animals. As we saw in Chapter 4, some viruses that have RNA as their genetic material are able to convert viral RNA into a DNA molecule in infected cells. In this way it is possible for viral information to become integrated into the cell's chromosomal DNA and possibly change the expression of the cell's genetic information. (Integration of viral DNA into chromosomes may explain how certain normal cells are converted into cancer cells—see Chapter 8.) However, with this single exception genetic information in cells always flows from DNA to RNA. The absence of a reverse arrow between protein and RNA is significant: information is never transferred from protein back into RNA, or from protein into DNA. (You might consider how this particular aspect of the central dogma bears on Lamarck's idea of the inheritance of acquired traits, discussed in Chapter 3.)

Finally, the diagram shows the permitted information transfers between identical kinds of molecules. Information in DNA is readily transferred to other DNA molecules during rep-

Archaebacteria:
Are They the "Missing Link" Between Procaryotes and Eucaryotes?

BOX 5-1

To the casual observer of nature, it is the differences among organisms that are most noteworthy. To the biochemist, however, it is the similarities among the molecules—especially the informational macromolecules, DNA, RNA, and protein—that are most significant. The processes of transcription and translation are remarkably similar in all cells, implying that these processes arose in primitive cells billions of years ago and stayed relatively unchanged as organisms evolved. Once the genetic code had become established (Chapter 6), any change in the basic processes of information transfer among molecules would probably have resulted in lethal changes for organisms.

New techniques of biochemical analysis have enabled biochemists to determine the base sequences of RNA and DNA molecules. Because ribosomes play such a crucial role in protein synthesis, sequence analysis of ribosomal RNA molecules provides a particularly reliable indicator of the relatedness between different species of organisms. That is, the more closely related species are, the more nearly identical are the sequences of bases in their rRNAs. The rRNAs are identical in the cells of all humans, for example, and the rRNAs from different strains of *E. coli* are also identical. Mutations that did not affect ribosome functions may have introduced a few base changes, but long stretches of the rRNAs from closely related organisms prove to be almost identical when analyzed. As would be expected, however, the base sequences for rRNA molecules from bacteria are quite different from the sequences for rRNA molecules in plant or animal cells, which presumably evolved from bacteria billions of years ago.

But a puzzling pattern emerges when the base sequences of rRNAs from certain unusual kinds of bacteria are examined. These organisms include certain anaerobic bacteria that survive only deep within rotting vegetation or inside a cow's stomach (they cannot tolerate exposure to oxygen), thermophilic bacteria ("heat lovers") that grow in hot springs where temperatures are hot enough to kill any other kind of microorganism, and halophilic bacteria ("salt lovers") that are obliged to live in concentrated brine, where few other organisms can survive.

When biochemists determined the base sequences of the rRNAs from these unusual bacteria, they made a surprising discovery: the sequences were more similar to those in the rRNAs of plant and animal cells than to those in the rRNAs of other bacterial species. This finding suggests that these unusual bacteria, previously believed to be close relatives to other bacteria, may actually be the distant ancestors of eucaryotic cells. Biochemist Carl Woese has proposed the name *archaebacteria* for these organisms and has suggested further that they are the ancestors of both other procaryotes and the eucaryotes.

Some of the answers to the mystery of the origin of life may emerge from further analysis of other informational macromolecules in archaebacteria. As DNA and RNA molecules from various organisms are sequenced, it may be possible to reconstruct the genealogy of cells. Woese believes that "biology is now on the threshold of a quieter revolution, one in which man will come to understand the roots of all life and thereby gain a deeper understanding of the evolutionary process."

Whether Woese's hypothesis is the correct interpretation of the sequence differences among rRNAs from various organisms, base sequences of RNA and DNA do provide a glimpse into biological history.

Reading: Carl Woese, "Archaebacteria." *Scientific American*, June 1981.

lication, as symbolized by the long curved arrow around DNA. Similarly, the short curved arrow under RNA indicates that in the special case of virus-infected cells information may be transferred from one viral RNA molecule to another. In uninfected cells this kind of exchange between RNA molecules never occurs. Information exchange between proteins does not occur under any conditions in cells.

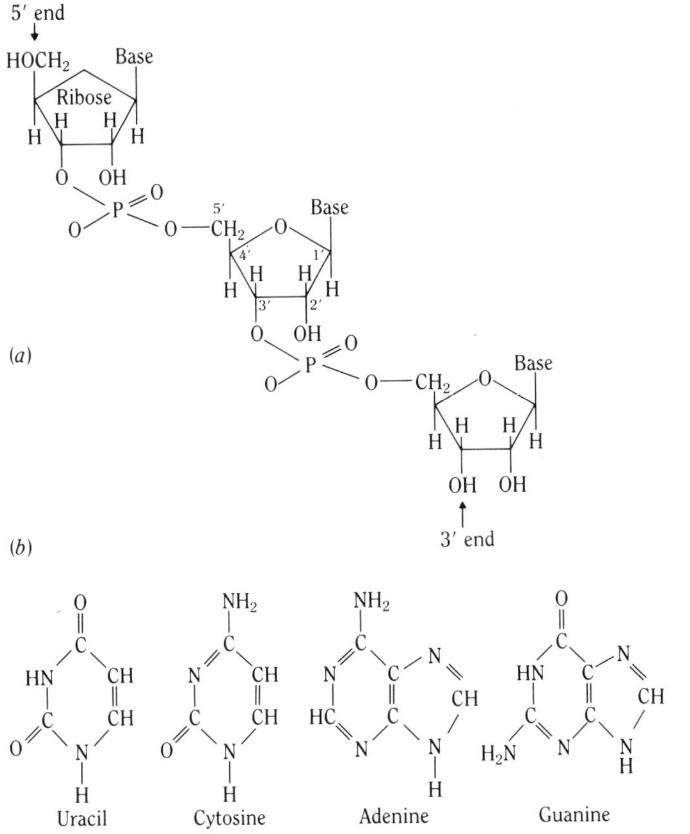

Figure 5-2 The structure of RNA. (a) The sugar (ribose)–phosphate backbone of RNA molecules. (b) The four RNA bases. The base sequence is determined by the base sequence in the strand of DNA from which the RNA is transcribed.

These are strong rules. Perhaps now you can understand why molecular biologists are more impressed by the great fundamental similarities among cells than by their more easily observed differences. Furthermore, analyzing the informational molecules in different kinds of organisms provides important clues to the origin and evolutionary relatedness of cells and organisms (see Box 5-1).

Transcription

Transcription is the process by which RNA molecules are synthesized from DNA in cells. The molecular details of both the transcription and translation processes are understood most thoroughly in bacteria, so the description that follows is derived from bacterial experiments. The important similarities and the relatively slight differences between these processes in bacterial and animal cells are discussed later.

TYPES OF RNA

Although all RNA molecules are composed of the same few chemical subunits (Figure 5-2), three functionally different

Table 5-2. Approximate Sizes and Functions of the Three Types of RNA
Found in Bacteria

Type of RNA	Approximate size (number of bases)	Function
Transfer (tRNA)	80	Carries each amino acid to its correct position in a growing polypeptide chain
Ribosomal (rRNA)	Three different kinds: 100, 1,500, and 3,000	Provides the framework for the construction of ribosomes
Messenger (mRNA)	Varies between 100 and 5,000	Carries the genetic information in a sequence of codons that determine the sequence of amino acids in proteins

types of RNA molecules—all transcribed from DNA—are found in all cells. **Messenger RNA (mRNA)** molecules carry information originating in DNA and are used subsequently in the translation process to produce proteins, many of which are enzymes. The other two types of RNA are **ribosomal RNA (rRNA)** and **transfer RNA (tRNA)**; although they are not translated into proteins, they do perform other vital functions in the translation process.

The different types of RNA in cells can be distinguished by both size and function (Table 5-2). The sizes of the three bacterial ribosomal RNAs are known with considerable precision. The largest of these, called 23S (the S value refers to its sedimentation rate during centrifugation), contains about 2,900 bases. The 16S rRNA is smaller, with about 1,500 bases. The smallest is 5S rRNA, which has about 120 bases. These three rRNAs are transcribed from genes in DNA and become part of the protein-synthesizing particles called **ribosomes** (to be discussed shortly). None of these three rRNAs contains genetic information that is translated into proteins, but all are required for protein synthesis and play both structural and functional roles in ribosomes. While a bacterium contains many thousands of ribosomes, a human cell may contain millions of them, each slightly larger than a bacterial ribosome.

The smaller transfer RNA (tRNA) molecules contain only about 80 bases, but they have a complex three-dimensional shape that enables them to perform a crucial function in the process of translation (Figure 5-3). Transfer RNA molecules are sometimes called "adapter" molecules because they convert the information in the mRNA molecules (sequence of bases) into the information in proteins (sequence of amino acids).

Each transfer RNA molecule can chemically join to one of the twenty different amino acids and also to a particular sequence of three bases (called a **codon**) in any messenger RNA molecule. Transfer RNAs deliver the amino acids they carry to a site on ribosomes where the amino acids are joined together to form proteins. After releasing the amino acid at the ribosome, tRNAs are released and are free to attach to another amino acid of the same kind—a cycle that is endlessly repeated throughout the life of the cell. In all organisms, from bacteria to humans, only thirty to forty different kinds of tRNA molecules are used to "read" the various codons in the different mRNAs. It is the particular sequence of codons in each mRNA that constitutes the "message" in messenger RNA and that directs the sequence of amino acids that are linked together during protein synthesis (the process of translation).

The messenger RNA (mRNA) molecules are the most diverse of the three classes of RNA found in cells. The mRNAs are transcribed from the thousands of different genes in the DNA and are subsequently translated into the thousands of different proteins that carry out the cells' functions. Each mRNA transcribed from a gene is "read" by each ribosome that attaches to it and moves along its length; thus each mRNA directs the synthesis of hundreds or thousands of identical proteins depending on the number of ribosomes that become attached. Different proteins are synthesized from different mRNAs, which is why so many different kinds of mRNAs are present in a cell.

All three classes of RNA molecules—rRNA, tRNA, and mRNA—are manufactured by the process of transcription. All are required to manufacture proteins in the process of translation. However, the mRNAs actually carry the genetic information for making proteins that is encoded in the sequence of bases in the DNA. The tRNAs and the rRNAs are essential in processing the information carried in the mRNAs and converting that information into proteins.

MECHANISM OF TRANSCRIPTION

DNA molecules can be thought of as enormously long tapes of information. The process of transcription is the first step in retrieving the information from specific regions (genes) of the DNA tape. The problems of transcription are similar to those that might be encountered when you first insert a cassette into an audiotape machine. If you want to listen to a particular song that is located somewhere on the tape, either you have to know precisely where the song is located or you have to search for it by running the tape backward and forward. To transcribe a gene or a group of genes from a DNA tape, the enzyme that synthesizes RNA molecules, called **RNA polymerase** (the suffix *ase* generally means that the substance is an enzyme), must be able to recog-

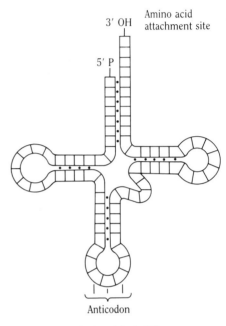

Cloverleaf model of tRNA

Figure 5-3 Transfer RNA. This small RNA molecule consists of about eighty bases that are hooked together to give a complex three-dimensional structure. An amino acid is attached at one end, and the anticodon is located in a loop at the other end of the tRNA.

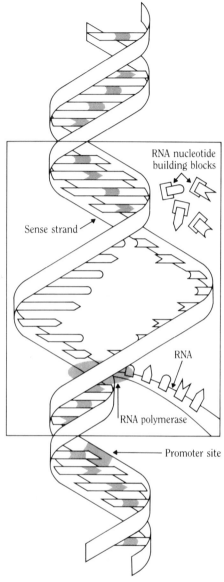

RNA nucleotide building blocks

Sense strand

RNA

RNA polymerase

Promoter site

Figure 5-4 The process of transcription. RNA polymerases attach to promoter sites at the beginning of genes and transcribe an RNA molecule. The RNA nucleotides that are used to construct the RNA molecules are synthesized by other enzymes in the cytoplasm of cells. The correct nucleotide is paired up with its complementary base in the sense strand of the DNA and joined to the RNA by RNA polymerase. Each RNA polymerase begins transcription at a promoter site, copies the sense strand of the DNA and finishes transcribing at the terminator. When transcription is completed, the RNA molecule is released from the DNA and the RNA polymerase is free to start transcribing another gene.

nize where to begin transcription of the DNA and also where to terminate it. That is, transcriptional "start" and "stop" signals are needed.

Two basic mechanisms of the transcription process allow the information carried in any group of genes to be selected and transcribed from any location in the DNA. At the beginning of genes in the DNA are sites, called **promoters**, that serve as start signals. These sites consist of a short sequence of bases (about thirty-five in bacteria) that is recognized by RNA polymerase, allowing it to initiate transcription of an RNA molecule there. At the end of the gene or genes is another sequence of bases, the **terminator site**, that signals the RNA polymerase to stop.

Because the two strands of the DNA molecule contain different, but complementary, base sequences, the genetic information contained in each strand is different. The promoter region determines not only where transcription is to begin but also its direction—and hence the strand ("sense" strand) that is to be transcribed. The RNA polymerase attaches to the promoter on the sense strand and uses the bases in that strand as the template for synthesis of the complementary strand of RNA (Figure 5-4). For different genes in the DNA the promoter site may be situated on either of the two DNA strands.

Every cell has many thousands of RNA polymerase molecules, so many genes can be transcribed from the DNA at the same time. Like other enzymes, RNA polymerase molecules can be recycled because they catalyze the chemical reactions used to synthesize RNA molecules but are not themselves used up or destroyed in those reactions. As we will see later, only those genes are transcribed that are needed by the cell at a given time. That is, gene expression is regulated in all cells.

Not all kinds of RNA molecules are synthesized in the same quantities; their transcription from DNA depends in part on the kind of promoter sequence providing access to the gene. For example, in *E. coli* several clusters of ribosomal RNA genes are scattered around its DNA molecule, and the rate at which these rRNA genes are transcribed is generally much greater than the rates observed for other genes whose products are used in lesser amounts. Furthermore, the rate of rRNA transcription will vary with the compositions of the mediums in which the bacteria are grown. The rate of transcription for other genes also depends on specific genetic regulatory mechanisms, which are discussed in Chapter 9.

STABILITY OF RNA MOLECULES

What happens to the various RNA molecules—rRNAs, tRNAs, and mRNAs—inside the bacteria once they have been transcribed? The ribosomal RNAs are immediately incorporated into ribosomes, where they remain functional as long as the bacteria survive. Transfer RNA molecules are also stable in the bacteria and function over and over again in the process of protein synthesis. In contrast, messenger RNA molecules in bacteria are quite unstable; they generally last for only a few minutes before they are broken down and their bases recycled into other, newly synthesized RNA molecules.

The instability of bacterial mRNAs created a puzzle for biochemists who were trying to isolate and analyze the various kinds of RNA molecules. At any moment about 90 percent of all the RNA being synthesized in bacteria is mRNA, yet when all of the RNA is isolated about 98 percent turns out to be a mixture of only ribosomal and transfer RNA. The explanation for the disappearance of mRNA became clear when researchers discovered that only rRNA and tRNA are stable molecules that, once synthesized, continue to accumulate in bacteria. Although mRNA is actually synthesized in the greatest quantity, it is broken down so rapidly that only a small percentage of the mRNAs can be found in the cell at any moment. Why is the life of mRNA molecules in bacteria so fleeting?

In Chapter 2 it was pointed out that bacteria can adapt rapidly to changes in their environment by expressing—switching on or off—certain genes, depending on which proteins are needed for growth. If mRNA molecules remained functional in bacteria for an indefinite period of time, the same kinds of proteins would continue to be made even if they were no longer needed. Thus, bacteria needed to evolve mechanisms enabling them to synthesize only those proteins they require for optimal growth and reproduction in the environmental conditions that exist at any particular moment. The mechanisms in bacteria that rapidly degrade the bacteria's mRNA without also degrading the chemically identical rRNAs and tRNAs are still unknown.

In plant and animal cells all three types of RNA molecules are generally quite stable and may persist in cells for many days or months. However, in developing embryos, where the functions of cells are changing, the RNA molecules may be less stable. In the cells of mature plants and animals, mRNA does not need to be rapidly broken down, because the internal cellular environment remains more or less constant. These cells are continuously supplied with the necessary nutrients and so do not need to respond to changes in their external environment. Therefore, transcription in plant and animal cells produces stable mRNA molecules that function for extended periods of time.

Translation

Translation is the process by which the information in mRNA molecules is decoded and converted into the information carried by the different sequences of amino acids that make up different

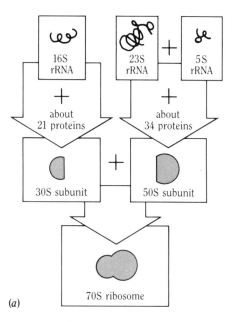

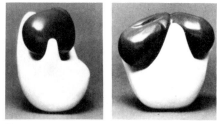

Figure 5-5 Ribosomes. (a) Each subunit of a ribosome consists of at least one RNA molecule and many different proteins. The 30 S and 50 S subunits join to make the 70 S ribosome, which is active in protein synthesis. The sedimentation (S) values are not exactly additive when molecules of different sizes are centrifuged. (b) The structure of the ribosome. The model shown in side and front views was constructed from many electron micrograph pictures of ribosomes extracted from cells. (Photographs courtesy of Miloslaw Boublik, Roche Institute of Molecular Biology.)

proteins. The translation process is basically quite similar in all cells, but once again the mechanism has been worked out in greatest detail in bacteria, so that is the one we shall describe.

Translation is much more complex than transcription because in this process the genetic information must be *decoded,* that is, information encoded in one molecular form (the sequence of bases) must be converted into another molecular form (the sequence of amino acids) that assigns to proteins their particular functions. Translation, then, is the cellular process that makes the genetic information accessible to the cell in the form of molecules that can carry out a variety of chemical reactions.

The translation process is analogous to that of converting information contained in the dots and dashes of the Morse Code into the information contained in the sequence of letters and words in sentences. In the Morse Code, the conversion is from a code of two elements (dots and dashes) to the code of twenty-six letters in the alphabet. In the process of translation, the conversion is from a genetic code of four elements (the four bases) to a protein alphabet of twenty different amino acids. (The genetic code is discussed in Chapter 6.)

The translation process consists of several components: mRNAs, ribosomes, tRNAs, activating enzymes, amino acids, and many other enzymes. The construction of even a single protein in a cell requires a coordinated effort by all of these component parts of the translation system. We will first examine the properties and functions of each of the components, then we will look at how they function together.

RIBOSOMES

Ribosomes are the biochemical machines in cells that are designed exclusively to carry out mRNA-directed protein synthesis. They are complex particles consisting of three different classes of ribosomal RNAs—23S, 16S, and 5S—together with about fifty-five different *ribosomal proteins* (Figure 5-5a). These particular proteins are found only in ribosomes and together perform special functions in the process of translation. (Remember that we are here describing bacterial translation as our example of the translation process. The rRNAs of eucaryotic organisms have various other S values, depending on their size.)

Because so many ribosomal proteins and other enzymes are required to synthesize even a single protein in a cell, we are faced with the classic chicken or egg question: Which came first, ribosomes or proteins? No one really has any idea how the first proteins came to be synthesized in primitive cells, although scientists speculate that short polypeptide chains may have been synthesized by quite simple enzymes, which themselves may have consisted of only a dozen or so amino acids joined together.

During millions of years of evolution, cells became ever more complex, eventually requiring thousands of different proteins to carry out their highly specialized functions. Even the simplest and smallest bacteria contain ribosomes and all of the other components of the translation machinery that are also found in more or less the same form in human cells.

Bacterial ribosomes are constructed from two subunits that are easily separated into their component parts, called the 30S and 50S subunits (Figure 5-5*b*). To begin translation of an mRNA molecule, the two ribosomal subunits must first recognize and bind to a particular sequence of three bases located near the 5′ end of the mRNA called the *translation initiation site*. The mRNA chain is somehow threaded through each ribosome that attaches to the initiation site, and then the ribosomes move along the mRNA much like beads strung on a thread. As each ribosome moves down the mRNA, other ribosomes can attach to the newly exposed initiation site. Thus, each mRNA is capable of making as many polypeptides as the number of ribosomes that attach to it. An mRNA continues to synthesize proteins until it is eventually destroyed. Any ribosome can attach to any mRNA molecule; therefore, the kind of protein that is synthesized is determined solely by the particular mRNA molecule, whose information is, in turn, determined by the sequence of bases in the gene from which it has been transcribed.

TRANSFER RNAS: ADAPTER MOLECULES

Ribosomes cannot, by themselves, accomplish the decoding of an mRNA molecule. It is the transfer RNA molecules, often called *adapter molecules,* that actually convert the information from one coding system to the other. As shown in Figure 5-6, one end of a tRNA molecule is covalently bonded to a particular amino acid—in this instance, phenylalanine (phe). At the other end of the phe-tRNA molecule are three bases called the **anti-codon**, which can form hydrogen bonds with the corresponding three bases in the mRNA (the *codon*). In this way, each mRNA codon specifies a particular amino acid and the tRNA molecule provides that amino acid at the correct position.

Each of the thirty to forty tRNAs has a unique three-dimensional structure that is recognized by one of twenty different **activating enzymes** in the cell (biochemists call these enzymes aminoacyl-tRNA synthetases). The activating enzymes attach each of the different amino acids to their corresponding tRNAs. For example, the phenylalanine activating enzyme recognizes both the phe-tRNA and the amino acid phenylalanine and joins the two together by a covalent bond. This group is now ready to attach to an mRNA-ribosome complex (Figure 5-7). The phenylalanine codon in an mRNA is recognized by the phe-tRNA anticodon and the codon and anticodon are joined by hydrogen

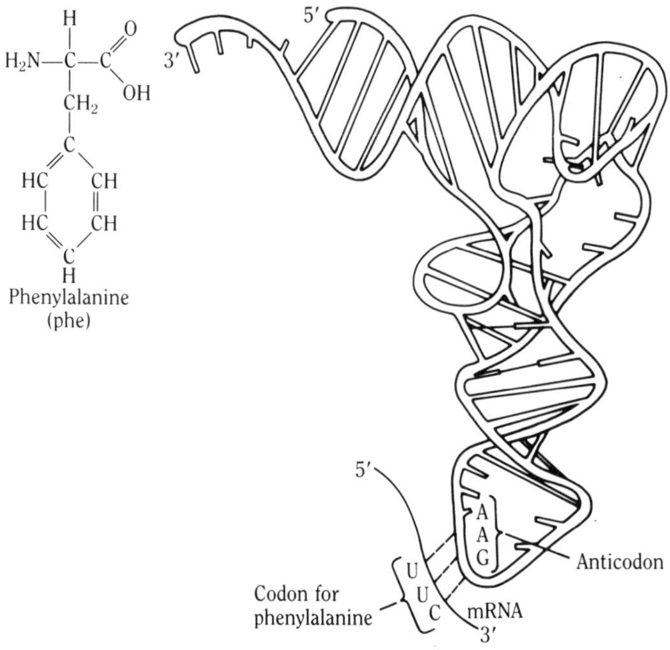

Phenylalanine
(phe)

Figure 5-6 Transfer RNA. Amino acid activating enzymes attach a particular amino acid (in this case phenylalanine) to its corresponding tRNA. The three-base anticodon of the tRNA recognizes the three-base codon in the mRNA that corresponds to the amino acid carried by the tRNA.

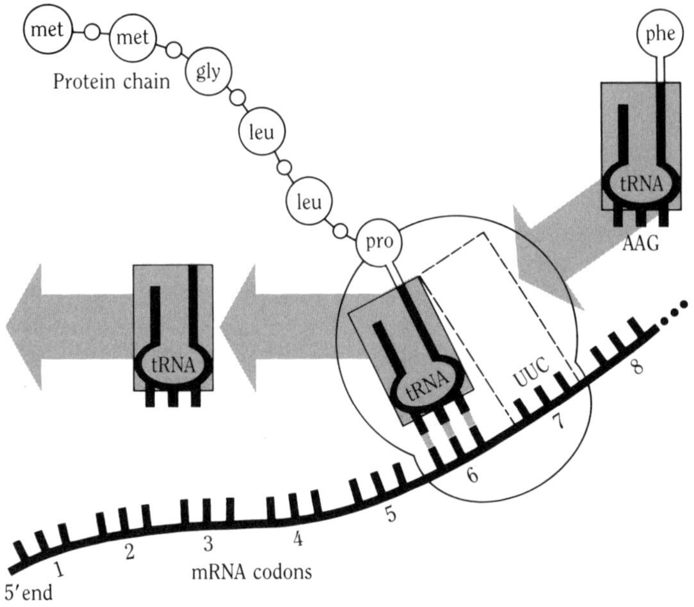

Anticodon

Codon for
phenylalanine

mRNA

Figure 5-7 The function of tRNA. The phe-tRNA anticodon (AAG) hydrogen bonds to the phenylalanine codon (UUC) in mRNA. Attached to the other end of the tRNA molecule is the amino acid phenylalanine, ready to be joined to the growing protein chain. See Figure 5-8.

bonds. At the same time, a ribosome positions the phenylalanine so that it can be connected to the chain of amino acids that is being synthesized by that particular ribosome.

Figure 5-8 summarizes the overall series of steps in the process of translation. In the cell cytoplasm, amino acids are attached to tRNAs by the activating enzymes. The amino acid-

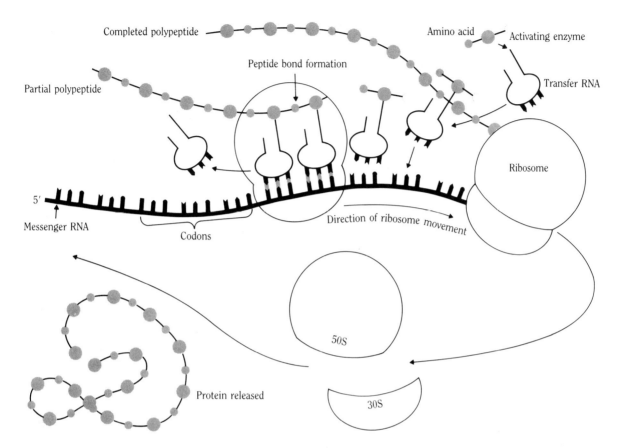

Figure 5-8 The process of translation. In the cytoplasm of cells amino acids are attached to their appropriate tRNA by one of the different activating enzymes. The tRNAs recognize specific codons in mRNAs and line up amino acids in the proper sequence. Ribosomes move along the mRNA forming a peptide bond between amino acids until a complete polypeptide is synthesized. The protein is released, and the ribosomes separate into their two subunits and reattach to another mRNA.

carrying tRNAs move about in the cell until a three-base anti-codon in the tRNA is paired with its complementary codon in an mRNA. Ribosomes position adjacent tRNAs along the mRNA so that enzymes can catalyze the formation of a bond between the amino acids. As each ribosome moves along an mRNA, a protein is synthesized.

Proteins are synthesized at a remarkably fast pace in cells. In *E. coli* each ribosome can chemically link amino acids together at a rate of about 15 per second. If we make the reasonable assumptions that an average protein contains about 225 amino acids and that there are about 15,000 ribosomes in each bacterium, we can calculate that about 1,000 protein molecules are synthesized every second, or 60,000 proteins per minute, per bacterium. Because proteins are needed to perform all enzymatic cellular functions, the process of translation, even at this high speed, must proceed with very few mistakes if the cell is to survive and function normally.

THE PEPTIDE BOND

The names and chemical structures of the twenty different amino acids used to synthesize proteins were listed in Chapter 1.

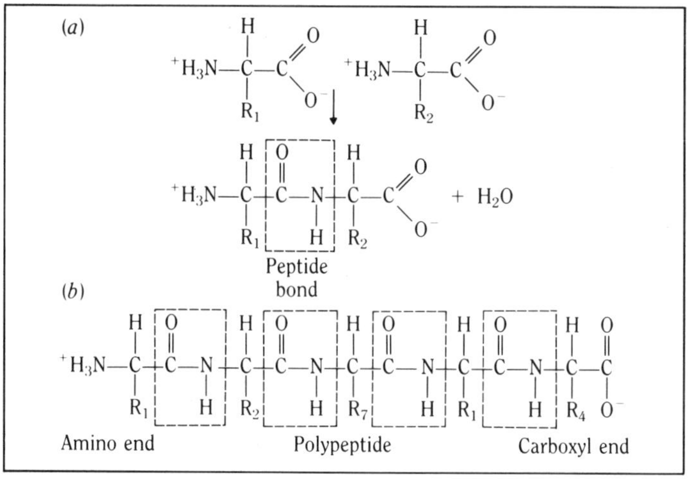

Figure 5-9 The peptide bond. Any two amino acids can be joined together in the process of translation. (a) The amino end ($-NH_3^+$) of one amino acid is covalently linked to the carboxyl end ($-COO^-$) of another amino acid by formation of a peptide bond and the release of a molecule of water in the enzymatically catalyzed reaction. (b) When peptide bonds link many amino acids together, a polypeptide results. The side groups (R_1, R_2, R_7, and so on) represent the different amino acids. For example, the amino acid with R_1 might be methionine, the one with R_2 might be leucine, and so on.

The chemical structures of the amino acids are not as complicated as they might seem, since all amino acids share a common structure and (with one exception) differ only in their *side groups* (review Figure 1-8). Each amino acid can be chemically joined to any other amino acid by a **peptide bond**. This special chemical bond joins the amino end ($|NH_3^+$) of one amino acid to the carboxyl end ($|COO^-$) of the adjacent amino acid. Because amino acids are linked by peptide bonds, a chain of amino acids is called a **polypeptide** (Figure 5-9). Each polypeptide reflects the information carried in the sequence of bases in the mRNA, which, of course, is determined by the sequence of bases in the sense strand of the DNA. Thus, polypeptides reflect the expression of the genetic information just as music reflects the information carried on a cassette tape.

The proteins that carry out specific functions in cells may consist of a single polypeptide chain or, as is commonly the case, several polypeptides that are interlocked to produce a specific configuration. For example, the bacterial enzyme β-galactosidase consists of four identical polypeptides, each composed of an identical sequence of 1,173 amino acids. Human hemoglobin, the protein in red blood cells that binds and transports oxygen, also consists of four polypeptide chains. However, these four chains are not identical. Rather, there are two pairs of polypeptides called alpha (α) and beta (β) chains. The identical α-chains have 141 amino acids, while the β-chains have 146 amino acids. Every human red blood cell contains 200–300 million hemoglobin molecules (represented as $\alpha_2\,\beta_2$).

Protein Structure

Most of the genetic information that is expressed by DNA is used to direct the synthesis of proteins. In many respects, proteins are the most essential molecules in living cells because enzymes (which are proteins) catalyze all cellular chemical reactions. The term *protein* itself is derived from the Greek *proteios,* meaning "of the first rank." Proteins perform many other functions in addition to catalyzing chemical reactions. They transport nutrients and other kinds of small molecules into and out of cells. Hemoglobin, for example, transports oxygen to body cells and removes carbon dioxide. Proteins are required for all physical movement—muscles are made of proteins. Skin is made of protein. Bone, which provides the body's mechanical support, is made up of proteins and mineral salts. Bacteria swim by means of a special tail-like structure—a *flagellum*—that is composed of protein. The regulation of genetic expression in DNA is accomplished by proteins, and the growth and differentiation of plant and animal cells are regulated by proteins.

Proteins perform their different functions with great speed and efficiency. For example, an enzyme may catalyze a chemical reaction in 1/1,000 of a second (one millisecond); proteins in the eye can change their shape and send an electrical signal to the brain in 1/1,000,000 of a second (one microsecond).

The unique and practically unlimited variety of shapes that proteins can assume gives them enormous functional versatility. Protein structure is characterized by four levels of complexity called primary, secondary, tertiary, and quaternary structure (Figure 5-10). The *primary structure* of a protein is the sequence of amino acids carried in its polypeptide chains. The amino acid sequence also determines the protein's *secondary structure,* which is the portions of the polypeptide chain that are folded or helical in configuration. The particular kind of secondary structure is determined by the hydrogen bonds that form between various amino acids in the polypeptide chain. The *tertiary* structure is the three-dimensional configuration of the polypeptide chain—that is, the way it is twisted or coiled around itself. For example, the three-dimensional structure of polypeptide chains that contain cysteine amino acids is partly determined by the chemical bonds called **disulfide bridges** that always form between the sulfur atoms of two particular cysteine amino acids. If you reexamine the structure of cysteine (Figure 1-8) you will notice that this amino acid has a sulfur atom that is free to form a covalent bond with another sulfur atom, as shown in Figure 5-11. To appreciate the importance of cysteine and disulfide bridges, consider what you are actually paying for when you go to a hairdresser for a permanent: You are paying for the destruction and rearrangement of disulfide bonds in the proteins of your hair fibers (see Box 5-2).

Implications of Protein Structures

The structure of proteins may seem far removed from any important social consequence or relevance to our lives. But knowledge of the protein structures in cells may be used by society to our advantage or disadvantage. Biochemical individuality may

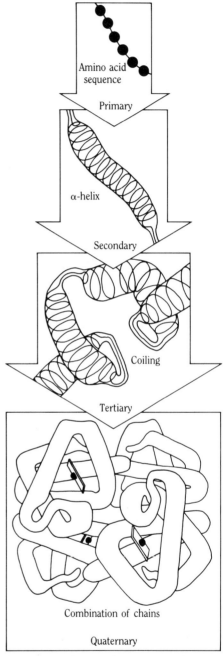

Figure 5-10 The structure of proteins. The primary structure is determined by the sequence of amino acids in a polypeptide chain. This sequence also determines how many folded or helical regions (the secondary structure) the polypeptide will have. The tertiary structure refers to the specific three-dimensional shape of the polypeptide. The quaternary structure refers to the shape of a protein such as hemoglobin that contains intertwined polypeptide chains.

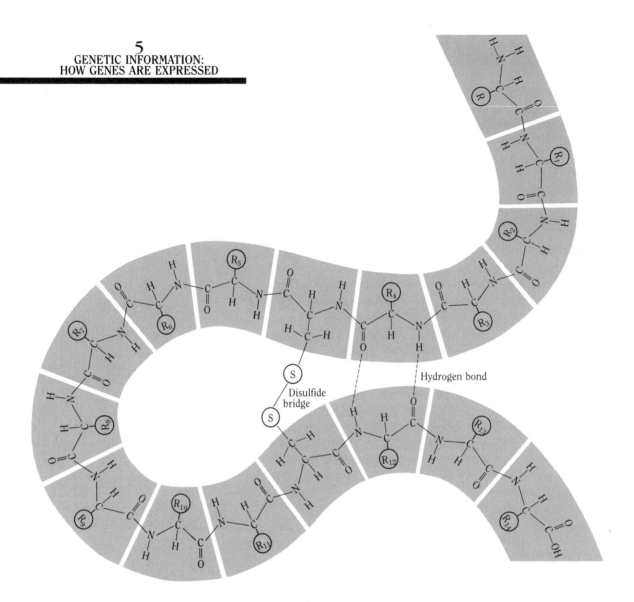

Figure 5-11 Disulfide bridges and hydrogen bonds determine the secondary and tertiary structures of proteins. The sequence of the amino acids (primary structure) determines which pairs can hydrogen bond to one another to twist the polypeptide into a specific three-dimensional configuration. The sulfur atoms in cysteine form covalent bonds and also help determine the protein's tertiary structure.

be used to estimate the chances that certain persons will suffer from genetic diseases (discussed in Chapter 16), to exclude individuals from certain jobs, or even to design chemical or biological weapons that selectively destroy certain populations of people.

Carl A. Larson, a Swedish biochemist, has suggested how biochemical weapons could be developed and used in the future. Such weapons would be possible because people carry different alleles and thus they differ from one another biochemically, and individuals of some ethnic groups tend to have certain alleles in

Hair Styles: Burning Your Disulfide Bridges

BOX 5-2

People spend vast amounts of time and money trying to improve the appearance of their hair—an effort that is actually an applied exercise in biochemistry. Human hair, like the hair of other animals, consists of structural proteins called *keratins*—long protein fibers twisted into helices or other regular structures that are linked to one another by the sulfur–sulfur bonds ($-S-S-$) known as disulfide bridges. The stiffness of the keratin protein matrix is determined by the number of disulfide bridges that connect the proteins. The many disulfide bridges in the keratins in animal horns, hooves, and claws make these protein structures extremely hard and inflexible; hair has fewer disulfide bridges, making it stiff yet flexible; and the low-sulfur proteins in skin make it quite soft and pliable.

A human hair is assembled from protein fibers that are wound together in a regular helical arrangement. When hair is brushed or combed, the helix in each fibril is stretched, temporarily breaking some of the hydrogen bonds holding the helix in position. When the tension is released, the protein helices reform, and the disulfide bridges that connect the various strands reorient the hair fibers into the normal hair pattern.

A permanent wave involves breaking the disulfide bridges with heat and chemicals and then allowing them to reform in different arrangements after the hair is shaped into the desired style by the hairdresser. As long as these new disulfide bridges remain, the hair keeps its new curl. Of course, as new hair grows out, normal hair appearance returns as the disulfide bridges in the new hair assume their natural configuration. An average human hair grows about 6 inches per year. This might not seem like very much, but it means that each hair follicle cell must spin out about ten turns of a helical keratin protein every second.

The proteins that make up hair are synthesized in special cells of the skin and scalp. Thus, adding proteins to hair cannot affect the "health" of hair; the quality of skin and hair proteins is determined by the cells that synthesize these structural proteins by the process of translation. In short, the claims of cosmetics and shampoo manufacturers that their products can improve skin or hair are actually intended to improve their profits.

Reading: Tom Conry (ed.), *Consumer's Guide to Cosmetics*. New York: Anchor Press/Doubleday, 1980.

---

higher frequency than individuals of other ethnic groups. Larson writes

Sometimes, gene frequencies agree fairly well between populations, but more often there are great differences. . . . Although the study of drug-metabolizing enzymes is only beginning, observed variations in drug response have pointed to the possibility of great innate differences in vulnerability to chemical agents between different populations. . . . When new enzyme varieties are discovered, some of them are likely to overstep the prevalence limits so far observed, both high and low, in different populations. . . . Thus, the functions of life lie bare to attack.

A well-known example of human biochemical differences is found in the ABO blood types. Human red blood cells have proteins on their cell surfaces that may be of four different structural types: A, B, AB, or O. A person's ABO blood type is determined by his or her alleles for a particular gene (Chapter 16). The frequency of the four ABO blood types can be measured in any human population. Sometimes the differences in blood types between populations are large. In central Asia, for example, the gene for type B blood is present in 30 percent of the population, whereas among American Indians type B blood is completely absent.

Such large differences in the occurrence of certain proteins can be found for almost any protein that can be analyzed. Because enzymes interact with specific chemicals, it is also relatively easy to find chemicals that can interfere with an enzyme's chemical function. For example, *acetylcholinesterase* is an enzyme that is vital to nerve-cell function in humans. If this enzyme is deactivated, electrical transmission in the brain and nervous system stops, causing death. Certain phosphorus-containing compounds known as nerve gases interfere with acetylcholinesterase activity; thus, exposure to these gases can cause death within minutes.

Now, suppose that a small group of soldiers could be identified who, because of their genetic makeup, had an altered acetylcholinesterase enzyme that was not affected by nerve gas. These soldiers could infiltrate enemy positions and release the gas, which would then kill everyone else but leave them unharmed. Or imagine a chemical that would prove deadly to Caucasians because of their particular alleles and enzymes but would not affect Asiatic or African people. Most people would find such ideas abhorrent, but there are others who believe that this kind of military biochemical research is vital to our nation's security. They believe that if we do not develop such weapons, the "other side" will.

Perhaps the real possibility of ethnic warfare will make you more aware of the ethical and social implications of fundamental genetic and biochemical research. Scientific knowledge is neither good nor bad in itself, but applications of that knowledge may have quite unexpected consequences. The only defense against the abuse of any scientific knowledge is as complete an understanding by as many people as possible of the individual and social implications of scientific discoveries.

Key Terms

transcription   the process of synthesizing RNA molecules from specific segments of DNA molecules.

translation   the process of converting the information in the sequence of bases in a messenger RNA molecule into a sequence of amino acids in a polypeptide.

central dogma   the rules that govern the exchange of information among DNA, RNA, and protein molecules.

messenger RNA (mRNA)   an RNA molecule whose sequence of bases is translated into a specific sequence of amino acids (a polypeptide).

ribosomal RNA (rRNA)   the RNA molecule that is the main structural component of ribosomes.

transfer RNA (tRNA)   a special type of small RNA molecule that helps line up amino acids in the proper sequence by serving as a link between an mRNA codon and the amino acid it codes for.

ribosomes   structures in cells on which protein synthesis occurs. Ribosomes contain three types of rRNA molecules as well as about fifty-five different proteins.

codon   a group of three bases in mRNA that specifies one of the twenty different amino acids.

RNA polymerase   the enzyme used to synthesize rRNAs, tRNAs, and mRNAs from genes in DNA.

promoter site   a sequence of bases in DNA where transcription of a gene or genes begins; the region of the DNA where RNA polymerase attaches.

terminator site   a sequence of bases in DNA where transcription stops; the RNA polymerase detaches from DNA at this site.

anticodon   the particular three bases in a tRNA molecule that recognize the codon corresponding to the amino acid attached to the end of the tRNA.

activating enzymes   the twenty different enzymes responsible for attaching tRNAs to their corresponding amino acids.

peptide bond   the specific covalent bond that joins amino acids together in polypeptide chains.

polypeptide   a chain of amino acids joined together; one or several polypeptide chains make up a protein.

disulfide bridges   a covalent bond between two sulfur atoms in different cysteine amino acids in a protein.

Additional Reading

Larson, C. A. "Ethnic Weapons." *Military Review*, November 1970.

Pederson, T. "Messenger RNA Biosynthesis and Nuclear Structure." *American Scientist*, January/February 1981.

Schmidt, O. and D. Söll. "Biosynthesis of Eukaryotic Transfer RNA." *BioScience*, June 1981.

Smith, M. "The First Complete Nucleotide Sequencing of an Organism's DNA." *American Scientist*, January/February 1979.

Temin, H. M. "RNA-Directed DNA Synthesis." *Scientific American*, January 1972.

Yanofsky, C. "Gene Structure and Protein Structure." *Scientific American*, May 1967.

Review Questions

1. What is the name of the process by which information is transferred from RNA to protein?

2. Which of the three classes of RNA is the smallest?

3. Where does transcription of a gene begin?

4. Where does translation of an mRNA begin?

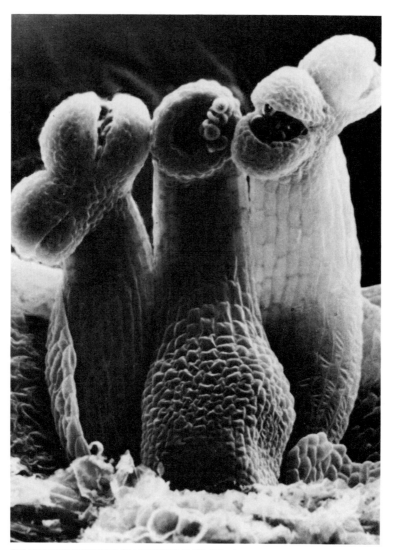

THE MALE ORGANS (STAMENS) OF A DUCKWEED PLANT. (180 ×)

6

THE
GENETIC
CODE

The
Dictionary
of Life

Nature is trying very hard to make us
succeed, but nature does not depend on us.
We are not the only experiment.
R. BUCKMINSTER FULLER

By the early 1960s great strides had been made in understanding the chemistry of DNA and how information in genes is expressed. Following the announcement of the structure of DNA in 1953 by Watson and Crick, scientists from many fields increased their efforts to figure out precisely how genetic information is stored in DNA and how that information is transferred to other molecules in cells. By 1961, the existence of messenger RNAs had been confirmed. Prior to that time these information-containing RNA molecules had been difficult to isolate and analyze in bacteria because of their biological instability. Along with the discovery of mRNA molecules came the first concrete idea of how hereditary information is extracted from genes in DNA and how genetic information is regulated. Also early in the 1960s, some ingenious genetic experiments were performed by Jacques Monod, Francois Jacob, and others at the Pasteur Institute in Paris that indicated how bacterial genes are switched on or off by specific molecules that regulate the transcription of RNA from DNA (their work is discussed in Chapter 9).

By 1961 Francis Crick and Sidney Brenner in England had proved that the genetic code is a *triplet code*—in other words, that a sequence of three bases in mRNA molecules codes for each amino acid. However, the most fundamental question of molecular genetics still remained unanswered: Precisely how does DNA direct the synthesis of proteins? More specifically, what *are* the codons—the three-base combinations that code for each different amino acid?

It began to appear, after the initial discovery of the triplet nature of the genetic code, that the actual deciphering of the code might prove to be an exasperatingly slow task. Efforts to "crack" the genetic code were hampered by a lack of suitable experimental techniques. But finally, toward the end of 1961, a couple of unexpected experimental breakthroughs were made, and the exciting scientific race to discover the base sequence of each of the codons began. By the middle of the 1960s, the complete genetic code had been cracked.

It may be difficult for most people not involved in scientific research to envisage serious, methodical scientists racing against one another to make discoveries and to announce them first.* But cracking the genetic code would mean fame and fortune for the successful scientists—academic promotions, trips to international scientific meetings, and, for the most fortunate individuals, such as Marshall Nirenberg, a Nobel prize. As codon after codon became known, Francis Crick arranged them into a meaningful order that eventually led to the standard genetic code dictionary—the code that is used by every living organism. This chapter explains how the genetic code dictionary

---

*For an engrossing account of such a race, read James Watson's controversial little book *The Double Helix*, which recounts Watson and Crick's race to be the first to discover the molecular structure of DNA.

Second base in codon

First base in codon	U	C	A	G	Third base in codon
U	UUU, UUC } Phe · UUA, UUG } Leu	UCU, UCC, UCA, UCG } Ser	UAU, UAC } Tyr · UAA, UAG } Stop	UGU, UGC } Cys · UGA Stop · UGG Trp	U C A G
C	CUU, CUC, CUA, CUG } Leu	CCU, CCC, CCA, CCG } Pro	CAU, CAC } His · CAA, CAG } Gln	CGU, CGC, CGA, CGG } Arg	U C A G
A	AUU, AUC, AUA } Ile · AUG Met (start)	ACU, ACC, ACA, ACG } Thr	AAU, AAC } Asn · AAA, AAG } Lys	AGU, AGC } Ser · AGA, AGG } Arg	U C A G
G	GUU, GUC, GUA, GUG } Val	GCU, GCC, GCA, GCG } Ala	GAU, GAC } Asp · GAA, GAG } Glu	GGU, GGC, GGA, GGG } Gly	U C A G

Figure 6-1 The genetic code. The code shows the correspondence between each three-base codon in mRNA and the amino acid for which it codes. The first base in each codon is closest to the 5′ end of the mRNA and the third base is closest to the 3′ end. U in the third position often translates the same as C; A in the third position often translates the same as G. The codon most commonly used to start synthesis of a polypeptide is AUG. Three codons (UAA, UGA, and UAG) are used to terminate polypeptide synthesis; these "stop" codons do not code for any amino acid. Figure 1-8 gives the complete names and structures for the amino acids.

is read and what insights can be drawn from the information carried by the code itself. It also describes some of the critical biochemical and genetic experiments that led scientists to formulate how genetic information is stored, expressed, and regulated within cells.

What Is the Genetic Code?

The information in the genetic code (Figure 6-1) is similar to a translation dictionary (say, English–German): it allows one molecular language—the sequences of bases in RNA—to be translated into the other molecular language: the sequence of amino acids in proteins. The genetic code is a triplet code consisting of the four different bases arranged in groups of three. Because proteins are synthesized from mRNAs, the genetic code is listed in Figure 6-1 in terms of messenger RNA bases. However, it should be remembered that DNA is the permanent storehouse of all genetic information and that the bases in mRNAs are synthesized from the complementary bases in DNA (Figure 6-2). What facts and important inferences can be drawn from the complete genetic code shown in Figure 6-1?

1. The genetic code has a well-defined structure; that is, the codons are not randomly assigned to the various amino acids. For those amino acids that are specified by more than one codon, the first two bases are usually the same; base differences occur only in the third position of the codon for most amino acids (the exceptions are leucine, serine, and arginine).

2. In the third position of each codon, the bases U and C

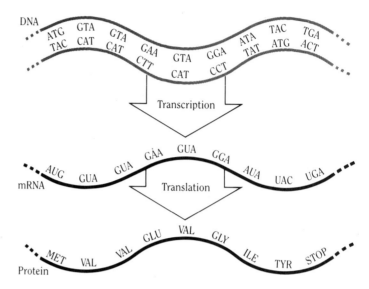

Figure 6-2 The flow of information from DNA to mRNA to protein. The bases in one of the two DNA strands are transcribed into complementary bases in an mRNA molecule. The mRNA codons are translated into the sequence of amino acids in a protein.

carry the same meaning; for example, UUU and UUC code for phenylalanine, and AGU and AGC code for serine.

3. The genetic code is **degenerate**. This means that most of the amino acids are coded for by several different codons. For example, phenylalanine has two codons, UUU and UUC, and leucine has six codons, UUA, UUG, CUU, CUC, CUA, and CUG. An important consequence of code degeneracy is that the base composition—that is, the relative amounts of the four bases—in the DNA of different organisms can be quite variable, yet the amino acid composition of their proteins may be virtually identical.

4. The genetic code is **universal**. This means that in every living organism each codon specifies the same amino acid. For example, the codon ACG specifies threonine in cells of bacteria, wheat, whales, and humans. Thus, mRNA molecules from any organism can, in principle, be accurately translated by the protein-synthesizing machinery of cells from any other organism. For example, mRNAs carrying genetic information for human hemoglobin have been inserted into frog cells and into bacteria, and in each instance human hemoglobin polypeptides have been synthesized. To date, only minor exceptions to the universality of the genetic code have been found (see Box 6-1).

5. The genetic code contains codons that are used to signal the beginning and end of genes, just as punctuation marks demarcate sentences. Four codons serve as start and stop signals for the translation process. The AUG codon, which codes for methionine, initiates the synthesis of most proteins in bacteria and probably does the same in eucaryotic cells. In principle, every bacterial polypeptide should have methionine as the first amino acid. However, it turns out that the initiating methionine is usually removed from newly synthesized proteins by enzymes.

The Genetic Code Is Universal,
But...

Box 6-1

When presented with a general rule by a teacher, some students feel challenged to find an experiment or fact that will contradict the generality of the principle that is being taught. This questioning attitude is healthy and shows that students are trying to solve things for themselves. Some teachers enjoy being questioned by students; others bristle at what seems to be a threat to their expertise or classroom authority. The maxim "Exceptions prove the rule" is often used to stifle further questions or debate. But, in fact, many scientific breakthroughs derive from paying attention to exceptions to the rule. Exceptions and unexplained observations create new insights, stimulate new hypotheses, and eventually result in new and more accurate theories.

Mitochondria are one exception proving the rule that the "universal" genetic code, while not quite universal, is very nearly so. They also provide some insight into the code's origins. All eucaryotic cells contain mitochondria—organelles that, as we noted in Chapter 4, provide the chemical energy (ATP molecules) necessary to carry out cellular chemical reactions. Mitochondria have several biochemical properties that, in some respects, more closely resemble biochemical properties of primitive procaryotic

cells than of eucaryote cells. Some scientists suggest that millions of years ago, eucaryotelike cells "swallowed" some procaryotic cells and that eventually the procaryote and the eucaryote established a symbiotic relationship. Ultimately, the "ingested" procaryotic cells may have evolved into mitochondria performing specialized functions. However, while it is true that mitochondria have biochemical properties characteristic of procaryotic cells, it has by no means been proven that these organelles arose from bacterialike organisms.

Each mitochondrion contains a small circular DNA molecule that codes for ribosomal RNAs, some tRNAs, and a few proteins. All mitochondrial DNA is very small compared to chromosomal DNA but varies greatly in size; human mitochondrial DNA has only 15,569 base pairs—enough for a dozen or so genes. The extremely small size of mitochondrial DNA in all organisms relative to the DNA in the nucleus has apparently forced some biochemical economies as well. Mitochondria take a few liberties in reading the genetic code in their small DNAs. In human mitochondria the stop

codon UGA is read as tryptophan, AGA and AGG mean "stop" instead of coding for arginine, and the codon AUA is read as methionine instead of isoleucine. The few tRNAs that are transcribed from genes in the mitochondrial DNA are much smaller than normal, and they have lost some of their codon specificity. In fact, some of the mitochondrial tRNAs are able to recognize four different codons instead of only one.

Natural selection may have changed the genetic code in mitochondrial DNA for reasons that have yet to be discovered. It is curious that one of the smallest and least complex of eucaryotic cells, yeast, has the largest known mitochondrial DNA (78,000 base pairs), whereas human cells have the smallest—a paradox that has yet to be explained. Comparing mitochondrial DNAs from different organisms and determining the extent to which the genetic code is interpreted differently by different mitochondria provide another powerful means of analyzing the genetic mechanisms that all cells have in common as well as the evolutionary mechanisms that have resulted in biological diversity.

Reading: Robert Reid, "Genes Break the Rules in Mitochondria." *New Scientist,* December 11, 1980.

Three stop codons, UAA, UAG, and UGA, are reserved exclusively for termination of translation—that is, for stopping protein synthesis; none of these three codons specifies any amino acid. Normally, no tRNA molecules recognize the stop codons, and therefore no amino acid can be inserted into the growing polypeptide chain when these codons appear in the mRNA.

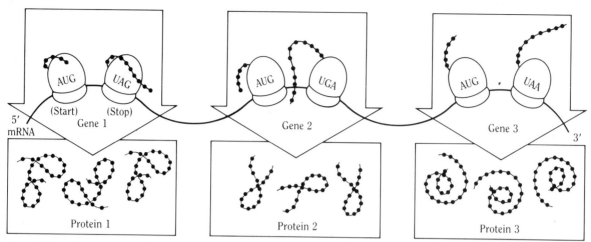

Figure 6-3 Polygenic mRNAs. These mRNAs, found in bacteria, carry information for the synthesis of several different proteins. Synthesis of each protein is initiated at a particular AUG codon and terminated at either UAA, UAG, or UGA (stop) codons. Thousands of proteins are synthesized from each polygenic mRNA molecule.

In bacteria, many mRNA molecules carry genetic information from several genes, which means that these **polygenic mRNAs**, as they are called, code for the translation of several different polypeptides. Because bacteria have polygenic mRNAs, the ribosomes must be able to attach correctly to the start of each gene (which is the function of the AUG codon) and to disengage correctly at the end of each gene (which is the function of the UAA, UAG, and UGA codons). Although any of the three stop codons can terminate the synthesis of a polypeptide chain, sometimes more than one stop codon is used to prevent attachment of additional amino acids and to ensure release of the finished protein from the ribosome (Figure 6-3).

In bacteria, mRNAs carrying information from as many as ten different genes have been observed. Even larger mRNAs are carried in RNA viruses, which inject a molecule of RNA (containing all of their genetic information) into living cells. The polio virus that infects human cells (and that crippled tens of thousands of people before the development of the polio vaccines) has an RNA molecule that contains 6,621 bases. This RNA carries genetic information coding for twelve different proteins and is translated without a stop from one end to the other. After the enormously long polypeptide is synthesized, the twelve proteins are separated by the action of enzymes.

Why the Code Must Be a Triplet Code

Now that we have reviewed the important features of the genetic code as it was eventually deciphered, we will go back in time to examine some of the ideas and experiments that led to the complete cracking of the code. Early theoretical arguments showed that the genetic code could not possibly consist of just one or even two bases. Because only four different bases are used—A; C, G, and T in DNA and A, C, G, and U in RNA—taken singly these four letters could only code for four amino acids. And if the four letters were used two at a time, still only sixteen different amino acids could be specified (4^2). Because there are twenty different amino acids, the four bases must be used in groups of at least three—and, in fact, four bases taken three at a time can code for as many as 64 amino acids (4^3). This realization reduced the problem to determining which amino acid each triplet signifies and accounting for the extra coding information since, in principle, only twenty triplet codons are needed yet sixty-four are available.

It is one thing to hypothesize that the genetic code is a triplet code consisting of sixty-four codons, but it is quite another matter to prove it experimentally. In 1961 Francis Crick, Sidney Brenner, and their coworkers devised a series of experiments with which they were able to demonstrate that bases in mRNA are indeed read in groups of three (or possibly six or nine, which is highly unlikely). They reasoned that if bases are read three at a time when the code is deciphered, then adding or removing one or two bases will cause the reading of the bases to be shifted so that the triplet codons no longer make any sense.

To understand what happens when a reading frame is shifted, notice what happens to a sentence if a letter is removed and the same grouping of letters in the words is maintained. For example, do ouu nderstandn oww hati mean? The addition or deletion of one or two bases in mRNA can be accomplished by the insertion or removal of bases in the DNA causing **frameshift mutations** (Figure 6-4).

Crick, Brenner, and their coworkers were able to produce a large number of different frameshift mutations by exposing the DNA of T4 phages to a particular chemical that causes the deletion or addition of base pairs. Although it is impossible to observe such a small change in DNA directly, these mutations can be detected indirectly by observing whether or not the mutant T-phages are able to grow in bacteria (Figure 6-5). By employing some ingenious genetic manipulations, the researchers were able to determine the exact number of base additions or deletions in each of the mutant T-phages. The key discovery in the experiments was that the addition or deletion of any three base pairs in the DNA would still permit the T-phages to grow. As a result of the base changes, a small region of the mRNA transcribed from the gene with the three mutations was usually misread, but some of the mRNA was translated correctly and produced some functional protein so that the T-phages were able to grow. These results, complicated as they may seem, were the first experimental confirmation of the triplet code. For genes to be expressed correctly, precise groups of three bases must be read sequentially from one end of the mRNA to the other just as words in a sentence are read from beginning to end.

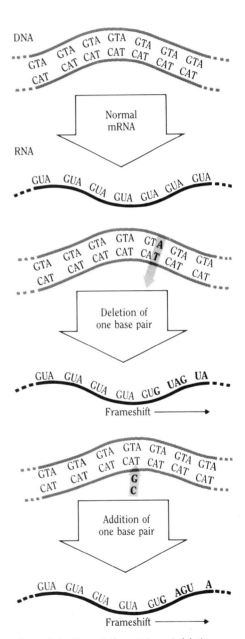

Figure 6-4 Frameshift mutations. A deletion or addition of one (or two) bases in a gene causes the mRNA made from that gene to be misread from the point of the insertion or deletion of the base pair.

6
THE GENETIC CODE:
THE DICTIONARY OF LIFE

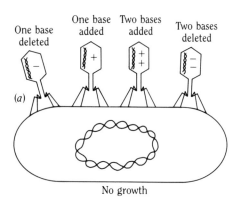

One base deleted · One base added · Two bases added · Two bases deleted

(a)

No growth

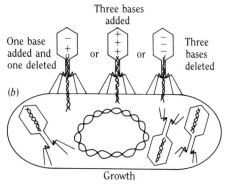

One base added and one deleted · or · Three bases added · or · Three bases deleted

(b)

Growth

Figure 6-5 Effects of addition or deletion of bases in a gene of a T-phage. (a) The addition or deletion of one or two base pairs in a particular gene prevents the growth of T-phages because the gene is inactivated by the frameshift mutation. (b) The correct reading frame can be restored if one base pair is deleted but another is added close by. The reading frame is also unchanged if groups of three bases are added or deleted. Each of these multiple mutations result in growth of the T-phages.

Cracking the Code

Knowing that the genetic code is read in groups of three bases still does not permit assignment of the correct three bases (codon) for each amino acid. An important technical breakthrough made by Marshall Nirenberg and Heinrich Matthei in the early 1960s finally led to the cracking of the genetic code—the assignment of codons to amino acids. Nirenberg and Matthei showed that an artificial, chemically constructed mRNA molecule consisting only of uracil bases (polyuridine) could be successfully translated into a polypeptide in an *in vitro* protein-synthesizing system. In such a system the necessary cellular components for carrying out the synthesis of proteins—ribosomes, tRNAs, activating enzymes, amino acids, and other essential enzymes—are first isolated from cells. Then the individual components are combined in a test tube, where, if messenger RNAs are added, they carry out the process of translation using the same chemical reactions that normally occur inside the cell.

Nirenberg and Matthei isolated the cellular components of the translation process, then carefully removed all remaining cellular mRNA molecules from the preparation in order to test the capacity of the artificial mRNA to direct the synthesis of a protein. When the polyuridine mRNA was added to the *in vitro* protein-synthesizing system, a polypeptide chain was produced. It proved to be polyphenylalanine, a polypeptide made up solely of the amino acid phenylalanine. The researchers thus concluded that UUU codons were being read as phenylalanine.

Artificial mRNAs composed of other bases were constructed and used to establish two other codons: AAA was found to code for lysine and CCC for proline. However, synthesis of artificial mRNAs was a difficult task for chemists at that time and, in general, only gave unambiguous codon assignments when all of the bases in the artificial mRNAs were identical.

As a result of another experimental breakthrough made by Nirenberg in 1964, within a year or so the complete genetic code was cracked—one of the more remarkable accomplishments of molecular genetics. Nirenberg discovered that specific codons—just three bases that could be constructed chemically—would function as mini-mRNAs in the *in vitro* protein-synthesizing system. In a series of experiments Nirenberg and scientists in other laboratories were able to show that each three-base codon would attach to one, and only one, tRNA carrying a particular radioactively labeled amino acid. Because the codon–tRNA–radioactive-amino-acid complex became attached to a ribosome, its presence was easily detected by pouring the contents of each experimental test tube (shown in Figure 6-6) through a filter that trapped the complex of molecules and the radioactivity. (The principle is the same as using a coffee filter to trap the

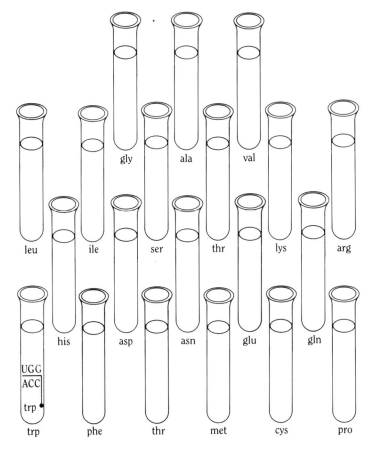

Figure 6-6 Nirenberg's tRNA-codon binding experiment led to the cracking of the genetic code. Each tube contains all of the components required for *in vitro* protein synthesis, including all twenty different amino acids. (No mRNAs are present, however.) One specific codon (UGG in this experiment) is added to each tube and functions as a mini-mRNA. A different radioactively labeled amino acid is then added to each tube, as indicated below each tube. In only one of the twenty tubes will a radioactively labeled amino acid become attached to its corresponding codon via the appropriate tRNA (UGG in this experiment). The one tube in which attachment occurs can be detected by pouring the contents of each tube through paperlike filters that trap the ribosome-codon-tRNA complexes. Because the experimenter knows which codon and which radioactively labeled amino acid have been added to each tube, he or she can assign the codon to a particular amino acid.

coffee grounds while letting the coffee-flavored water run through.) Because each different chemically constructed three-base codon would attach to only one particular radioactive amino acid in each tube, each codon could be assigned unambiguously to each amino acid. Thus, the code was rapidly cracked and led to the codon–amino acid assignments shown in Figure 6-1.

Proving Codons Are Recognized by tRNAs

The crucial step in protein synthesis is the recognition of the codons in mRNA molecules by the tRNAs that carry the corresponding amino acids. So far in our discussion we have taken this crucial step in the translation process for granted. Molecular biologists, however, had to demonstrate that tRNAs do, in fact, recognize codons through binding of anticodons. A definitive experiment showing that tRNAs play this role in interpreting the genetic code was finally devised in 1968. This important experiment also utilized an *in vitro* protein-synthesizing system, in this case "primed" with mRNA molecules that coded for the synthesis of rabbit hemoglobin. This experiment showed that it

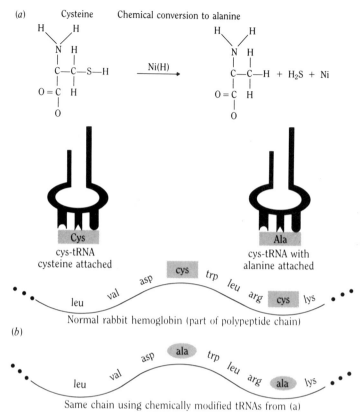

Figure 6-7 Experiment showing that tRNAs recognize codons regardless of what amino acids are attached to them. (*a*) A mixture of tRNAs carrying the correct amino acids is chemically treated with a nickel catalyst that converts cysteine to alanine. (*b*) The modified tRNA-amino acid mixture is used in an *in vitro* hemoglobin-synthesizing system. Alanine is incorporated into newly synthesized hemoglobin polypeptides in the position where cysteine is normally found.

is the tRNAs—not the amino acids—that actually recognize codons in mRNAs.

Ribosomes with hemoglobin mRNAs attached were extracted from rabbit red blood cells. These hemoglobin mRNA–ribosome complexes were then incubated in a test tube along with all of the other cellular components needed for *in vitro* protein synthesis—with one critical modification. In a separate step a mixture of tRNAs with their amino acids attached was treated with a metallic nickel catalyst that converted the amino acid cysteine to alanine (Figure 6-7a). As a result of this chemical modification, cys-tRNA was attached (incorrectly) to alanine. The researchers predicted that when this modified tRNA–amino acid mixture was added to the *in vitro* system, wherever a cysteine codon appeared in the mRNA it would be recognized by the cys-tRNA but that alanine would be inserted into the hemoglobin instead of cysteine. They further suggested that this would occur only if the tRNAs are responsible for recognizing codons and are indifferent to the amino acids that they carry.

The tRNA–amino acid mixture was then added to the other cellular components capable of synthesizing rabbit hemoglobin *in vitro*. Would alanine appear in the completed hemoglobin proteins at positions where cysteine normally is found? It did

(Figure 6-7*b*). Thus, the experiment showed that the tRNAs function as predicted in the process of translation: Specific tRNAs recognize their corresponding codons in mRNAs regardless of the amino acids attached to them, and the tRNAs are the molecules that actually decipher the genetic code. This experiment also demonstrated the crucial role of the activating enzymes in the process of protein synthesis: if they fail to attach the correct amino acid to the corresponding tRNA, it is possible for an inactive protein to be produced, since the wrong amino acid would be inserted at certain positions in the protein.

Errors in the Code: Altered Human Hemoglobins

Insight into the important interrelationship of the genetic code, the processes of transcription and translation, and protein activity in cells can be gained by examining the effect of altering a single base pair in the gene that codes for human hemoglobin. As noted in Chapter 5, the human hemoglobin protein consists of four polypeptide chains: two identical alpha (α) and two identical beta (β) chains. Each hemoglobin molecule picks up as many as four oxygen (O_2) molecules as it passes through blood vessels in the lungs and transports them to other body tissues, where the oxygen is exchanged for carbon dioxide, a by-product of other cellular chemical reactions.

Two different genes are involved in the synthesis of adult human hemoglobin: one to direct the synthesis of α-polypeptide chains, and one to direct the synthesis of the β-chains. By the middle of the 1960s the complete amino acid sequence for the β-chain had been determined by biochemical techniques. The first seven amino acids of the β-chain of normal human hemoglobin are shown in Figure 6-8, together with the corresponding amino acid sequences of hemoglobins isolated from the blood of anemic patients—people whose blood is deficient in functional red blood cells. In each abnormal hemoglobin protein, a single amino acid substitution occurs at position 6 of the β-chain. People with hemoglobin S have valine inserted at position 6 instead of glutamic acid; they suffer from **sickle-cell anemia,** named after the shape taken by oxygen-deprived red blood cells (Figure 6-9). People with hemoglobin C have a lysine substitution at position 6 and also suffer from severe anemia. These single amino acid substitutions in the β-chain of hemoglobin reduce the hemoglobin's ability to bind oxygen and thereby cause serious anemia. Several hundred variations of human hemoglobin caused by mutations have been detected; however, only a few cause any symptoms of disease. One interesting aspect of many of these mutant hemoglobins is that they have been preserved in human populations, in some instances at least, because of natural selection (see Box 6-2).

By referring to the genetic code (Figure 6-1) it is possible to

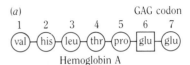

Figure 6-8   The normal function of human hemoglobins is lost by substituting one amino acid for another in the β-chain. (*a*) Normal adult hemoglobin has glutamic acid at position 6 of the β-chain. (*b*) Sickle-cell hemoglobin results from the changing of glutamic acid to valine at position 6. Note that this amino acid substitution can result from changing only one base in the codon—a change that results from a mutation of just one base pair in the DNA. (*c*) Substituting lysine at position 6 causes another form of severe anemia.

(a)   (b)

Figure 6-9 Normal blood cells (*a*) compared to sickled cells (*b*) from patients with sickle-cell anemia. The characteristic sickle shape is produced when the supply of oxygen in the blood is low. People with this blood disease are unable to transport sufficient oxygen to other body tissues because of the altered hemoglobin's inability to function normally. (Courtesy George Brewer, University of Michigan Medical School, and Marion I. Barnhart, Wayne State University Medical School.)

determine which codon or codons specify each amino acid in the normal hemoglobin β-chain. Because some amino acids can be specified by more than one codon (degeneracy of the genetic code), the codon assignments corresponding to each amino acid in hemoglobin are ambiguous. However, when mutant codons are also assigned to the substituted amino acids (Figure 6-8), it is often possible to make an unambiguous codon assignment for each amino acid at a particular position in the polypeptide. In a normal hemoglobin β-chain, GAG must be the codon (not GAA) that specifies glutamic acid at position 6, since single base-pair changes in the DNA will generate the codons for valine (GUG) or lysine (AAG). You will discover (try it) that it is not possible to assign the GAA codon to glutamic acid in such a way that it can generate the altered amino acids by single base changes.

The chemical differences between hemoglobin S, hemoglobin C, and normal hemoglobin were first demonstrated by Nobel laureate chemist Linus Pauling in 1949. Pauling used an analytical technique called **electrophoresis**, which separates similar protein molecules by exploiting the fact that different proteins migrate at different rates on a moist piece of paper or in a gel-like material when an electrical current is passed through it (Figure 6-10). The more electrically charged the proteins are (charges form on amino acids in the proteins when they are placed in solutions of salts), the more rapidly the proteins move in the electrical field created by the current. Figure 6-10 shows characteristic electrophoretic separation of hemoglobin molecules from an individual with normal hemoglobin-coding genes and from individuals with abnormal hemoglobins. Individuals who are **homozygous** (from the Greek *homos,* "one and the same") for normal hemoglobin genes produce only one kind of hemoglobin (A, S, or C in Figure 6-10). That is, they have identical alleles at the same genetic locus on a pair of homolo-

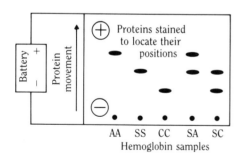

Figure 6-10 The use of electrophoresis to separate and identify different hemoglobins. Six various hemoglobin samples differ from one another by single amino acid substitutions that change the electrical charge on the hemoglobin molecules, causing them to move at different velocities in the electrical field. Individuals that are homozygous for hemoglobin genes produce only one kind of hemoglobin. Normal and mutant alleles can be distinguished by the different electrophoretic patterns of the hemoglobin proteins.

Are Mutations Always Harmful?

Box 6-2

The effects of mutations depend on the chemical environment in which they are expressed, which in turn depends on the environment in which the organism lives. Because most organisms are well adapted to their particular environments, any mutation is likely to change the organism in a way that makes it less well adapted to its environment. Thus, *most* mutations are harmful.

Natural selection usually works to eliminate from a population those genes that cause serious disease, because individuals carrying these mutant genes often do not survive long enough to reproduce. There are exceptions, however. Certain complex gene-environment interactions sometimes permit the maintenance of otherwise lethal genes in human populations. Sickle-cell anemia and β-thalassemia (Cooley's anemia) are good examples. These serious human diseases are caused by mutations that reduce the oxygen-carrying capacity of hemoglobin molecules in blood cells, yet they are maintained in certain human populations because they are caused by mutations that also enhance the survival and reproduction of certain individuals in those populations.

Sickle-cell anemia and β-thalassemia are both caused by different mutations arising in the gene that codes for the β-chain of human hemoglobin. This gene is carried on human chromosome 11; if both chromosomes carry mutant alleles, the individual is homozygous for the mutation and will suffer from severe anemia. Without medical help, the homozygous individual is likely to die quite young. However, if the individual is heterozygous for that allele—that is, if the cells have one normal and one mutant gene—then half the hemoglobin molecules will function normally, and only half will be defective. Because of the amount of normal hemoglobin, heterozygous persons rarely suffer from anemia. Now, in some regions of the world, especially in those portions of Africa where sickle-cell anemia is prevalent and along the southern coast of the Mediterranean Sea where β-thalassemia is common, malaria is responsible for up to half of all childhood deaths. Interestingly enough, however, children with one mutant hemoglobin gene tend to be among the survivors—they are less affected by malaria than children with normal hemoglobin. What does malaria have to do with hemoglobin?

Malaria is caused by a protozoan that reproduces in human red blood cells (rbc's). When this parasite grows and reproduces in human rbc's, the rbc's are destroyed, causing anemia and even death. Recent research has shown that the malaria protozoa cannot grow nearly as well in rbc's with mutant hemoglobin molecules—such as those that occur in people with sickle-cell anemia—as they can in rbc's whose hemoglobin is normal. Apparently, the altered shape of the rbc reduces the amount of potassium in the cell, thereby preventing reproduction of the parasite. In people who are heterozygous, then, the effects of malaria infections are reduced without their experiencing a change in their red blood cell functions serious enough to cause symptoms of anemia.

Thus, a deleterious mutation is maintained in the DNA of certain human populations because it allows people to survive in an environment that otherwise might kill them. Of course, in areas of the world (such as the United States) where malaria has been eradicated, there is no selective advantage in possessing sickle-cell genes. But neither is there a strong selection against them, and so these mutant genes persist generation after generation.

The advantage of having defective hemoglobins in certain environments shows why it is difficult to predict with confidence what the consequences of a particular genetic change will be. The interaction of genes and the environment is very complex; today's lethal mutation may be tomorrow's salvation.

Reading: Anthony C. Allison, "Sickle Cells and Evolution." *Scientific American.* August 1956.
Milton J. Friedman and William Trager, "The Biochemistry of Resistance to Malaria." *Scientific American.* March 1981.

Inborn Errors of Metabolism

Box 6-3

Archibald Garrod, an English physician who practiced medicine at the turn of this century, made a major contribution to our understanding of inherited human metabolic diseases. But as happened with Mendel's experiments (see Chapter 11), the significance of Garrod's discoveries was generally unappreciated for more than thirty years.

In 1909 Garrod published a book in which he described four human diseases: *alkaptonuria, pentosuria, cysteinuria,* and *albinism.* He diagnosed the first three diseases on the basis of the presence of unusual chemical substances excreted in the person's urine. Alkaptonuria, in particular, is easily diagnosed because on exposure to air the urine turns black. Albinism is characterized by a defect in cellular pigmentation—the absence of the substance called *melanin*—that is especially noticeable in the skin, hair, and eyes. (See photo.)

Garrod realized that albinism was similar to the other three diseases because people with albinism lacked skin pigments whose synthesis depends on enzymes. He deduced that the missing pigments must be synthesized by the action of enzymes that were also genetically determined. Garrod proposed three possible explanations for albinism: "We might suppose that the cells which usually contain pigment fail to take up melanins formed elsewhere; or that the albino has an unusual power of destroying these pigments; or again that he fails to form them." He went on to argue that the last explanation was correct and that the lack of melanins resulted from a missing enzyme.

Garrod was familiar with Mendel's discoveries concerning the patterns of inherited traits in peas. He recognized that the four human diseases he had been studying were caused by inherited, recessive genes because the pattern of inheritance behaved according to Mendel's laws (discussed in Chapter 11). With remarkable insight, he proposed that all four human diseases were the result of a missing or inactive enzyme, the absence or malfunction of which was due to a defective gene that had been inherited.

Garrod proposed the general term *inborn errors of metabolism* to describe this group of diseases. Today more than 150 of such inherited *metabolic diseases* (as they are usually called today) have been identified in humans.

In 1982 physicians at the University of California Medical School in San Francisco for the first time were able to detect an inherited metabolic disease in a fetus *in utero* and begin treatment even before the infant was born. Using the technique of amniocentesis (discussed in Chapter 14), they determined that the fetus's cells were unable to synthesize biotin, an essential vitamin, because of a mutant gene that had been inherited from its parents. Giving the mother large doses of biotin during her pregnancy ensured that the fetus received enough of the vitamin to develop properly while still in the womb. After the infant was born, its diet could be supplemented with the vitamin it was unable to synthesize because of the inherited defect.

Albinism in a Hopi child (center). (Courtesy of Field Museum of Natural History.)

gous chromosomes. Individuals who are **heterozygous** (from the Greek *heteros,* "different") have different alleles at the locus and thus produce two different proteins.

Suppression: Misreading the Code

The selection and persistence of defective hemoglobin genes and proteins in human populations (Box 6-2) attests to the unpredictable consequences of mutations. Almost a half century before the hemoglobin changes responsible for the various kinds of anemia diseases were discovered, an English physician, Sir Archibald Garrod, proposed that certain human diseases were caused by defective enzymes. Garrod also suggested that these human diseases were hereditary and that the defects could be transmitted from generation to generation—even though he knew nothing about mutations or DNA at the time (see Box 6-3).

Most changes in DNA result in mutations that do harm the organism in some way. Yet mutations may occur that not only are not harmful but, in fact, may actually correct otherwise lethal defects caused by other mutations. Such mutations, called **suppressor mutations,** are one mechanism by which organisms can correct the effects of harmful mutations.

Suppression of mutations has been studied extensively in bacteria and in simple eucaryotic cells such as yeast. Because the experiments that are used to demonstrate mechanisms of suppression are more difficult to perform with higher organisms, it has not been proven that suppression of mutations occurs in human cells—although it would be surprising if it did not. Moreover, the study of various kinds of suppressor mutations in bacteria has been enormously helpful in uncovering how the genetic code is normally read during the process of translation. Individual components of the translation machinery (most commonly the tRNA) are responsible for suppressing the effects of mutations and they do so by preventing their expression—that is, by preventing the incorrect codons from substituting the insertion of one amino acid for another in proteins.

Unlike frameshift mutations, which add or delete a base pair (or in some cases two base pairs) in the DNA, most mutations simply *substitute* one base pair for another (discussed in Chapter 7). Changing a base pair in a gene in DNA changes a codon in the mRNA transcribed from that gene, and the result is an amino acid substitution in the protein (see Figure 6-11a). In order to suppress the phenotypic effects of the original mutation, the suppressor mutation must change some component of the translation machinery in such a way that the altered codon can be misread occasionally, thereby restoring the original amino acid and activity of the protein. Thus, either the original amino acid must be inserted at the position in the protein dic-

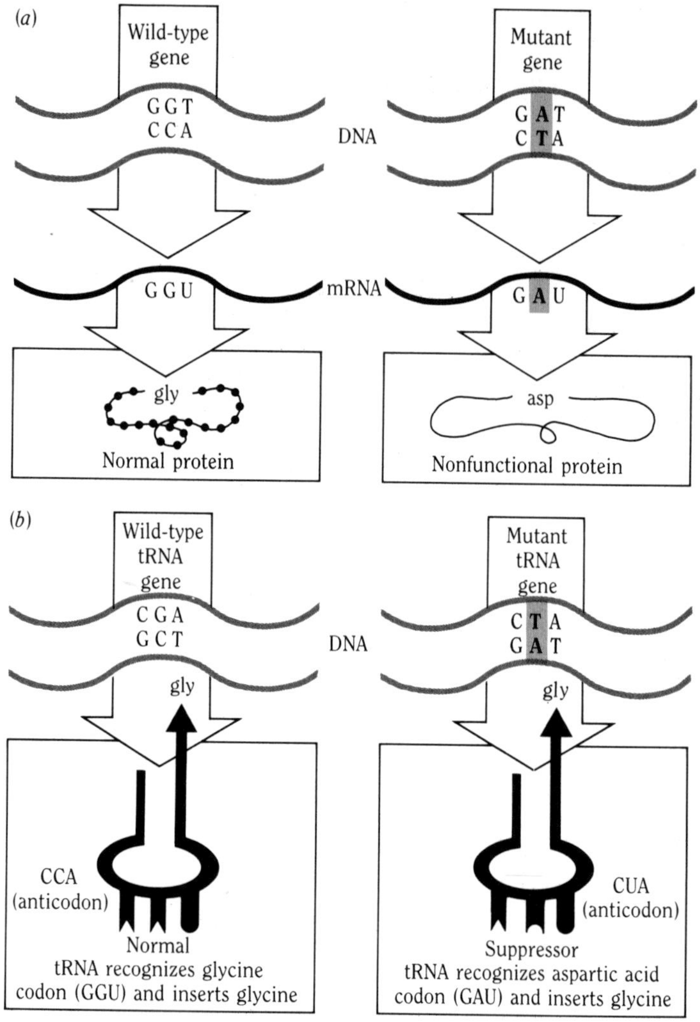

Figure 6-11 Suppression of mutations by tRNAs that have altered anticodons. (a) A wild-type gene contains information for a glycine codon; a mutation in that gene changes one base pair and changes the codon in the mRNA to one that directs insertion of aspartic acid instead of glycine in the protein. (b) Another chance, independent mutation in the gene that produces the glycine-tRNA changes the anticodon of the gly-tRNA so that it recognizes the codon for aspartic acid in mRNA but still inserts glycine.

tated by the changed codon, or some other amino acid can be inserted as long as it restores the protein's activity.

It is the tRNA molecules that are primarily responsible for "reading" the codons correctly. If a base in the anticodon of the tRNA is changed, certain codons in mRNAs will be "misread" (Figure 6-11b). Thus, when suppression of a mutation occurs, two "wrongs" do make a "right." By mistranslating the defective codon, the mutant (suppressing) tRNA inadvertently inserts the correct amino acid into the polypeptide chain. Protein activity is thereby restored, and the cell carries out normal functions even though its DNA now contains two mutations.

Many mutations that would otherwise be lethal in bacteria can be suppressed by altered tRNAs and, less frequently, by altered ribosomes. Often the original amino acid need not be restored, since some proteins will function normally with a vari-

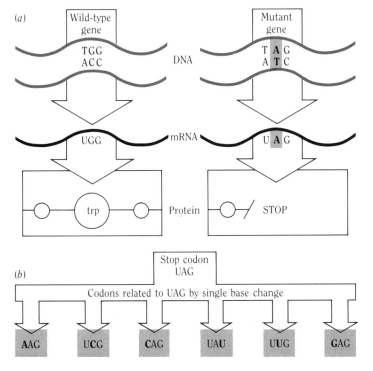

Figure 6-12 Suppression of a stop codon by mutant tRNAs. (a) A mutation in DNA changes a tryptophan (trp) codon in the mRNA into a UAG stop codon. (b) Because six other codons differ from the stop codon by a single base, mutations in any of six different tRNA genes can suppress the effect of mutations that give rise to stop codons in any other gene. Thus, the lethal effect of one mutation may be suppressed by a second mutation that also arises by chance.

ety of amino acid substitutions. Thus, any one of several altered tRNAs may be effective in suppressing mutations that arise in almost any gene. For example, a mutation may change the tryptophan codon (UGG) into a "stop" codon (UAG) within any gene in the organism. Such a mutation invariably destroys the protein's activity, because polypeptide synthesis cannot proceed beyond a stop codon. However, any one of six different tRNA genes may mutate in such a way that their anticodon can now recognize the UAG codon and insert an amino acid. Suppression of the harmful mutation will occur if any one of six different amino acids inserted into the protein is able to restore its activity (Figure 6-12).

Suppression of mutations by changing tRNA molecules is not quite as innocuous to the cell as our description might make it appear. Perhaps it has occurred to you that if tRNA molecules can insert an amino acid at a mutant UAG ("stop") codon, they might also prevent normal polypeptide chain termination at the necessary UAG codons in other genes. Fortunately, suppression by mutant tRNAs is usually rather inefficient. What generally happens is that a small amount of the protein requiring a suppressor mutation is synthesized, while the accuracy of reading of other mRNAs and synthesis of other proteins is somewhat reduced. The net result is that the cell survives with the two mutations and is able to grow and reproduce.

The mechanism of suppression by mutant tRNAs illustrates

the delicate balance between the information carried in DNA and its eventual expression as proteins. Although the genetic code itself is both fixed and universal, the way in which individual codons are read during translation is somewhat flexible. At any one of several steps in the transfer of information from DNA to protein, the fidelity of the information transfer may be modified or lost all together.

All hereditary changes are ultimately caused by mutations. A mutation may cause a lethal defect in an organism, or it may suppress an otherwise lethal change. Mutations are the source of new species and the cause of the extinction of others. Extinction of a species is not a sign of failure of that species, any more than a person's death signifies that his or her life has been a failure. Deaths of individuals and extinction of species are part of the normal and necessary processes of biology and evolution. Nobody can predict with any assurance what organisms will inhabit earth a million years from now. It is quite conceivable—even likely, judging from the past history of life forms on earth—that the human species may not be the most complex or even the most intelligent form of life that will ultimately inhabit the earth. However, it is quite safe to predict that whatever forms of life arise in the distant future, they will be composed of cells that will utilize this same genetic code to store genetic information and that the same cellular mechanisms of transcription and translation will be used to express it.

_____ Key Terms _____

code degeneracy   the property of the genetic code by which some amino acids are coded for by more than one codon.

code universality   the fact that in every kind of organism each codon specifies the same amino acid.

polygenic mRNA   an mRNA molecule that carries information from more than one gene and that is translated into different polypeptides.

frameshift mutation   a mutation that results from the insertion or deletion of one or two base pairs in DNA.

sickle-cell anemia   a serious blood disease caused by a mutant hemoglobin. One amino acid is changed at position 6 of the β-chain as a result of a mutation.

electrophoresis   a technique used for the separation of nearly identical proteins based on differences in their electrical charge.

homozygous   having identical alleles at the same locus on homologous chromosomes.

heterozygous   having different alleles at the same locus on homologous chromosomes.

suppressor mutation   a mutation that suppresses the effects of another mutation. The two mutations usually occur in different genes, although this is not always the case.

Additional Reading

Allfrey, V. G. and A. E. Mirsky. "How Cells Make Molecules." *Scientific American,* September 1961.

Gorini, L. "Antibiotics and the Genetic Code." *Scientific American,* April 1966.

Lagerkvist, V. "Codon Misreading: A Restriction Operative in the Evolution of the Genetic Code." *American Scientist,* March/April 1980.

Rich, A. "Bits of Life." *The Sciences,* October 1980.

Review Questions

1. How many codons are there in the genetic code?

2. What kind of mutation results when a base pair is deleted from DNA?

3. What is the function of hemoglobin proteins in the blood?

4. What class of molecules is usually changed by suppressor mutations?

5. What three codons stop protein synthesis?

6. What kind of molecules contain anticodons?

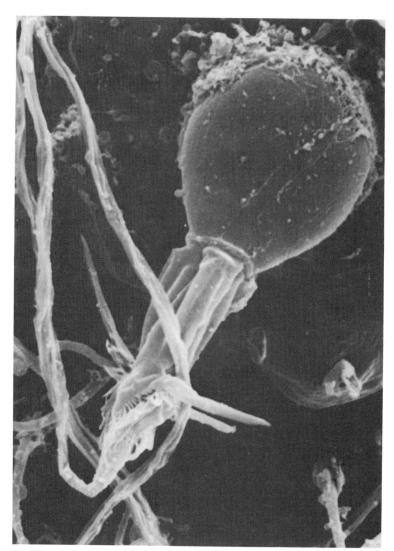

A NEMATOCYST (POISONOUS DART-THROWING ORGAN) IN A HYDRA. (8,120 ×)

7

MUTATIONS

*Changes
in DNA*

*Whoever is led to believe that species are
mutable will do good service by
conscientiously expressing his convictions;
for thus only can the load of prejudice by
which this subject is overwhelmed
be removed.*
CHARLES DARWIN

For many people the word *mutation* conjures up images of physical deformities or late-show movie monsters. But a **mutation** is *any* change that arises in the genetic information of cells. As we pointed out in Chapter 6, the precise phenotypic consequences of any mutation are impossible to predict because the expression of any mutation depends on the organism's environment. Mutations are chance events and arise spontaneously in all DNA molecules. Mutations are neither good nor bad; they are a natural consequence of normal cellular processes. There is no basis for the belief, still held by some people, that human genetic defects represent some sort of punishment or retribution. Mutations are not caused by immorality or unethical conduct; rather, they are a natural consequence of physical and biological processes.

The occurrence of mutations in cells is analogous to automobile accidents. It is impossible to predict which individuals will be involved in the 15 million or so automobile accidents that occur each year and whether or not they will be hurt or killed. However, it *is* possible to identify persons who are at increased risk of being involved in an accident. For example, many studies show that drivers who consistently speed or who drink substantial amounts of alcohol are more likely to be involved in an accident. Speeding and drinking do not directly cause accidents, but they do increase the chance of an accident occurring. Similarly, while the overall frequency of mutations can be estimated, it is impossible to predict which DNA molecule will change or what the effects of the mutations will be. As with auto accidents, certain factors increase the chance that a mutation will occur. These factors, called **mutagens**, include such environmental agents as x-rays, radioactive materials, ultraviolet light, and chemicals that interact with molecules of DNA and increase the frequency with which bases are changed. The greater the exposure to mutagenic agents, the more likely it is that one or more mutations will occur.

Just as most random changes in a fine watch mechanism would be far more likely to cause malfunctioning than to improve functioning, so most mutations are far more likely to impair cellular functions than improve them. Thus, mutations are generally regarded as harmful, and exposure to highly mutagenic substances should be avoided.

Even with minimal exposure to highly mutagenic agents such as x-rays or weak mutagenic agents such as ultraviolet radiation in sunlight there is still some probability for any gene to mutate. Such changes in DNA are called *spontaneous mutations*. These mutations appear to be unavoidable and arise from occasional errors in base pairing during DNA replication or from unavoidable exposure to naturally occurring background radiation (such as cosmic rays) or to naturally occurring chemicals in our environment and food. The frequencies of spontaneous

Table 7-1. Rates of Spontaneous Mutations of Particular Genes in Various Organisms

Organism, and traits	Mutation Rate (mutations per genome per generation)
T-even phages	
Bacterial host range	3×10^{-9}
Rapid-lysis of bacteria	1×10^{-8}
E. coli	
Resistance to streptomycin	4×10^{-10}
Resistance to infection by T-even phages	1×10^{-9}
Inability to metabolize lactose	2×10^{-7}
Neurospora crassa (bread mold)	
Ability to synthesize adenine	4×10^{-8}
Zea mays (corn)	
Shrunken seeds	1×10^{-6}
Purple seeds	1×10^{-5}
Drosophila melanogaster (fruit fly)	
White eye	4×10^{-5}
Yellow body	1×10^{-4}
Homo sapiens (humans)	
Hemophilia type A (bleeding disease)	3×10^{-5}
Achondroplasia (dwarfism)	4×10^{-5}
Retinoblastoma (tumor of eye)	1×10^{-5}
Huntington's chorea (nerve degeneration)	1×10^{-6}

mutations vary enormously from gene to gene and from organism to organism (Table 7-1). When the frequency of mutations rises significantly above these spontaneous mutation rates, the additional mutations are assumed to be caused by some agent—radioactivity, x-rays, chemical mutagens, and so on. However, the kinds of mutations are the same whether the mutations are spontaneous (unattributable to specific causes) or induced (caused by an environmental agent).

In order to evaluate the effects of mutations in causing disease, in producing genetic defects, and in creating new varieties of organisms, it is first necessary to understand the various kinds of mutations that can occur.

Point Mutations

Mutations are classified according to the kinds of changes that occur in DNA molecules. The simplest sort of mutation, called a **point mutation**, involves changing of a single base pair. This change may or may not affect the functions in cells, depending on whether the activity of a protein is changed or whether the regulation of a gene is modified.

When a purine base in one strand of the DNA is changed

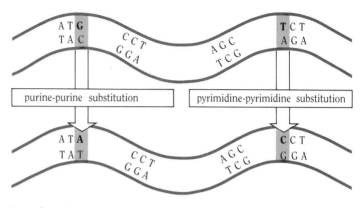

Figure 7-1 Transition mutations. Point mutations are called transition mutations if one purine is substituted for the other or one pyrimidine for the other in a strand of DNA. These kinds of point mutations occur by copying errors when DNA is replicated.

into the other purine by mutation (adenine ↔ guanine) or when one pyrimidine base is changed into the other (cytosine ↔ thymine), the change is called a **transition mutation** (Figure 7-1). It is also possible to convert a purine into a pyrimidine or a pyrimidine into a purine, but these changes are much less common.

Point mutations can also be classified by determining the effect that particular codon changes will have on the protein that is synthesized. Point mutations causing codon changes in mRNA that result in amino acid substitutions in proteins are called **missense mutations**. Quite often the function of the protein will be changed by the amino acid substitution (recall from Chapter 6 the example of sickle-cell anemia caused by hemoglobin S). Other point mutations may change a codon from one specifying an amino acid to any one of the three terminating or stop codons. These kinds of point mutations are called **nonsense mutations**. Examples of both missense and nonsense point mutations are shown in Figure 7-2. Mutations can also occur that change the codon but that do not change the amino acid in the protein or the organism's phenotype. For example, the codon UUU could be changed to UUC by a point mutation, but

Figure 7-2 Missense and nonsense mutations. Missense mutations result from base-pair changes that substitute one amino acid for another in proteins. Nonsense mutations cause the polypeptide to terminate prematurely.

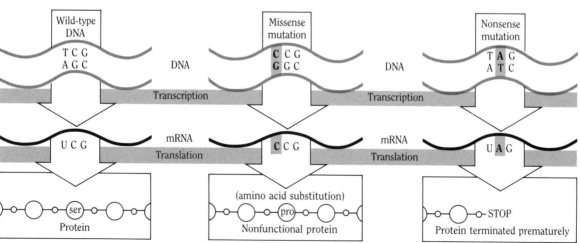

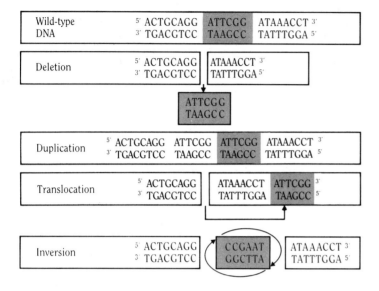

Figure 7-3 Types of mutations. Deletions involve the loss of one or many base pairs; duplications the addition of one or many base pairs. When a segment of DNA is moved from one region to another (or from one chromosome to another), the mutation is called a translocation. When a segment of DNA is removed, rotated end for end, and reinserted into the same location, the mutation is called an inversion.

since both codons specify phenylalanine the protein would be unchanged. Such mutations are said to be silent or **neutral mutations.**

Mutations Involving More Than One Base Pair

Many mutations cause more than just a change in a single base pair. These kinds of mutations may change many bases in DNA or they may even cause the addition, removal, or relocation of parts of chromosomes or even whole chromosomes. The four categories of these more extensive kinds of mutations are deletions, duplications, translocations, and inversions (Figure 7-3).

Deletions result from the loss of at least one base pair in DNA and generally involve the loss of hundreds or thousands of bases. If many bases are lost, there is virtually no chance that they all will be replaced correctly; therefore, deletion mutations are unable to revert to the wild-type sequence of the DNA. Point mutations, on the other hand, may revert; that is, the original base pair may be restored in the DNA by a second mutation.

Duplications result when a segment of DNA is added that is identical to one that already exists. Geneticists believe that duplication of genes has played a particularly important role in the evolution of genetic information. By duplicating genes and having extra copies present in a chromosome, the organism would, so to speak, have "spares" of the genes. Thus, if a gene on one copy were to mutate (sometimes to produce a useful new protein, sometimes not), the unchanged gene would remain intact and continue to function in its essential role. Under certain environmental conditions, the duplicated genes might enhance the survival of the organism and, in this way, be preserved and passed on to progeny.

Table 7-2. Analogies for the Various Kinds of Mutations that Occur in DNA

DNA	Kind of mutation
This is an accurate statement	Wild-type
This is a*m* accurate statement	Point
This is *not* an accurate statement	Insertion
This is *an an* accurate statement	Gene duplication
This is an *in*accurate statement	Gene duplication followed by a point mutation
This is an accurate statement	Chromosome duplication
This is an accurate statement	
This *statement* is an accurate	Translocation
This is an *etarucca* statement	Inversion
Thi sisa nac cur ates tatem ent	Frameshift

When a segment of DNA is moved from its normal location in the chromosome to a new site, the mutation is called a **transloca-tion.** Segments of DNA can translocate to other sites in the same chromosome, or, in the case of cells with many chromosomes, segments of DNA can translocate from one chromosome to another.

If a segment of DNA is removed, rotated end for end so as to maintain the correct chemical polarity, and then reinserted into the DNA, the mutation is called an **inversion.** In such a case bases have not been added to or deleted from the DNA, but the genetic information has been changed because the sequence of bases is different. Inversions have evolutionary significance because they help keep particularly useful groups of genes intact when they are passed on to progeny. Inversions provide a mechanism by which an organism can preserve essential groups of genes over many generations. This is because, during the formation of eggs and sperm when recombination occurs during meiosis, inverted seg-ments of DNA do not recombine since they cannot pair up with the corresponding sequence of bases in the other chromosome (this is discussed further in Chapter 11).

Individual base pairs, the arrangements of large segments of DNA, and even the number of chromosomes in cells may be changed by mutations. To help you understand the various kinds of mutations that arise in DNA, Table 7-2 provides analogies in the form of sentences. You can think of each sentence as a molecule of DNA, each word as a gene, and letters as base pairs. Keep in mind that in a cell the change caused by a mutation first arises in the DNA, is transcribed into the mRNA, then appears as an amino acid change in the protein, and finally results in a change in the phenotype. The organism may look different, function differently, or exhibit different behavior as a consequence of a mutation, just as the sentence in Table 7-2 undergoes alterations in meaning with the various changes.

Dominant and Recessive Mutations

The phenotypic effects of mutations in diploid organisms (those having pairs of chromosomes) may be different from those in haploid organisms (those having only a single chromosome). Because a bacterium has only one chromosome, every mutation in its DNA can be detected in an environment in which the gene carrying the mutation is expressed. For example, if a mutation inactivates a bacterial gene that codes for an essential enzyme, the bacterium cannot survive. All eucaryotic organisms, however—from yeasts to trees to people—are diploid and carry two copies of each chromosome and, therefore, two alleles for every gene. Having two copies of each gene protects organisms from otherwise lethal mutations. In fact, nature seems to design most essential functions and organ in pairs: people have two arms, two legs, two eyes, two ears, two lungs, two kidneys—even two hemispheres of the brain.

Human cells contain twenty-two pairs of chromosomes that are called **autosomes**; the other pair, the **sex chromosomes**, may be identical (XX in females) or different (XY in males). Therefore, all human genes, except for those on the sex chromosomes in males, occur in pairs. As we noted in our discussion of anemia and hemoglobin in Chapter 6, if the pair of genes (alleles) are identical—that is, have the same base sequence—the organism is homozygous for that pair of alleles. If the genes have a different base sequence, the organism is heterozygous for that pair of alleles.

Both alleles may be expressed, although to different degrees, because the amino acid sequences of the proteins coded for by the different alleles are slightly different. In the heterozygous state, the expression of one allele (activity of the protein) may dominate the expression of the other allele. The allele whose expression is reflected in the phenotype is called the **dominant allele**; the nonexpressed gene is called the **recessive allele** (Figure 7-4). Since mutations change genes, they may also change the dominant or recessive relationship of the alleles (this is discussed in more detail in Chapter 11).

For any diploid organism, each gene can be represented by only two different alleles (ignoring gene duplications). However, the alleles may vary among individuals, so that a large number of different alleles may be present in a population. (The evolutionary significance of different alleles is discussed in Chapter 17.) For some human genes the number of alleles at a particular genetic locus is quite large. More than eighty alleles have been detected for the gene that codes for the enzyme glucose-6-phosphate dehydrogenase (G6PD) in various human populations. Because

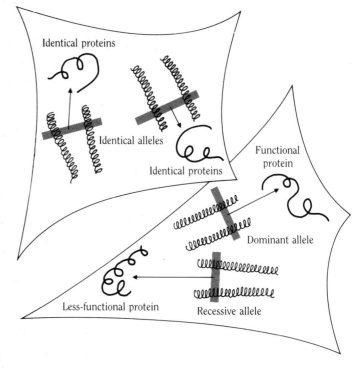

Figure 7-4 Dominant and recessive alleles. Whereas homozygous individuals have identical alleles that produce identical functional proteins, heterozygous individuals have two different alleles. If only one protein is functional, the allele that produces it is dominant and the other allele is recessive.

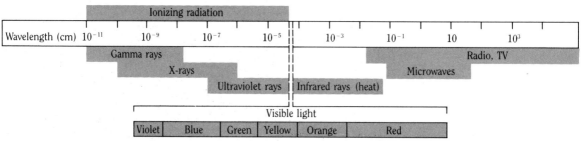

Figure 7-5 The electromagnetic spectrum. Energy is inversely related to wavelength. Ionizing radiation has the shortest wavelengths and the greatest energy. It penetrates cells, damages DNA, and produces mutations. Ionizing radiation is measured in roentgens; one roentgen (pronounced "renkin") produces 2×10^9 ion pairs (positively and negatively charged atoms) in a cubic centimeter of air. Ionizing radiation in cells is expressed in *rems* (roentgen equivalent in *man*), and for most purposes a roentgen and a rem are equivalent.

this gene is located on the X chromosome, females can be either homozygous or heterozygous for it; males, however, have only one copy of the gene. Most of the mutant G6PD alleles function normally; however, some alleles cause *favism,* a disease that results in a severe anemia that can be fatal if the individual eats fava beans or takes certain antimalarial drugs such as primaquine. Some speculations as to why these potentially lethal mutant genes persist in human populations are discussed at length in Box 7-1.

Radiation as a Mutagen

Spontaneous mutations occur at low rates under normal conditions, but certain environmental agents can interact with DNA and dramatically increase mutation rates. For example, **ionizing radiation** (gamma rays, x-rays, and ultraviolet rays) and particles emitted from radioactive materials damage DNA and thereby

Favism:
When It Is Dangerous to
Eat Beans

BOX 7-1

Most mutations that arise in organisms nowadays are likely to be harmful or at best to have little or no effect on survival. The widely accepted explanation for this is that natural selection has kept all the beneficial mutations that arose in the past. The plant or animal surviving today is the one that is "fittest" in its present environment. But a mutation that is adaptive in one environment may be quite maladapted in another. A case in point is a group of mutant alleles that persist in human populations because they enhance people's survival in one kind of environment yet may also cause a lethal disease under a different set of environmental conditions.

These mutant alleles are of a gene on the human X chromosome that codes for the enzyme glucose-6-phosphate dehydrogenase (G6PD), which is involved in metabolism of glucose. More than eighty different mutant alleles of this gene have been detected in human populations. About one person in forty carries one of the mutant G6PD alleles, and cells of these persons show some deficiency in G6PD enzymatic activity. For most individuals, the altered enzymes function well enough that no disease is detectable, and they are unaware that they carry the mutation.

However, for certain individuals the generally innocuous presence of a mutant G6PD gene can abruptly prove fatal. If these individuals eat fava beans, a popular food in Mediterranean countries, they may suffer from *hemolytic anemia*—destruction of red blood cells. If a sufficiently large fraction of their red blood cells is destroyed, they die. In people susceptible to this disease, called *favism,* substances from the digested fava beans alter the chemistry of their red blood cells. If the G6PD enzyme level is sufficiently low, the chemical changes may be sufficient to destroy the red blood cells. If these individuals never eat fava beans, they are usually without symptoms because their red blood cells maintain some G6PD activity.

Why is such a potentially harmful mutation maintained in human populations? As in the case of the altered hemoglobins discussed in Chapter 6, the G6PD mutations make people more resistant to malaria infections. As might be expected, most of the mutant G6PD genes—and most people suffering from favism—are found in malaria-infested countries, which include those around the Mediterranean Sea. It has been possible to demonstrate that the malaria-causing protozoa have difficulty growing in red blood cells taken from people with favism or from people with other G6PD mutations. Thus, the resistance to malaria conferred by G6PD mutations more than compensates for the lowering of cellular G6PD activity in terms of survival.

Favism demonstrates once again why it is impossible to predict the consequences of any mutation without a detailed understanding of the environment in which the mutation is expressed. Do you think that persons with mutant alleles of G6PD should be classified as genetically defective?

Reading: Milton J. Friedman and William Trager, "The Biochemistry of Resistance to Malaria." *Scientific American*, March 1981.

---

increase the frequency of mutations. As Figure 7-5 shows, the full electromagnetic radiation spectrum includes gamma rays, x-rays, ultraviolet radiation (UV), visible light, microwaves, and radio waves. As the wavelength of electromagnetic radiation becomes shorter, the energy of the radiation increases, and above a certain energy level the radiation is able to penetrate cells, damage molecules, and affect chemical reactions. The photons in ionizing radiation have sufficient energy to strip the outer electrons from atoms, thereby causing atoms and molecules to become electrically charged. (Recall from Chapter 1 that in this electrical condition they are called *ions.*)

UV rays have enough energy to penetrate the skin and dam-

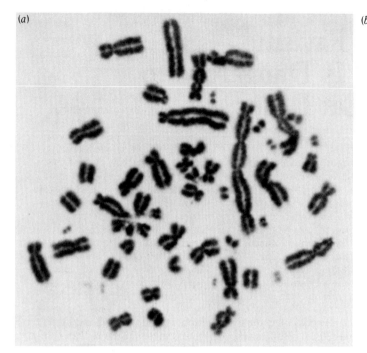

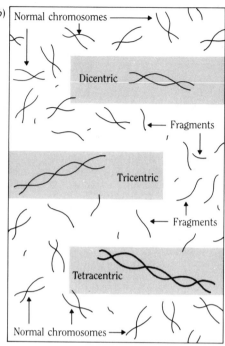

Figure 7-6 Human chromosomes damaged by x-rays. (*a*) Photograph of metaphase chromosomes showing abnormalities caused by x-rays. (Courtesy of Judy Bodycote and Sheldon Wolff, University of California, San Francisco.) (*b*) Diagram illustrating the various kinds of abnormal chromosomes.

age skin cells. It is UV radiation in sunlight that causes tanning—and, in some individuals who are out in the sun a great deal, skin cancer. X-rays are considerably more energetic than UV radiation and can pass completely through the body. That is why an x-ray–sensitive film placed behind the body forms an image when the body is exposed to x-rays. Denser structures such as bones absorb more of the x-rays than the softer body parts do, so an image of the skeleton appears on the film. However, in passing through the body tissues, the x-rays also damage DNA as well as other molecules and cells. If the x-rays damage the DNA of somatic (body) cells, mutations may occur in those cells that may eventually transform them into cancer cells (see Chapter 8). If the x-rays damage the DNA in sex cells, they may cause heritable mutations—that is, mutations that can be passed from generation to generation.

In 1927 H. J. Müller discovered that x-rays cause mutations in *Drosophila* (fruit flies). By irradiating the sperm of male flies, Müller was able to generate more than a hundredfold increase in the mutation rates for various *Drosophila* genes. The mutant genes were passed on to successive generations of flies, proving for the first time that x-rays do indeed cause hereditary mutations. Since Müller's original experiments, it has been shown that x-rays and other forms of ionizing radiation increase the frequency of mutations in all living organisms, including humans.

Subatomic particles (electrons, protons, neutrons) are emitted from radioactive atoms as their nuclei decay into other elements. These subatomic particles also have sufficient energy to penetrate cells, damage DNA molecules, and cause mutations. The physical damage produced in DNA by such radioactivity and by ionizing radiation can be observed by examining irradiated chromosomes under the microscope. Figure 7-6 shows the kind of physical chromosome damage that is caused by low doses of x-rays, for example. Since the chemistry of DNA molecules is the same in all organisms, it is logical to expect that the genetic damage caused by ionizing radiation would be the same in all cells. This expectation is borne out by data showing that the number of x-ray–induced mutations in various species is proportional to the amount of DNA in their cells (Figure 7-7).

While the evidence that x-rays and other forms of ionizing radiation cause mutations is undisputed, nobody knows just how harmful low-level ionizing radiation is to living tissue, and the question continues to be hotly disputed. At extremely low doses of radiation mutation rates are practically impossible to measure accurately. Thus, the dispute centers on the question of whether or not it is fair to extrapolate from the measurable mutagenic effects of high doses of radiation in order to estimate the effects of much lower doses. For example, if a certain dose of x-rays produces one hundred detectable mutations, will a dose 1/100 as strong produce just one mutation? The as yet unknown answer to this question has important consequences for anyone who ever undergoes an x-ray—and for radiologists, x-ray technicians, nuclear reactor technicians, and others who work with and are exposed to even low-level ionizing radiation.

The problem of low-level radiation doses is pointed up in two different interpretations of how mutation rates vary as a function of dose, as shown in Figure 7-8. In both interpretations the number of mutations produced is proportional to the radiation dose. However, the mutagenic effects predicted for low doses of radiation are quite different. Line A extrapolates to zero, which means that regardless of how low the radiation dose is, some mutations will still be produced. Line B, on the other hand, has a *threshold* below which no additional mutations arise. Because accurate experimental data on the number of mutations produced by very low doses of radiation are difficult to obtain, especially for humans, it is not possible to determine which relationship—the one indicated by line A or the one indicated by line B—is the one that actually occurs in nature. If, in fact, there is no threshold, then any exposure to ionizing radiation is potentially harmful.

Periodically, a committee of experts of the U.S. National Academy of Sciences surveys all the data pertaining to radiation and publishes reports on the biological effects of ionizing radia-

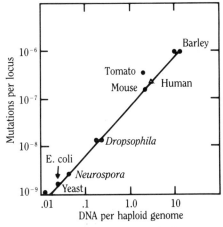

Figure 7-7 Relationship between the frequency of mutations produced by ionizing radiation and the amount of DNA in various organisms. All of the points except for human have been determined experimentally. The human value is estimated from the DNA content of human cells. (Adapted from S. Abrahamson, M. A. Bender, A. D. Conger, and S. Wolff, "Uniformity of Radiation-Induced Mutation Rates Among Different Species," *Nature* 245 (1973) p. 460.)

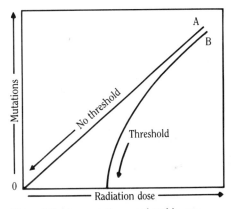

Figure 7-8 Are mutations produced by any amount of radiation? Line A shows no threshold, so some mutations will occur down to zero dose. Line B shows a threshold; below a certain dose no mutations occur. It is virtually impossible to determine which of these radiation response curves applies to humans.

tion (BEIR Reports). At this writing, the BEIR guideline for individual exposure to ionizing radiation recommends that such exposure not exceed 5 rems over a thirty-year period. (A rem is a standard measure of radiation and stands for *r*oentgen *e*quivalent in *m*an.) Natural sources of radiation such as cosmic rays and radioactivity contribute about 2.5 rems to the average person over a thirty-year span, and dental and medical x-rays provide an additional 2 rems. Thus, 5 rems is generally regarded as the amount of unavoidable exposure to ionizing radiation.

Now, this guideline does not say or mean that this amount of radiation is safe; it is simply what the committee feels constitutes an acceptable and unavoidable risk to the public based on current data. It is worth noting that over the past thirty years the maximum recommended dose of radiation considered acceptable has been steadily reduced. Although the effects of small doses of radiation are still being debated, the trend in recent years has been to reduce human exposure to radiation and to assume that all ionizing radiation—whatever the dose—is potentially harmful. As with other aspects of life, the risks and benefits of radiation exposure must continually be weighed by each of us, just as risk-benefit tradeoffs must be considered when determining our exposure to chemicals that are known, or suspected to be, mutagenic.

Chemicals as Mutagens

Many chemicals can penetrate cells and interact with the DNA molecules within them, thereby increasing the frequency with which mutations occur. Of the many thousands of chemicals that can cause mutations, some are highly mutagenic, whereas others are only slightly mutagenic. Therefore, it is important to know not only which chemicals are mutagenic but also the degree to which they increase mutations—and the extent to which humans and other animals are exposed to them.

Our modern lifestyle is intimately linked to the use of manufactured chemicals. Almost everybody uses several—be they paints, plastics, pesticides, or pharmaceuticals—every day. We depend on styrofoam cups for carrying hot coffee, we sit and sleep on polyurethane foam cushions, we wear nylon, rayon, and polyester clothes, we use plastic combs, cups, and pens. And many of us are exposed to some of the thousands of different chemicals that are used in agriculture and industry. Furthermore, hundreds of chemicals are added to our foods and beverages. Since World War II, when the boom in the use of manufactured chemicals really began, about 70,000 chemicals have come into common use in the United States. Only a small fraction of these chemicals—a few thousand at most—have been adequately tested for their mutagenic potential, despite the fact, established by numerous tests, that a strong positive correlation exists between environmental agents that increase the likelihood of mutations (*mutagens*) and agents that cause cancer (*carcinogens*). A chemical mutagen not only increases the frequency of mutations in DNA, it also causes cancers to develop in animals exposed to the chemical (see Chapter 8).

Phage and bacteria have been particularly useful in testing the mutagenicity (the capacity to induce mutations) of chemicals. Because millions, even billions, of microorganisms can be tested for the occurrence of mutations, accurate mutation rates are easily obtained for a wide variety of mutagenic agents. Moreover, since DNA is chemically identical in all organisms, any chemical that produces mutations in bacteria is also likely to cause mutations in human cells. For these reasons, microorganisms are usually chosen for the initial screening of chemicals.

An example of a simple, inorganic chemical that is quite mutagenic and that occurs naturally in many plants, including vegetables such as spinach, is sodium nitrite ($NaNO_2$). This chemical is added to processed meats such as hot dogs, bacon, sausage, and salami to retard spoilage and to enhance the red color. Sodium nitrite is readily converted to nitrous acid (HNO_2) in the stomach by reactions with other stomach acids (HCl) normally present. Nitrous acid not only kills cells but is also highly mutagenic.

The phage $\emptyset$X174 has been used to demonstrate the biological effects of nitrous acid. The advantage of using $\emptyset$X174 is that it contains only a single strand of DNA, so that all mutations resulting from nitrous acid treatment of the phage are expressed and can be detected. Even a brief exposure of phage particles to nitrous acid may cause a mutation in the phage DNA.

The chemical mechanism by which nitrous acid causes mutations and increases their frequency in DNA is shown in Figure 7-9. The amino groups (NH_2) are removed from adenine and cytosine bases in the DNA by nitrous acid. As a result, the bases are converted into other bases that have different hydrogen bonding properties, in turn causing changes in the base pairs after the DNA is replicated. Other chemicals similar to nitrous acid, such as nitrosamines, are also highly mutagenic. These related chemicals are found in pesticides, herbicides, cigarette smoke, and the drinking water of some communities.

Forward and Reverse Mutations

As mentioned in Chapter 3, organisms typically found in natural environments are said to be *wild-type*. Any change from the wild-type must be a mutant if the altered phenotype is caused by a genetic change. Mutations that cause organisms to change from the wild-type to a mutant phenotype are known as **forward mutations**. Generally, whenever mutations are discussed, forward mutations are meant. However, it is also possible to start

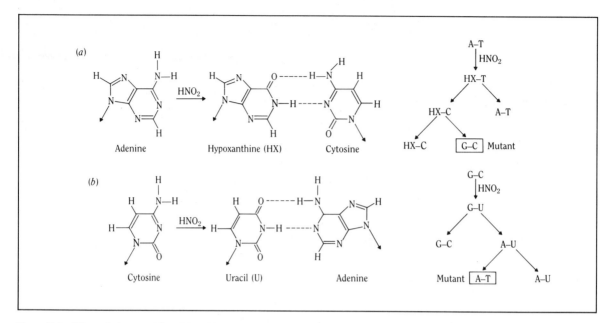

Figure 7-9 Effect of nitrous acid on DNA. (*a*) Adenine is converted to hypoxanthine by nitrous acid. After two replications, one of the DNA molecules contains a G-C base pair instead of the original A-T base pair. (*b*) Cytosine is converted to uracil by nitrous acid. After two replications one of the DNA molecules contains an A-T base pair instead of the original G-C base pair.

with a population of mutant organisms and search for mutations that cause reversion to wild-type. A **reverse mutation** is one that causes the organism to revert to the wild-type phenotype.

Wild-type *Salmonella* bacteria, like *E. coli,* can grow in a liquid medium containing certain minerals and the sugar glucose. From this medium the wild-type *Salmonella* cells are able to synthesize all the other molecules they need for growth—amino acids, vitamins, bases, and so on. For example, histidine is one of the twenty amino acids that wild-type *Salmonella* synthesizes. A mutation arising in any one of the nine genes that code for the enzymes required for histidine biosynthesis creates a histidine mutant, a bacterium that can no longer synthesize histidine. Histidine-requiring *Salmonella* mutants grow only if histidine is added to the medium. However, if a reverse mutation occurs in the DNA that restores the enzyme activity, these bacterial revertants are again able to grow without added histidine (Figure 7-10). Some of the revertant bacteria are true revertants—that is, the wild-type base sequence has been restored in the DNA. Other revertant bacteria grow without histidine because of suppressor mutations and are actually double mutants. **Pseudorevertants** (as such double mutants are called) and true bacterial revertants cannot usually be distinguished from each other phenotypically since they grow under the same conditions.

In the early 1970s Bruce N. Ames, professor of biochemistry at the University of California, Berkeley, devised a simple test for chemical mutagens using histidine-requiring mutants of *Salmonella.* His hope at the time, which has been borne out to a

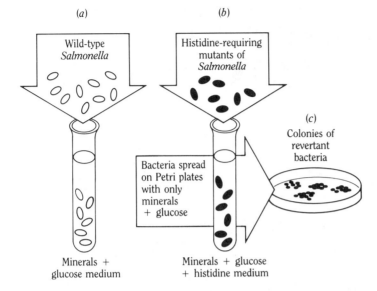

(a) Wild-type *Salmonella*

(b) Histidine-requiring mutants of *Salmonella*

(c) Colonies of revertant bacteria

Bacteria spread on Petri plates with only minerals + glucose

Minerals + glucose medium

Minerals + glucose + histidine medium

Figure 7-10 Detection of revertant bacteria. (a) Wild-type *Salmonella* bacteria grow in a glucose-mineral medium. (b) Histidine-requiring mutants will grow in the same medium if histidine is added. (c) Revertants of the mutant bacteria can be selected by plating large numbers of bacteria on a solid medium that lacks histidine. The histidine-requiring mutants cannot grow there, but any mutation that allows the bacteria to synthesize histidine will grow and form a colony. Some of these colonies are true reversions: the wild-type base pair has been restored in the DNA. Other revertant colonies (pseudorevertants) result from suppressor mutations.

considerable degree, was that testing of a mutagenic chemical by bacteria would provide an indication of the chemical's carcinogenic potential. In fact, about 85 percent of the chemicals that mutate DNA in *Salmonella* are also able to induce cancers in laboratory test animals, making the Ames test both useful and reliable for its intended purpose. Since the development of the Ames test, this simple bacterial system has been used to screen thousands of potentially dangerous chemicals (see Box 7-2).

In the standard Ames test, four different strains of histidine-requiring *Salmonella* are grown in the presence of varying amounts of the suspected chemical mutagen. Each *Salmonella* strain has a particular kind of mutation (frameshift or missense) in one of the genes in the histidine biosynthetic pathway. None of these four mutant strains is able to grow unless the medium is supplemented with histidine. However, in the presence of a mutagenic chemical, both true revertants and pseudorevertants are generated and grow without added histidine. The increase in the number of histidine revertant bacteria after exposure to the chemical indicates the mutagenic potency of the chemical (Figure 7-11).

Several genetic modifications of the *Salmonella* strains and use of a special growth medium have greatly increased the sensitivity of the Ames test for detecting mutagenic (and possibly carcinogenic) chemicals. Specific mutations have been introduced into each of the histidine-requiring *Salmonella* strains that make it easier for chemicals to enter the cell and that also prevent the DNA from being repaired after it has been mutated by the chemical. The bacteria are then grown on Petri plates containing a mixture of rat liver enzymes. The presence of these

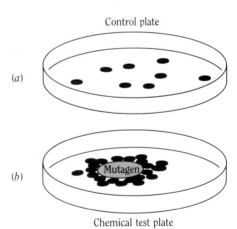

Control plate

(a)

(b)

Mutagen

Chemical test plate

Figure 7-11 The Ames test for the mutagenic potency of chemicals. (a) Millions of histidine-requiring mutant bacteria are spread on a medium lacking the amino acid histidine. A few colonies appear because spontaneous mutations produce a few revertant or pseudorevertant bacteria. (b) Approximately the same number of histidine-requiring mutant bacteria are spread on a medium lacking histidine but also containing a suspected chemical mutagen. If more revertant colonies appear on the chemical test plate than on the control plate, the chemical is considered a mutagen. The extent of the increase in revertant colony numbers indicates the chemical's mutagenic potency; some chemicals are millions of times more mutagenic than others (Table 7-3).

Dangerous Chemical Mutagens

CAUTION

BOX 7-2

Since the development of the Ames test and other biological tests for measuring the mutagenic potency of chemicals, it has become increasingly evident that we live in an environment full of mutagens. Millions of pounds of mutagenic chemicals have been widely disseminated in the soil, water, and air in the United States in the past thirty years, and in years to come people still will be exposed to them, like it or not. Bruce Ames, who devised the Ames test of chemical mutagenicity, describes the problem:

Since the late 1950s we have been exposed to a flood of chemicals—from flame retardants in our children's pajamas to pesticides accumulating in our body fat—that were not tested for carcinogenicity or mutagenicity before their use. A few of these chemicals are now being tested in animals, but for most of them the human population is serving as the test animal. We are exposed to a very large number of chemicals that are mutagens and carcinogens, many of them quite useful to society, and it is clearly impractical to ban them all, yet foolish to ignore their potential danger. We must have some way of setting priorities for regulation of these chemicals, and this requires an assessment of human risk.

Exposure to mutagens cannot be avoided entirely, but there are many substances that *can* be avoided. For example, Tris-BP (2,3-dibromopropyl phosphate) is a flame-retardant that was routinely added to all children's sleepwear until its mutagenic activity was discovered. Children absorbed the chemical through their skin while they were asleep, and the chemical was detectable in urine samples collected from them in the morning. About 50 million children were exposed to this chemical before it was finally banned by the Federal Drug Administration (FDA) in 1977. Other chemicals are still added to many fabrics as fire retardants, and not all of them have been tested for mutagenicity. In some cases, 10–20 percent of the weight of the fabric may be due to the chemical additive.

About 20 million men and women dye their hair in this country. Many hair dyes have been found to be mutagenic through Ames testing. In a study reported in 1975, 150 out of 169 commercial hair-dye preparations were mutagenic to some degree. While many of the hair dyes are not as strongly mutagenic as Tris-BP and certain other chemicals, they are used in large quantities by millions of people over many years. The mutagenic chemicals in hair dyes are absorbed through the skin of the scalp, and once they enter the bloodstream they are carried to all parts of the

enzymes increases the likelihood of detecting a chemical that is mutagenic in animals because even harmless chemicals that are ingested by animals are modified in the liver by enzymes and can be converted in the liver to mutagenic compounds.

The sensitivity of the Ames test in detecting mutagens is quite extraordinary (Table 7-3). It has shown that some chemicals, such as epoxybutane and benzyl chloride, are scarcely mutagenic at all, whereas chemicals such as aflatoxin B and furylfuramide are potent mutagens. Because millions of mutant bacteria can be plated onto Petri plates and because even a few revertants can be detected, the Ames test measures chemical mutagenicity over a millionfold range.

In addition to the Ames test, numerous other biological systems for evaluating mutagens have been devised that incorporate "guinea pigs" ranging from *Drosophila* to yeast to human cells grown *in vitro*. The identification of chemical mutagens is now a relatively easy task; however, the social, political, and

body, where they may damage the DNA in cells. Although manufacturers have removed some of these chemicals from their products, the replacement chemicals may be untested ones.

Not all of the chemicals that prove to be mutagenic in the Ames test can be tested in animals to determine whether they also cause cancer; the cost in dollars and time would be prohibitive. Yet of the more than 300 chemicals that have been carefully tested both in the Ames test and in animals, there exists a strong correlation between mutagenicity and carcinogenicity. About 85 percent of the chemicals that damage DNA in bacteria (mutagens) also cause cancer in animals (carcinogens). Thus, bacteria do provide an early warning system, albeit an imperfect one, for dangerous chemicals.

A final word of caution. Cigarette smoke also causes mutations in bacteria and is a proven carcinogen in humans. If cigarettes were being introduced for the first time today, they would have to be banned by the FDA to comply with existing laws. Why do you think the sale of cigarettes is still permitted while the sale of other proven carcinogenic substances is prohibited?

Reading: Stewart Brand, "Human Harm to Human DNA." *The CoEvolution Quarterly*, Spring 1979.

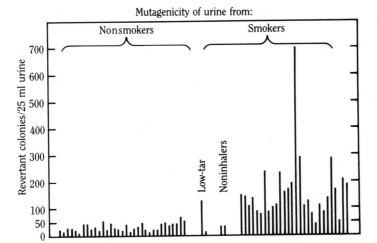

Mutagenicity of urine from:

Table 7-3. Range of Mutagen Potency in the *Salmonella* Ames Test

Chemical	Revertant ratio*
1,2-Epoxybutane	1 : 1
Benzyl chloride	3 : 1
Methyl methanesulfonate	105 : 1
2-Naphthylamine	1,400 : 1
2-Acetylaminofluorene	18,000 : 1
Aflatoxin B_1	1,200,000 : 1
Furylfuramide	3,500,000 : 1

Data from J. McCann and B. N. Ames in H. H. Hiatt, J. D. Watson, and J. A. Winsten (eds.), *Origins of Human Cancer*, Cold Spring Harbor Laboratory, *1977.*

*The revertant ratio refers to the number of revertant or pseudorevertant *Salmonella* colonies able to grow in the absence of the amino acid histidine divided by the number of *Salmonella* colonies observed in the presence of the chemical mutagen. For example, the chemical furylfuramide is 3.5 million times more mutagenic than 1,2-epoxybutane.

economic decisions involved in regulating the use of chemicals are invariably controversial and often difficult to implement.

Repair of DNA Damage

At this point you might be wondering how cells and organisms manage to survive the frequent onslaught to their DNA caused by radiation and mutagenic chemicals. Fortunately, cells have a variety of efficient DNA repair systems, and most mutations that arise are corrected immediately by enzymes that restore the correct base sequence in the DNA. Numerous DNA repair enzymes continuously monitor the synthesis of DNA molecules in order to detect and repair mutations as they occur. Some enzymes specifically repair radiation-damaged DNA; other enzymes are designed to detect and correct the specific kinds of genetic damage caused by mutagenic chemicals. Two well-understood biological mechanisms involved in preventing mutations are the repair of UV-damaged DNA and the repair of chemically damaged bases in DNA.

One effect of UV light on DNA is to cause formation of thymine dimers, the covalent linking together of adjacent thymine bases in a strand of DNA (Figure 7-12a). The DNA replication machinery cannot cope with the replication of thymine dimers, and a mutation will arise at the site of the thymine dimer unless it is removed by DNA repair enzymes before the strands of DNA are replicated.

The first step in the removal of the thymine dimer is to break the DNA strand at or near the site of UV-induced damage. This is the function of an **endonuclease**, an enzyme that breaks the sugar-phosphate backbone of DNA at specific locations (Figure 7-12b). The free end of the DNA strand is then recognized by an **exonuclease**, an enzyme that removes the damaged thymine bases plus a few others, creating a gap in the DNA strand (Figure 7-12c). The single-strand gap is filled in by DNA polymerase I, an enzyme that repairs gaps by reinserting the correct bases using the opposite strand of DNA as a template (Figure 7-12d). Finally, the sugar-phosphate backbone of the DNA strand is again sealed by the enzyme DNA ligase (Figure 7-12e). Repair of radiation-damaged DNA is a continuous process in bacteria as well as in human cells. Loss of any one of the repair enzymes by mutation can be lethal to bacteria that are exposed to UV light, demonstrating that repair of radiation damage is essential to the survival of cells.

Evidence of the vital need for repair of radiation damage in humans is provided by the hereditary human disease *xeroderma pigmentosum*. This disease is caused by mutations that result in loss of essential UV repair enzymes that are similar to the DNA repair enzymes studied in bacteria. People with this disease develop skin cancers early in life because their skin cells are

Adjacent thymines in a strand are chemically linked by the energy in the UV light.

An endonuclease breaks the DNA strand near the dimer.

An exonuclease removes several bases, including the damaged ones.

DNA polymerase I repairs the damage by replacing the correct bases in the gap, using the undamaged strand as a template.

DNA ligase repairs the sugar-phosphate backbone by joining the ends of the strand.

Figure 7-12 Repair of mutations caused by UV light.

unable to repair damage to DNA caused by even brief exposure to sunlight.

As mentioned earlier, nitrous acid and other related chemicals can cause mutations by deaminating bases in DNA, which subsequently changes base pairs formed during replication. These chemically induced base changes result in base pair substitutions in the DNA unless DNA repair enzymes remove the chemically damaged bases and reinsert the correct ones. This is accomplished by a series of enzymatic steps similar to those that repair UV-damaged DNA. These steps are outlined in Figure 7-13. A specific enzyme removes the deaminated cytosine (uracil) in one of the DNA strands (Figure 7-13b), and the remaining steps are identical to the ones used to repair UV-damaged DNA.

The DNA in all cells continuously runs the risk of being damaged by radiation, chemicals, and other environmental agents. Most DNA damage is probably repaired before it causes harm to the cell or organism. However, if the exposure to mutagenic agents is too severe or if the DNA repair processes are defective, mutations do occur. In humans, DNA repair processes become less efficient with age, because of poor nutrition, hormone imbalances, or other (poorly understood) reasons.

Hereditary (Genetic) Versus Somatic Mutations

In bacteria and other single-celled microorganisms, all mutations that arise and are not repaired are inherited—that is, they are passed on to succeeding generations of cells. With these organisms the terms *inherited, hereditary,* and *genetic* are often used interchangeably in the sense that the mutation is passed on to progeny in the next generation. However, in higher organisms that reproduce by classical sexual processes (Chapters 11 and 12), it is important to understand the distinction between a *hereditary mutation* and a *somatic mutation.*

Mutations are hereditary in higher plants and animals only if they occur in sex cells—in **male gametes** (sperm in animals, pollen in plants) or in **female gametes** (eggs in plants and animals). It is the union of male and female gametes and the growth of a new individual that accounts for the transfer of genetic information from generation to generation in plants and animals. However, most of the billions of cells in plants and animals are **somatic cells,** which include all cells other than gametes.

All somatic cells in plants and animals contain the entire genome of the organism. That is, every cell in human tissues—heart, skin, eye, liver, and so on—contains the full complement of forty-six chromosomes. The DNA in the chromosomes of these human cells can mutate as they grow and divide, just as the DNA in sperm or egg cells can. However, mutations that arise in somatic cells are never passed on to progeny, and so somatic mutations are not hereditary. That is why lung cancer, for exam-

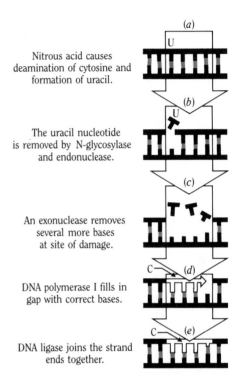

Nitrous acid causes deamination of cytosine and formation of uracil.

The uracil nucleotide is removed by N-glycosylase and endonuclease.

An exonuclease removes several more bases at site of damage.

DNA polymerase I fills in gap with correct bases.

DNA ligase joins the strand ends together.

Figure 7-13 Repair of mutations caused by nitrous acid.

ple, is not hereditary. Such mutations may, however, affect the health and physiological functioning of the person in which they occur—lung cancer often kills its victims.

Somatic mutations may alter a vital function of a particular cell, and as that cell grows and reproduces, all other cells derived from the mutant cell will also have the same altered functions. Because the cells grow in an unregulated manner, somatic mutations may initiate the growth of a tumor in a plant or animal. The accumulation of somatic mutations is also thought to be involved in aging processes. This is why mutations, cancer, and aging are biological processes that are inextricably linked.

Key Terms

mutation   any change in an organism's genetic information; a change in the DNA of cells.

mutagen   any environmental agent that increases the frequency of mutations.

point mutation   any change in one DNA base pair.

transition mutation   single base-pair substitutions resulting from adenine ↔ guanine or cytosine ↔ thymine exchanges.

missense mutation   a single base-pair change that causes the substitution of one amino acid for another in the protein.

nonsense mutation   a single base-pair change that generates a stop codon and terminates the polypeptide chain.

neutral mutation   a point mutation that does not change the amino acid in a protein and, hence, does not change the protein's function or the organism's phenotype.

deletion   loss of one or many base pairs in DNA.

duplication   duplication and insertion of a group of base pairs into DNA.

translocation   movement of a group of base pairs from one location to another.

inversion   clockwise or counterclockwise rotation of a group of base pairs so that the original sequence of bases is changed.

autosomes   all chromosomes in eucaryotes excluding the sex chromosomes.

sex chromosomes   the chromosomes that determine the sex of many organisms.

dominant allele   the allele in a heterozygous individual that determines the phenotype.

recessive allele   the allele in a heterozygous individual that is not observable in the phenotype.

ionizing radiation   electromagnetic radiation with energy sufficient to strip electrons from atoms, thus creating ions (electrically charged atoms).

forward mutation   a change from the wild-type to a mutant organism.

reverse mutation (revertant)   the restoration of an original base pair in DNA by a second mutation arising at the same site as the first mutation. A change from mutant to wild-type.

pseudorevertant   a bacterium that grows in the same conditions as wild-type but actually contains two (or more) mutations.

Ames test   a test for the mutagenic and carcinogenic potential of chemicals that involves measuring the reversion of histidine-requiring mutants of *Salmonella* bacteria.

thymine dimer   adjacent thymine bases in DNA that are covalently linked by UV light.

endonuclease   an enzyme that breaks and removes bases from DNA (or RNA) at specific sites within the molecule.

exonuclease   an enzyme that removes bases from the ends of DNA (or RNA) molecules.

male gametes   sperm (pollen in plants).

female gametes   eggs in all sexually reproducing organisms.

somatic cells   all cells other than the sex cells in plants and animals.

Additional Reading

Brand, S. "Human Harm to Human DNA." *The CoEvolution Quarterly*, Spring 1979.

Howard-Flanders, P. "Inducible Repair of DNA." *Scientific American*, May 1981.

Kohn, K. W. "DNA Damage in Mammalian Cells." *BioScience*, September 1981.

Upton, A. C. "The Biological Effects of Low-Level Ionizing Radiation." *Scientific American*, February 1982.

Review Questions

1. Which bases in DNA are altered by mutagens?

2. How many bases in DNA are changed by point mutations? by deletions?

3. Is a dominant or recessive mutation more likely to change the phenotype?

4. What is the principal difference between somatic and germinal mutations?

5. What kind of mutagen causes thymine dimers?

6. How are mutations in the DNA of cells corrected?

7. Which of the following is mutagenic: salt, sugar, fava beans, or cigarettes?

SNAKE SKIN IS COMPOSED OF OVERLAPPING HORNY SCALES. (30 ×)

8

CANCER

*Not a
Hereditary
Disease*

*I believe that the only promising strategy,
in the forseeable future, is to learn what
are the causes of cancer and then to learn
how to avoid them.*

JOHN CAIRNS

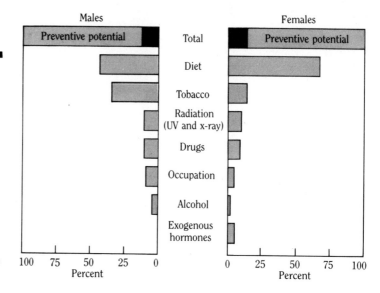

Figure 8-1 Percent of cancer incidence attributable to specific environmental factors. Reducing or eliminating exposure to environmental carcinogens would dramatically reduce the incidence of many cancers.

After heart disease, the second most frequent cause of death in the United States is cancer. One out of every four Americans will develop some form of cancer in his or her lifetime, and one in five will die from it. Many people worry about getting cancer, especially if some close family member has died from it. Although treatments for various cancers have improved, many cancers are still incurable.

With the exception of a very few rare forms, cancer is *not* a hereditary disease. Even if several members of the same family have died of it, that fact does not justify classifying cancer as a genetic disease, nor does it mean that other family members are necessarily predisposed to getting cancer. Certain individuals do have a genetic predisposition to cancer, and family members do share a significant number of their genes. But family members also share a common environment, and many environmental factors—ionizing radiation, exposure to mutagenic chemicals and viruses, nutritional excesses or deficiencies—contribute to increased risks of cancer formation.

The World Health Organization and the National Cancer Institute have estimated that as much as 90 percent of all forms of cancer is attributable to specific environmental factors (Figure 8-1). Because exposure to these environmental factors can, in principle, be controlled, most cancers could be prevented. Thus, the maxim "prevention is the best cure" is particularly appropriate for cancer. The mistaken but commonly held idea that cancer is a hereditary disease is due to a lack of understanding of how environmental agents cause mutations and initiate cancers. The mutations that lead to formation of cancer cells arise in somatic cells and, as pointed out in Chapter 7, mutations that occur in cells other than the sex cells are not inherited.

What Is Cancer?

Cancer can be defined as the unregulated growth and reproduction of cells in higher animals and in some plants. The term is used as a general description for more than a hundred kinds of human diseases that are caused by the accumulation of abnormal cells into masses (**tumors**) in various organs of the body. Cancer diseases can be subdivided into **leukemias**, cancer of white blood cells; **sarcomas**, cancer of bone and muscle cells; and **carcinomas**, cancer of skin and membrane cells. *Benign* tumors are noncancerous growths; they can usually be removed surgically. *Malignant* tumors, on the other hand, are cancerous and may spread throughout the body, causing numerous other tumors to appear in vital organs, threatening the patient's life.

Cancer Cells Differ From Normal Cells

The precise regulation of the growth, reproduction, and functions of cells in any animal is essential to its survival. Body organs grow to a certain size and then stop growing; at the appropriate point in the development of each organ, cells cease reproducing except to replace those that die. Your heart, lung, liver, brain, and other organs developed to a size that was genetically predetermined, just as your overall shape, size, and features were genetically programmed by the DNA in your cells. Environmental factors can modify an animal's cellular genetic instructions and change physiology (bodily functions) to some extent, but no amount of environmental manipulation will change a dachshund into a Great Dane or a house cat into a tiger.

Cancer cells, unlike normal cells, have been changed by one or more somatic mutations in a way that causes them to lose the capacity to respond to the chemical and environmental signals that normally regulate cell growth and reproduction. The abnormal growth of cancer cells can be demonstrated by comparing their growth and the growth of normal cells in a liquid medium containing all of the necessary nutrients. Unlike bacteria, which grow freely in a liquid medium, most types of normal animal cells only grow if they become attached to the surface of the container. For example, if normal human skin cells are placed in a plastic Petri dish or flask and are covered with liquid nutrient medium, they will grow until a layer just one cell in thickness uniformly covers the bottom of the dish (Figure 8-2*a*). If cancer cells are allowed to grow under the same conditions, however, they will rapidly cover the bottom of the dish and then continue to grow and reproduce, eventually piling up on top of one another and forming thick, uneven layers of cells (Figure 8-2*b*). Thus, a distinguishing characteristic of all cancer cells is their failure to correctly regulate cell growth.

There is now abundant experimental evidence to support the idea that most cancers begin with one or several genetic changes

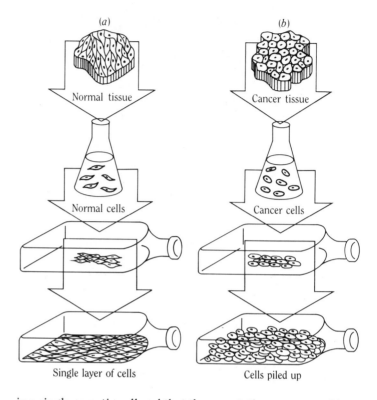

Figure 8-2 The growth of normal and cancer cells *in vitro*. (*a*) If normal tissue is broken up into single cells, and a few are placed in a flask containing nutrients, the cells will grow until a layer just one cell thick is formed. Then growth stops. (*b*) If cells from cancer tissue are treated in the same fashion, they continue to grow, piling up on top of one another to form many layers.

in a single somatic cell and that these mutations are caused by a variety of environmental agents. Mutations that might lead to formation of a cancer cell can be caused by exposure to ionizing radiation, to mutagenic chemicals, or to infection by tumor viruses. However, it is important to realize that not all mutations in animal cells result in cancer. Only if the mutation or mutations alter genes that regulate cell growth and reproduction will the cell begin to grow uncontrollably and possibly form a tumor. And it seems that most cells, if they are to become cancerous, must not only carry certain mutations but must be exposed to other substances that affect growth as well. An important confirmation of the idea that somatic mutations can cause cancer was made in 1982, when researchers showed that particular bladder cancer cells differed from normal bladder cells by a change in just one base pair in the DNA.

The smallest tumor already contains many millions of cells before it becomes detectable; large tumors contain hundreds of millions of cells. Cells from malignant tumors are often dislodged from the original mass of cells and carried by the bloodstream to other parts of the body by a process called **metastasis**. As the cancer cells migrate through the body, they penetrate other organs and develop into new tumors. It is the process of metastasis that causes cancer to spread (metastasize) and eventually destroy vital body organs.

Figure 8-3 Crown gall tumor on a turnip that was infected with bacteria carrying Ti plasmids. (Courtesy of C. I. Kado, University of California, Davis.)

Tumors develop in plants by a process different from that observed in animals. Some plant tissues can be infected by a bacterium called *Agrobacterium tumefaciens* and a few other closely related species. This bacterium has been shown to contain a self-replicating plasmid in addition to the bacterial chromosome, and it is this Ti (*Tumor-inducing*) plasmid that actually causes tumors to develop in plants (Figure 8-3). Since this discovery, Ti plasmids have become an important tool in moving genes from one species of plant to another and in the genetic engineering of new varieties of plants (see Chapter 10).

Carcinogens in the Environment

When laboratory animals such as mice or rabbits are exposed to x-rays or other forms of ionizing radiation, mutations arise in both somatic and sex cells, and the animals frequently develop tumors. The higher the dose of radiation, the greater the number of tumors and the number of animals affected. If the animals are

Table 8-1. Occupations and Cancer

Occupation	Carcinogen identified	Location of cancer
Chimneysweeps; manufacturers of coal gas	Polycylic hydrocarbons in soot, tar, oil	Scrotum; skin; bronchus
Chemical workers; rubber workers; manufacturers of coal gas	2-Naphthylamine; 1-naphthylamine	Bladder
Chemical workers	Benzidine; 4-aminobiphenyl	Bladder
Asbestos workers; shipyard and insulation workers	Asbestos	Bronchus; peritoneum
Sheep dip manufacturers; gold miners; some vineyard workers and ore smelters	Arsenic	Skin; bronchus
Makers of ion-exchange resins	Bis (chloromethyl) ether	Bronchus
Workers with glues, varnishes, etc.	Benzene	Bone marrow (leukemia)
Poison gas makers	Mustard gas	Bronchus; larynx; nasal sinuses
PVC manufacturers	Vinyl chloride	Liver
Chromate manufacturers	Chrome ores	Bronchus
Nickel refiners	Nickel ore	Bronchus; nasal sinuses
Isopropylene manufacturers	Isopropyl oil	Nasal sinuses

heavily irradiated, most of them die prematurely of cancer. That humans are also susceptible to the tumor-producing effects of ionizing radiation has been documented by the high incidence of leukemia among survivors of the Hiroshima and Nagasaki atomic bomb blasts. These laboratory and human results leave little doubt that exposure to ionizing radiation can cause cancer. The more important questions that remain to be answered are: How much radiation does it take to produce tumors, and how many individuals will be affected by a certain dose of radiation?

Evidence that certain chemicals are **carcinogens** (cause cancer) in humans is found in the increased incidence of certain types of cancer among workers in particular industries. Ever since an English physician, Sir Percival Pott, first observed an unusually high incidence of scrotal cancer among London chimneysweeps during the nineteenth century, workers in many industries have been found to have an increased risk of developing certain kinds of cancers because of their occupational exposure to certain substances (Table 8-1). For instance, the incidence of lung cancer is high among workers in asbestos industries, and exposure to vinyl chloride, which is used to manufacture polyvinylchloride (PVC) plastics, causes a type of liver cancer among plastic industry workers that is rarely observed among the general public.

Exposure to carcinogenic chemicals is not restricted to industrial workers. Deaths from lung cancer are almost directly proportional to the number of cigarettes smoked (Figure 8-4). As

Matthew Meselson, a biochemist at Harvard University, has observed, "The continued high consumption of cigarettes in spite of their being the cause of nearly all lung cancer (plus a large proportion of pulmonary heart disease, bronchitis and emphysema) results in part from ignorance or disbelief of the facts, especially among young people." That is why the facts about cancer need to be disseminated more widely.

Because many forms of human cancer are caused by environmental agents, they are, in principle, preventable (see Box 8-1). The solution to the cancer problem in modern industrial societies lies not in more effective cures (although cures are also desirable), but in preventing cancer from occurring in the first place. Prevention of disease is not, of course, a new or radical idea. Most lethal human infectious diseases—tuberculosis, smallpox, typhoid fever, plague—have been virtually eradicated not by curing people but by instituting health practices that prevent these diseases: vaccination, sanitation, and improved nutrition. The variation in cancer incidence from country to country is another indication that cancers are influenced more by environmental factors than by heredity.

Cancer and Heredity

Human cancers, except for a few rare exceptions, are not classified as genetic diseases because they do not conform to any of the criteria (discussed in Chapter 13) that are normally used to establish that a disorder is hereditary. Nor are cancers contagious—that is, they are not transmitted from person to person by physical contact or other means. Moreover, cancers, by and large, are species-specific; people do not contract cancer by eating plants with tumors, and they do not "catch" cancer from other animals. In the laboratory, cancers can be transferred between animals of the same species and sometimes between animals from different species or genera, but this process involves inoculating an animal with hundreds of thousands or millions of tumor cells, a few of which will become established in the test animal and eventually grow into new tumors.

Cells growing in malignant tumors have been genetically altered in one or more ways. The evidence is now overwhelming that somatic mutations of various kinds can initiate the conversion of a normal cell into a cancer cell, although the various mechanisms responsible for the genetic and cellular changes are not understood. Somatic cell mutations are not passed on to progeny, so it is extremely unlikely that most cancers can be inherited. The progeny of thousands of survivors of the Hiroshima and Nagasaki atomic bombs have been followed for nearly forty years. Among this group of Japanese who were exposed to large amounts of radiation and who experienced an increased incidence of cancer, no increase has been observed in the inci-

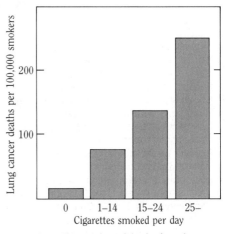

Figure 8-4 The number of deaths from lung cancer in relation to number of cigarettes smoked.

Epidemiological Studies Show Most Cancers Are Environmental in Origin

BOX 8-1

Epidemiology is that branch of medical science that investigates the frequency and geographical distribution of diseases and then tries to identify factors that cause them. Various kinds of epidemiological studies support the conclusion that most cancers are caused by environmental agents and are not hereditary. Such studies have shown that the frequency with which different cancers occur varies greatly from country to country. For example, cancer of the mouth makes up about 3 percent of all cancers in the United States, but in parts of Asia where people chew tobacco and betel nuts it can account for as much as 35 percent of all cancers diagnosed. Breast cancer varies by a factor of six from nation to nation, Japan having the lowest incidence and Holland the highest.

Comparison of the incidence of various cancers among Japanese citizens, American citizens, and Japanese who have emigrated to the United States shows quite convincingly that most cancers are attributable to environmental factors. The accompanying table shows that death rates from various cancers in Japan are dramatically different from those for Japanese who have moved to the United States. The incidence of cancer among Japanese Americans born and raised in the United States tends toward the frequencies found among U.S. whites, indicating that genetic differences are not significant in the development of these cancers.

Epidemiological studies have been carried out among workers in different industries as well as among people in different countries. In some industries workers are found who suffer from cancers that virtually never arise in the general population. For example, *mesothelioma* is a rare form of lung cancer that occurs only among workers involved in the mining and manufacturing of asbestos. It is now known that exposure to microscopic asbestos fibers from any asbestos-containing material increases the risk of this kind of lung cancer (and of other respiratory diseases as well.)

Vinyl chloride (VC) is an important industrial chemical that is the basic ingredient in a wide range of polyvinylchloride (PVC) plastics used in phonograph records, food containers, electrical insulation, garden hoses, plastic pipe, and many other products. The finished plastic products themselves do not cause cancer, but the starting material does. By the 1970s, it became apparent from numerous epidemiological and laboratory studies that unpolymerized vinyl chloride, which escapes from PVC products, is a carcinogen—it does increase the risk of developing cancer, particularly of *angiosarcoma,* an unusual form of liver cancer.

Two conclusions consistently emerge from epidemiological studies of cancer: (1) most cancers are environmentally caused, not hereditary, diseases; (2) most cancers are preventable.

Cancer location	Relative cancer mortality rates		
	Japanese	Offspring of migrants	U.S. whites
Stomach	100	38	17
Colon	100	288	489
Pancreas	100	167	274
Lung	100	166	316
Leukaemia	100	146	265

Source: W. Haenszel and M. Kurihara, "Studies of Japanese Migrants. I. Mortality from Cancer and Other Diseases Among Japanese in the United States." *Journal of the National Cancer Institute* 40 (1968), 43–68.

dence of genetic defects or genetic diseases. This does not mean that the radiation did not cause any mutations in their sex cells, it simply means that the increase in the number of mutations causing observable genetic defects is too small to be measured in a population of a few thousand individuals.

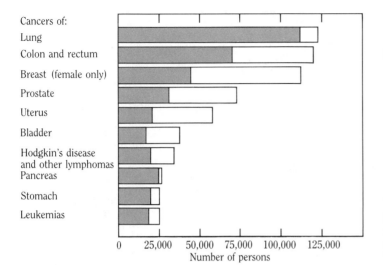

Cancers of:
Lung
Colon and rectum
Breast (female only)
Prostate
Uterus
Bladder
Hodgkin's disease and other lymphomas
Pancreas
Stomach
Leukemias

0   25,000   50,000   75,000   100,000   125,000
Number of persons

Figure 8-5 Incidence of the ten most common forms of cancer in the United States, 1981. The shaded areas indicate the estimated number who will die within five years of the diagnosis.

It has been speculated that under unusual circumstances tumor viruses that could increase the chance of developing cancer might be present in the germ line and transmitted from parent to child. However, to date no human cancer has been shown conclusively to be caused by viruses, and viruses are not regarded as causative agents for the most common forms of human cancer, which are listed in Figure 8-5.

It should be noted that there are some rare human diseases that *are* hereditary and that may predispose the affected individual to develop some form of cancer, even though the cancer itself is not inherited. According to John J. Mulvihill of the National Cancer Institute, there might be as many as 200 rare hereditary human diseases that predispose their victims toward development of some form of cancer, but taken all together they probably account for fewer than 5 percent of all human cancers. Saying that people are genetically predisposed to getting cancer often conveys a false idea about the relation of heredity and cancer. Perhaps the relation can be clarified with the example of fair skin. People with fair skin are more likely to get sunburn or skin cancer. Fair skin is inherited, but sunburn and skin cancer are caused by overexposure to sunlight, which can be avoided if people realize their increased susceptibility.

Two extremely rare inherited human diseases, xeroderma pigmentosum and ataxia telangiectasia, provide biochemical support for the idea that cancer begins with genetic changes and that these changes result in synthesis of defective enzymes that may alter cell growth. As mentioned in Chapter 7, the skin cells in persons suffering from xeroderma contain a mutant gene or genes whose consequence is a lack of one or more enzymes required to repair UV-damaged DNA. People with this hereditary disease who are exposed to even slight amounts of sunlight

167

Screening for "Cancer-Resistant" Workers

BOX 8-2

Certain occupations entail more risk of developing cancer than others. In some instances precautions can be taken to reduce this risk, such as shielding x-ray technicians or limiting the exposure of nuclear workers to radioactivity. Industry and government efforts have reduced the exposure of workers to proven carcinogens such as ionizing radiation, asbestos, vinyl chloride, and other agents. Yet there are limits to which this exposure can be controlled by industry or regulated by government agencies.

Are there other solutions to the problem of worker exposure to cancer-causing substances? The answer is yes—but some people have questioned whether the other solutions are ethical or fair. Human beings differ from one another biochemically because they carry different alleles for essential genes and therefore synthesize different enzymes. Even the levels of particular enzymes vary from person to person. So it is quite reasonable to expect that enzymes that repair damage to DNA will be present in greater amounts in some persons than in others. Simple biochemical tests could be used by employers to identify applicants who are more resistant to the effects of the carcinogens they would be exposed to in a particular job. Such genetic screening by industries could also be used to eliminate carcinogen "sensitive" workers from jobs.

Manufacturers are aware of the monetary benefits of identifying "cancer-resistant" workers. For one thing, such workers might be able to tolerate higher levels of harmful substances. For another, fewer workers might bring lawsuits claiming that their diseases were caused by occupational exposure to chemicals or claiming negligence on the part of their employer in informing them of the dangers.

The questions now are: Is it ethical or fair for workers to be denied certain jobs simply because of the biochemical and genetic constitution that they inherited? And should exposure of some workers to toxic or cancer-causing substances be permitted just because they are more resistant to biological damage than others?

Readings: Constance Holden, "Looking at Genes in the Workplace." *Science*, July 23, 1982.
Mary P. Lavine, "Industrial Screening Programs for Workers." *Environment*, June 1982.

invariably develop skin tumors because the UV-damaged DNA in their skin cells cannot be repaired. The UV light in sunlight causes the skin cancer because of the loss of an enzyme. People suffering from ataxia are particularly sensitive to x-rays. Because of an inherited mutation, the person's cells lack enzymes that normally repair x-ray damaged DNA. As a consequence, they are predisposed to developing leukemia. Thus, in both of these rare inherited diseases, a genetic change is indirectly responsible for the eventual cancer formation.

The fact that people vary genetically in their capacity to repair DNA damaged by various environmental agents means that their cells may be more or less susceptible to mutations and to the formation of cancer. This fact has prompted the controversial idea of screening for cancer-resistant workers in occupations where individuals are exposed to radiation or mutagenic chemicals (see Box 8-2).

Animals that develop cancers do not do so immediately following exposure to a carcinogen, even though the mutations

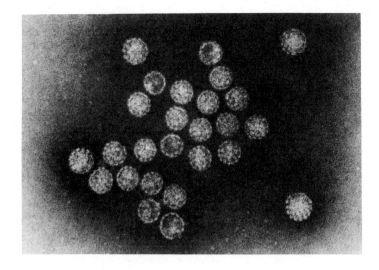

Figure 8-6 Electron micrograph of an oncogenic (tumor-producing) virus. (Courtesy of Robley Williams.)

presumably occur at the time of exposure. Many years pass before the cellular changes characteristic of cancer diseases become evident in the form of tumors. It may be that a significant number of mutations have to accumulate before cells actually begin to grow uncontrollably. It is also likely that other chemical substances called **cancer-promoting agents** trigger the development of cancer in those cells that already carry mutations. One such cancer-promoting substance, croton oil, need only be applied to the skin of mice and most of them that have been previously exposed to a carcinogen develop tumors rapidly.

Cancer is caused by a combination of factors, including genetic changes resulting from somatic mutations, and chemicals that affect the expression of normal or mutated genes. In addition to radiation, carcinogenic chemicals, and cancer-promoting agents, certain tumor viruses have been implicated as causal agents in the development of some animal cancers.

How Viruses Cause Cancer

The genetic information in animal viruses is carried by either RNA or DNA molecules. Either class of virus may cause the development of malignant tumors in animals, in which case the virus is referred to as an **oncogenic** (cancer-causing) **virus** (Figure 8-6). In 1911 Peyton Rous, working at the Rockefeller Institute (now Rockefeller University) in New York City showed that tumors could be induced in chickens by injecting them with an RNA virus, later named the Rous sarcoma virus after its discoverer. Since then, many other oncogenic RNA and DNA viruses have been isolated from animal tumors that will again produce tumors when the purified viruses are injected into susceptible animals.

In nature, oncogenic viruses tend to be species specific; that is, viruses that cause cancer in one species generally are harmless

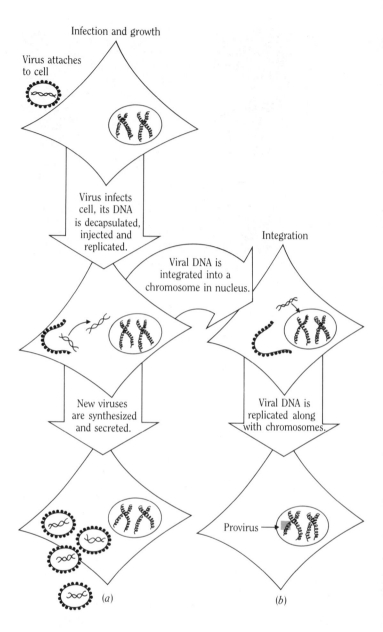

Figure 8-7 Infection of an animal cell by oncogenic viruses. (*a*) New virus particles may be synthesized, or (*b*) the viral DNA may be integrated into one of the cell's chromosomes. In its integrated state the viral DNA is called a provirus.

in another. For example, feline leukemia virus, which is quite commonly found in household cats, has never been shown to cause leukemia or tumors in other animals, including humans. Moreover, many cats whose saliva and blood contain great numbers of the feline leukemia virus have no symptoms of leukemia, indicating that oncogenic viruses by themselves do not necessarily produce cancer. In fact, many of these cats live a normal life-span and die of causes other than cancer.

Oncogenic viruses do not usually destroy the cells they infect, even if the viruses multiply in the cells. Both the cells and viruses continue to reproduce. When an animal cell is infected by oncogenic DNA viruses, two outcomes of the infection are possible. The viral genome may be replicated and newly synthesized virus particles released from the cell, whereupon they may infect other nearby cells and repeat the process (Figure 8-7). Or, the genetic information of the virus may become integrated into one of the cell's chromosomes. If it becomes integrated, the **provirus** DNA is replicated along with the rest of the chromosomal DNA and is transmitted to other cells in the same way as normal cellular DNA.

If animal cells are infected by an oncogenic DNA virus, the viral DNA can be directly integrated into one or more of the cell's chromosomes. However, if the infection is caused by an oncogenic RNA virus, the genetic information in the single-stranded viral RNA molecule must first be converted into a DNA molecule. This conversion is carried out by the enzyme **reverse transcriptase**, which is responsible for synthesizing a molecule of viral DNA by first synthesizing a single strand of DNA that is complementary to the infectious strand of RNA (Figure 8-8). Next, a double-stranded viral DNA is synthesized that can become integrated in one of the chromosomes. When animal cells are infected by oncogenic RNA viruses, conversion of the RNA to DNA and the integration of the viral DNA into the cell's chromosome are essential for continued RNA virus reproduction. Once the viral DNA has been integrated into a chromosome, the cell may or may not continue to actively synthesize RNA viruses, although in either case it retains its capacity to do so.

THE ONCOGENE HYPOTHESIS

There is reason to believe that viruses have existed on earth as long as cells have. In fact, most biologists believe that viruses must have evolved from cells, possibly in situations where a few genes became separated from a chromosome and somehow became encapsulated by proteins. One evolutionary consequence of such a process is that it would promote the transfer of genetic information among organisms by accelerating the spread of mutations from their point of origin in the population. In many instances the acquisition of new genes or rearrangement of genes caused by insertion of a provirus would be advantageous; on other occasions it might change the regulation of expression of genes, leading to uncontrolled cellular growth. If, on the average, the presence of the virus were to increase the adaptive variation of organisms, the maintenance of such viruses in the population would be ensured.

The **oncogene hypothesis** proposes that millions of years ago viral genes became integrated into the chromosomes of animal

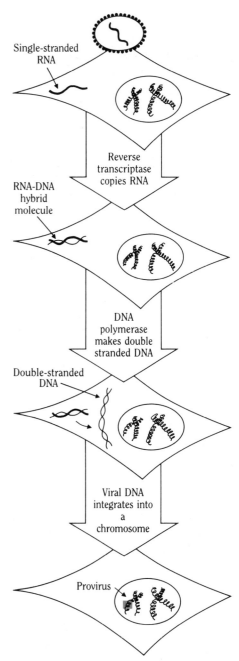

Single-stranded RNA

Reverse transcriptase copies RNA

RNA-DNA hybrid molecule

DNA polymerase makes double stranded DNA

Double-stranded DNA

Viral DNA integrates into a chromosome

Provirus

Figure 8-8 Infection of an animal cell by an oncogenic RNA virus. The infectious RNA is first copied by the enzyme reverse transcriptase into a complementary strand of DNA. DNA polymerase enzymes then make double-stranded viral DNA molecules. One or more of these DNAs may integrate into chromosomes and become inactive proviruses.

171

Figure 8-9 The single-stranded RNA molecule of the Rous sarcoma virus contains four genes: *gag* codes for a protein in the core of the virus; *pol* codes for the reverse transcriptase; *env* codes for the protein that encapsulates the RNA; and *src* codes for a protein that transforms normal cells into cancer cells. The R stands for an identical sequence of bases at each end of the RNA that are essential for replication.

reproductive cells and that the viral DNA was passed on, along with normal cellular genes, generation after generation. Like many other genes, these particular viral genes (the theory suggests) are generally not expressed in the cells that carry them. As long as the viral genes are not transcribed into mRNA or translated into proteins, no viral functions are expressed in the cell, and the provirus DNA causes the cell no harm. However, if ionizing radiation or some other mutagenic agent should damage the DNA and trigger the expression of the previously quiescent viral oncogenes, these cells may begin to grow in an unregulated manner and eventually develop into a tumor.

Several years ago it was shown that the *src* (pronounced "sark") gene carried by some proviruses is the gene that is responsible for changing a normal cell into a cancer cell. The genome of the Rous sarcoma virus contains only four genes, one of which is the *src* gene (Figure 8-9). When these four viral genes are not expressed, the cells carrying them function normally in all tissues. But if the *src* gene becomes expressed in a cell, many properties of the cell change, particularly the capacity to grow rapidly and to form tumors.

Until very recently the prevailing view was that existence of oncogenes in cells is abnormal, since cells infected by oncogenic viruses are often changed into tumor-producing cells. However, in the past several years it has been discovered that oncogenes such as *src* exist normally in the DNA of most animal cells even when no proviruses are present in the DNA.

This discovery means that oncogenes that were originally thought to be contained in viruses may, in fact, be a normal part of the cell's genome. The fact that *src* genes occur in animal viruses and are able to transform normal cells into cancer cells probably does not reflect their original origin or function. *Src* genes might have arisen in cells millions of years ago, possibly to regulate cell growth, and subsequently became associated with viruses. As was pointed out earlier in this chapter, the rate of cellular growth and the number of cells in a particular tissue must be genetically regulated. It may be that *src* genes originally regulated some of these processes. In most instances these genes may still function normally in cells; in other instances the oncogenes in viruses may cause cells to grow abnormally.

Genes that are essential for the growth and reproduction of organisms need to be shared if other organisms besides the ones in which they originated are to compete successfully. Viruses provide almost perfect vehicles for shuttling genes from cell to cell and from organism to organism. After packaging genes from one cell's chromosome, viruses can infect other cells and transfer the genes. In the case of the oncogenic viruses, they are also able to insert the genes they carry into a chromosome, making the genes a permanent part of the organism's genome.

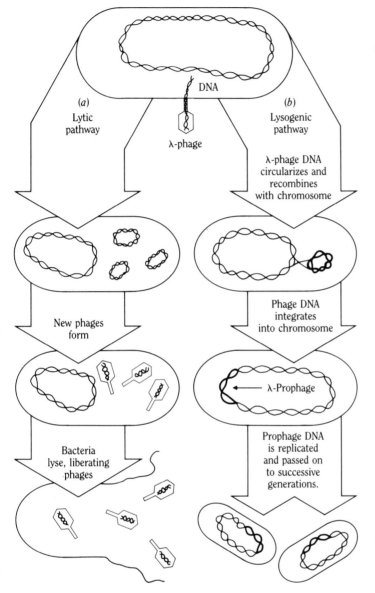

Figure 8-10 Infection of bacteria by λ-phages. Infection is followed by either a lytic or lysogenic response. (*a*) In the lytic pathway the injected phage DNA forms a circular molecule and is replicated hundreds of times. These DNA copies are assembled into new phages, the bacteria lyse (burst open), and the phages that are released can repeat the cycle. (*b*) In the lysogenic pathway the λ-phage DNA also forms a circle and may be replicated into a few copies. One of these copies is recombined into the bacterial chromosome, where the integrated DNA remains as an inactive prophage; that is, its genes are not expressed. The phage genome is passed on to successive generations.

LYSOGENIC BACTERIA

A process analogous to what occurs when animal cells are infected by oncogenic viruses takes place in bacteria. When certain phages such as lambda (λ) infect *E. coli,* they have the option of two developmental choices (Figure 8-10). The λ-phage can reproduce in the infected bacterium, leading to the production of hundreds of new phages and lysis of the bacterium, or the phage DNA can become integrated into the bacterial chromosome, causing the bacterium to become a *lysogen,* that is, a bacterium that carries a "prophage" in its genome. Superficially, the proc-

The Human E-B Virus: A Possible Cancer-Causing Agent

BOX 8-3

For more than seventy years researchers have known that malignant tumors can be produced in animals by injecting them with certain tumor viruses. For the most part, the tumors that have been studied are laboratory-induced diseases that are caused by deliberately inoculating animals with millions of virus particles. It is not yet known to what extent viruses actually cause tumors in animals living in their natural environments. Nor has it been shown that any viruses capable of infecting people actually cause human cancers, in the sense that the crippling disease poliomyelitis is caused by the polio virus or that the "flu" is caused by influenza viruses.

Because ethical considerations preclude deliberately infecting human experimental subjects with suspected cancer-causing viruses, a causal connection between viruses and human cancer must be deduced from other evidence, most of which is circumstantial. The most suggestive evidence so far has come from the association of the *Epstein-Barr* (E-B) *virus* with a particular form of human cancer, called *Burkitt's*

lymphoma, which is found primarily in certain areas of Africa.

Burkitt's lymphoma, which is characterized by a rapidly growing tumor that develops in the jaws of children, was first described by a British physician, Denis Burkitt, while he was working in Uganda. Since his discoveries in the 1950s, extensive studies have shown that the E-B virus is always present in the tumor cells that are removed from children with Burkitt's lymphoma. The viral DNA can integrate into the cell's chromosomes like the DNA of other tumor viruses and change the cell's growth. The purified E-B viruses also cause tumors when they are injected into laboratory animals. However, it has not yet been proven that the E-B virus by itself causes lymphomas, although it seems almost certain that its presence is necessary for development of the tumor.

It turns out that almost everyone, Americans as well as Africans, has been infected by E-B viruses, usually during childhood. As a result, most people develop antibodies against the E-B virus. It has been shown that among Americans E-B virus causes *mononucleosis,* a mild infectious disease that occurs frequently among college students. Gener-

ally, college students who were not exposed to E-B viruses as children and who did not develop immunity are the ones who catch "mono." However, students who develop mononucleosis are not at any greater risk of developing Burkitt's lymphoma. Since almost everybody appears to have been infected by E-B viruses at some time or another in their lives, there must be additional factors that lead to formation of Burkitt's lymphoma in Africa.

The lesson to be learned from the association between E-B viruses and Burkitt's lymphoma is that development of human cancer is not caused simply by infection by tumor viruses. Other environmental factors (as yet unknown) must be necessary for tumors to develop. In fact, as a general rule, infectious agents such as viruses are only partly the cause of infectious diseases among people. Immune responsiveness, nutritional state, drug use, degree of stress, and emotional condition all contribute to people's susceptibility to infections of any type.

Reading: W. Henle, G. Henle and E. T. Lennette, "The Epstein-Barr Virus." *Scientific American,* July 1979.

esses by which oncogenic viruses and λ-phages integrate their DNAs into the cell's chromosome appear quite similar. However, the mechanisms of the two processes are actually quite different.

The phage DNA that becomes integrated into the bacterial DNA is called a **prophage**, and most of its genes are unexpressed—its DNA is simply replicated along with the bacterial genes. No phage particles are produced. **Lysogenic bacteria** (those that carry prophages) are, in most respects, indistinguishable from normal bacteria unless the prophage genes are expressed—that is, unless the phages begin to grow in the bacteria.

How is the prophage DNA maintained in the bacterial DNA generation after generation without the genes in the prophage DNA becoming expressed? The answer requires understanding the mechanisms of gene regulation that are discussed in the next chapter. Briefly, however, a protein called a **repressor** determines whether the infecting phage DNA reproduces itself and generates more λ-phages or whether the phage DNA stays integrated and unexpressed in the bacterial DNA. One gene in the phage DNA codes for a particular repressor protein that determines whether the lysogenic bacterium survives or produces λ-phages and dies. When the repressor is synthesized quickly enough after infection and in sufficient amounts, the phage DNA integrates into the bacterial DNA. However, if the repressor is not made in the infected cells, new λ-phages are synthesized. Many environmental factors, such as temperature, kind of growth medium, and number of infecting phages, affect the synthesis of the repressor protein and, hence, the fate of the bacterium.

If lysogenic bacteria that contain the integrated prophage are exposed to UV light or to certain mutagenic chemicals, the repressor may be inactivated, in which case the phage DNA begins to replicate and new λ-phages are produced that destroy the bacteria. The repressor proteins may also be inactivated by mutations in the phage gene that codes for the synthesis of the repressor. Thus, the survival of lysogenic bacteria depends on the expression of particular phage genes.

Regulation of genetic expression is the key to bacterial survival in the changing environments that many bacteria experience. Regulation of genes is also the key to developmental processes in plants and animals. For example, how do plant cells regulate their genes so that cells that initially are identical are able to *differentiate* into various specialized cells that eventually become leaves, roots, stems, or flowers? How do animal cells regulate gene expression so that the cells can differentiate into the specialized tissues of lung, brain, ovary, and bladder? The regulation of gene expression is what gives each cell and each organism its unique phenotype. Elucidating some of the molecular mechanisms governing gene expression in bacteria, plants,

8
CANCER:
NOT A HEREDITARY DISEASE

and animals has been one of the remarkable achievements of molecular genetics. Genetic regulation in bacteria (the main subject of our next chapter) is quite well understood; a detailed understanding of how higher organisms regulate the expression of their genes and control development has yet to be achieved.

Key Terms

cancer   unregulated growth of plant or animal cells. The various diseases that result from the growth of masses of cells.

tumor   a mass of cells that accumulates at a particular site. If the cells spread to other parts of the body causing disease or death, the tumor is *malignant*. Benign tumor cells remain at the original site and do not usually cause disease.

leukemias   cancers of blood cells.

sarcomas   cancers of bone and muscle cells.

carcinomas   cancers of skin and membrane cells.

metastasis   the process by which tumor cells are carried to different parts of the body, where they grow into new tumors.

carcinogen   any agent that can cause cancer.

cancer-promoting agents   chemicals that, while not themselves carcinogens, promote formation of tumors by cells previously exposed to carcinogens.

oncogenic viruses   viruses capable of causing cancers in animals when the viral genes are expressed.

provirus   an unexpressed viral genome carried in one of the chromosomes of animal cells.

reverse transcriptase   an enzyme that synthesizes DNA strands by copying single-stranded RNA molecules.

oncogene hypothesis   the idea that viral genes became integrated into the chromosomes of animal cells millions of years ago. These genes are normally unexpressed but if induced may convert normal cells into tumor cells.

prophage   an unexpressed phage genome carried in the DNA of bacteria.

lysogenic bacteria   bacteria that carry a prophage in their DNA.

repressor   a protein that "turns off" or prevents a gene or a group of genes in DNA from being expressed.

Additional Reading

Bishop, J. M. "Oncogenes." *Scientific American*, March 1982.

Cairns, J. *Cancer: Science and Society*. San Francisco: W. H. Freeman, 1978.

Dulbecco, R. "The Nature of Cancer." *Endeavor* vol. 6 no. 2 (1982).

LaFond, R. E. (ed.). *Cancer—The Outlaw Cell*. Washington, D.C.: American Chemical Society, 1978.

Meselson, M. *Chemicals and Cancer*. Boulder, Colorado: Associated University Press, 1979.

Oppenheimer, S. F. "Causes of Cancer: Gene Alteration Versus Gene Activation." *American Laboratory*, November 1982.

Reif, A. E. "The Causes of Cancer." *American Scientist*, July/August 1981.

Richards, V. *Cancer, The Wayward Cell: Its Origins, Nature and Treatment*. Berkeley: University of California Press, 1978.

Shodell, M. "The Intimate Enemy." *Science 82*, September 1982.

1. What kinds of environmental agents cause cancer in animals?

2. What distinguishes cancer cells from all other kinds of cells?

3. What is the relationship between a mutagen and a carcinogen?

4. What carcinogen causes more human cancer deaths than any other?

5. About how many human cancers are hereditary?

6. What molecule is synthesized by reverse transcriptase?

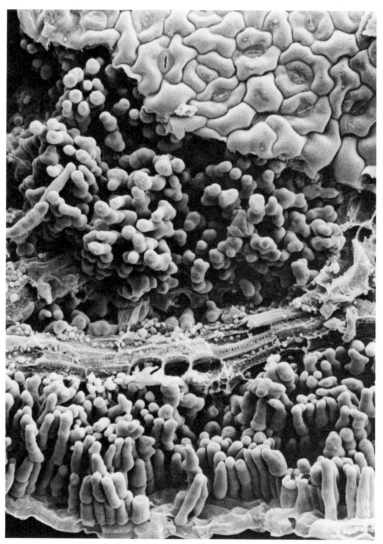

THE COMPLEX STRUCTURES OF DIFFERENT CELLS IN A BEAN LEAF. (600 ×)

9

GENE REGULATION

Switching
Genes
Off and On

Unpredictability is in the nature of the
scientific enterprise. If what is to be found
is really new, then it is by definition
unknown in advance.
FRANÇOIS JACOB

9
GENE REGULATION: SWITCHING GENES OFF AND ON

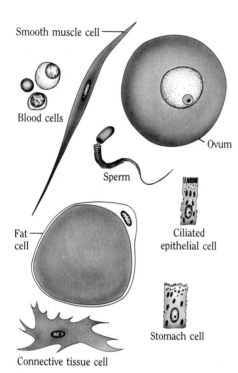

Smooth muscle cell

Blood cells

Ovum

Sperm

Fat cell

Ciliated epithelial cell

Stomach cell

Connective tissue cell

Figure 9-1 Various kinds of human cells. Cells in different tissues have different shapes and functions as a result of different genes being switched on or off.

Every moment, many thousands of chemical reactions are occurring in every cell of your body. The kinds of chemical reactions and the rate at which they occur must be regulated with exquisite precision if your tissues and organs are to function properly. The functions performed by each organ in the body are different, and it is unlikely that even identical cells in the same organ are performing precisely the same reactions at the same moment. How are all of the chemical reactions in all of the different kinds of cells regulated?

The particular biochemical activities of cells are determined by the particular genes that are being expressed in those cells. Most genes in a cell code for the enzymes that carry out the cell's chemical reactions. The enzyme-coding genes, in turn, are often regulated by other proteins or by regulatory sites in the DNA itself. During the development of an organism, its cells must differentiate; that is, particular genes must be switched on or off according to the functions the cells must perform during that particular stage of development. Cellular **differentiation** determines the sort of plant or animal that will be formed and is the process by which cells undergo a change (usually irreversible) from a relatively unspecialized state to more specialized functions as an organism develops.

The cells of a multicellular organism vary greatly in size, shape, and function (Figure 9-1). A human blood cell differs from a muscle cell in many respects, yet both cells contain precisely the same number of genes and the same number of chromosomes. What causes cells to be so different from one another is the ways their genes are regulated and expressed—which genes are switched on and which ones are switched off.

Even after cells have differentiated and have become integrated into a particular tissue, where they perform specific functions, they must still be able to adapt to changing conditions. For example, the tissue may become damaged or infected, the water or nutrient supply to the cells may change, or hormone or nervous system signals may require changes in gene expression. As with all dynamic processes, conditions both inside and outside the cell must be monitored, and expression of genes must be continually adjusted in response to stimuli.

To further appreciate the complexity of gene expression in animals, think of your body as a biological orchestra made up of billions of instruments (cells). The kind of music the orchestra plays will depend on which instruments are being played at any given moment as well as on the particular notes, the intensity or loudness of the sounds, and tone of the sounds. Clearly, in such a biological orchestra the possibilities for different kinds of sounds are limitless. To even begin to understand how such an orchestra functions, it is first necessary to understand in detail how each individual instrument plays and responds to the instructions of

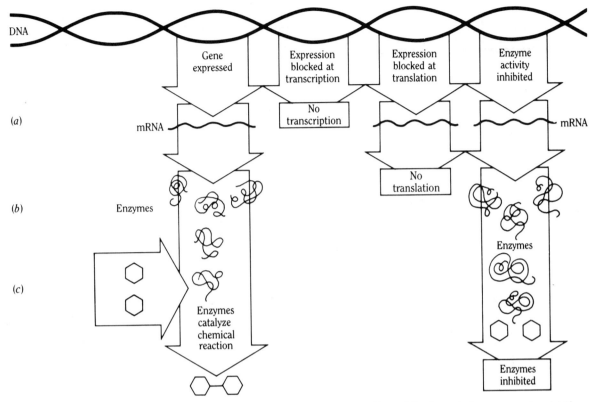

DNA

| (a) | Gene expressed | Expression blocked at transcription | Expression blocked at translation | Enzyme activity inhibited |

No transcription

mRNA — mRNA

No translation

(b) Enzymes

Enzymes

(c) Enzymes catalyze chemical reaction

Enzymes inhibited

Figure 9-2 Control of gene expression. (*a*) If transcription of the gene is prevented, no mRNA is made. (*b*) If transcription of the gene proceeds but the mRNA is not translated, no protein is synthesized. (*c*) If the enzymes cannot function because of chemical inhibitors present in the cell, gene expression is also prevented.

the conductor (the brain). Because this is such an enormous task, the simpler the instrument one chooses to study, the more likely it is that one may begin to understand how the biological orchestra is regulated.

Bacteria have been chosen as model systems for studying gene regulation because they are the simplest of all organisms. But even in these simple, single-celled organisms, the ways in which expression of even a single gene is regulated may be extremely complex. However, once one understands some of the details of gene regulation in bacteria, it is easier to formulate ideas about gene regulation in the complex cells in the human body.

Enzyme Activities Are Regulated at Three Cellular Levels

Enzymes catalyze most of the chemical reactions that occur in cells. As we saw in Chapter 5, genetic information flows from DNA into RNA into protein. So, in principle, at least three opportunities exist for regulating gene expression and enzyme activities in cells (Figure 9-2). Preventing transcription of genes is the most economical method of gene regulation because it saves the most cellular energy, since neither RNA nor proteins are made. This is

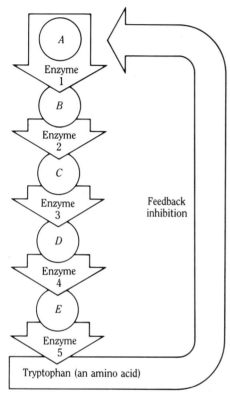

Figure 9-3 Feedback inhibition of amino acid biosynthetic pathways in bacteria. The final product of the pathway, in this example the amino acid tryptophan, can inhibit the activity of the first enzyme in the pathway (Enzyme 1). Thus, when tryptophan is present in the cell in excess, its further synthesis is prevented.

called **regulation of transcription**, and it occurs by preventing the synthesis of mRNAs. If a gene is not transcribed into a molecule of RNA, the information carried by that gene remains unexpressed in the cell. If, however, a molecule of mRNA is transcribed, expression can still be regulated by preventing translation; this is called **regulation of translation**. This kind of regulation might occur if the mRNA molecules are destroyed or inactivated before being translated, or inhibitory substances might prevent ribosomes from attaching to the mRNAs. Finally, even if a gene is transcribed into mRNA and the mRNA is translated, the enzyme that is synthesized may be inactive because of chemical inhibitors in the cell, or the enzyme may be prevented from being transported to the place in the cell where it normally functions. For example, enzymes may be chemically modified by other enzymes in such a way that their action is reduced or prevented.

For many bacterial enzymes, such as those that catalyze reactions leading to synthesis of amino acids, addition of excess amino acid to the medium will abruptly shut off the enzyme's activity by a process called **feedback inhibition**. The excess amino acid blocks the activity of the first enzyme in the biochemical pathway used to synthesize that particular amino acid, as shown in Figure 9-3.

Feedback inhibition is quite similar to other feedback mechanisms used in many assembly-line processes such as automobile production. If more automobiles roll off the assembly line than can be handled at the end of the line, information is fed back to the beginning of the assembly line to slow or halt the process until the excess products at the end have been removed.

As has already been pointed out, the mechanism of gene regulation that conserves the most energy for the cell is the regulation of transcription by blocking the synthesis of mRNAs. The discussion that follows in this chapter describes only those genetic regulatory mechanisms that affect transcription, since these are the most fundamental mechanisms by which the expression of genes in DNA is regulated. Remember, though, that even if the gene is expressed, regulation can still occur at other locations in the cell.

Genes Regulate Bacterial Growth

Escherichia coli normally grows in human intestines and assists in the digestion of food. Outside of the human digestive system, this bacterium can grow in a wide range of conditions; it does so by expressing only the particular genes that are needed for optimal growth in each habitat.

In the early part of this century, a simple experiment was performed that showed how *E. coli* bacteria are able to grow in a nutrient medium consisting only of essential minerals and two different sugars, a mixture of glucose and lactose. Not until almost fifty years later, however, did researchers figure out the mechanisms of gene regulation that produce this particular pattern of growth. If bacteria are grown in a solution containing a small amount of glucose and a large amount of lactose, they grow by utilizing only the glucose; the lactose in the medium is not touched until all the glucose is used up (Figure 9-4). When the glucose is depleted, the cells abruptly stop growing for a brief period of time. After a short delay, they resume growing by metabolizing the lactose, which up to this point has remained unused.

An enzyme called **beta-galactosidase** must be synthesized by the bacteria in order to break down the lactose into the two simpler sugars glucose and galactose (Figure 9-5). (Other enzymes are needed to convert the galactose into glucose, but these enzymes will not be considered here.) Most plant and animal cells use glucose to provide energy and the carbon atoms that are needed to synthesize other molecules. Ultimately, all sugars must be converted into glucose before their atoms can be used. What is not explained by the growth pattern shown in Figure 9-4 is why β-galactosidase is synthesized by the bacteria only after all the glucose has been used up. Since much more lactose than glucose is available, why wouldn't at least some of the lactose be used along with the glucose? Pursuing the answer to this question eventually led scientists to discover how synthesis of the enzyme β-galactosidase is regulated in bacteria.

THE *lac* OPERON: NEGATIVE GENE CONTROL

In 1961 Jacques Monod and François Jacob at the Pasteur Institute in Paris proposed an unusual model to explain how the expression of genes is regulated in bacteria. These researchers had isolated numerous bacterial mutants in which the normal synthesis of β-galactosidase was changed. In particular, they isolated mutants in which synthesis of the enzyme could not be switched on under any condition and also mutants in which the

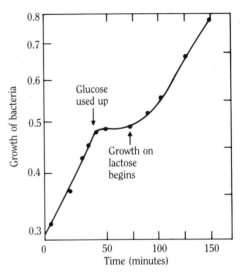

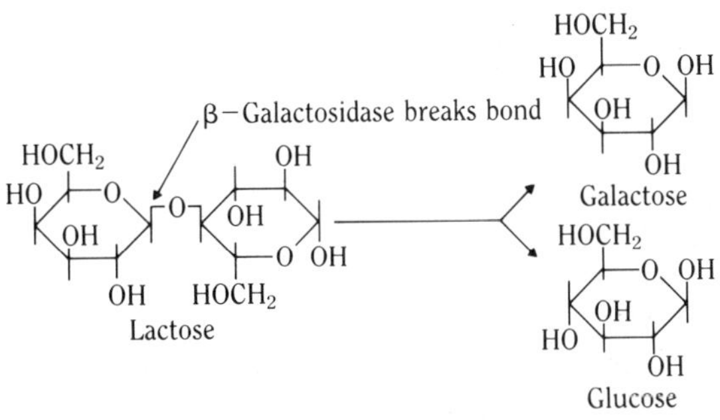

Figure 9-4 Growth pattern in *E. coli*. In a medium containing both glucose and lactose, bacteria grow until they use up all the glucose. At that point genes are switched on and the enzyme β-galactosidase is synthesized. Bacteria need this enzyme in order to utilize the lactose. When enough β-galactosidase has accumulated in the bacteria, they begin to metabolize the lactose and grow again.

HOCH$_2$
HO — O OH
β−Galactosidase breaks bond OH
OH
HOCH$_2$ OH Galactose
HO — O HOCH$_2$
OH OH — O OH
OH — O OH
OH HOCH$_2$ OH
OH HO
Lactose OH
Glucose

Figure 9-5 The chemical reaction carried out by the enzyme β-galactosidase. Lactose is broken into the simple sugars glucose and galactose.

183

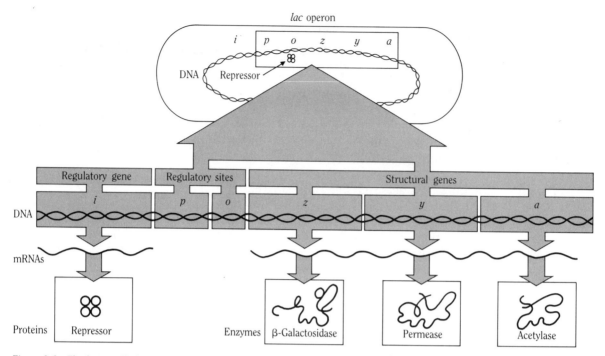

Figure 9-6 The lactose (*lac*) operon in bacteria. In order to utilize lactose, bacteria use three genes that synthesize enzymes and three genetic regulatory elements in the DNA. The genetic elements of the *lac* operon are the *z* gene (structural gene coding for the synthesis of β-galactosidase), the *y* gene (structural gene coding for the synthesis of permease), the *a* gene (structural gene coding for the synthesis of acetylase), the *i* gene (regulatory gene coding for the synthesis of repressor proteins), the *o* site (operator site in the DNA to which repressor proteins attach), and the *p* site (promoter site in the DNA to which RNA polymerase attaches).

enzyme was continuously synthesized and could not be switched off. The most novel feature of the Monod-Jacob model for regulation of β-galactosidase synthesis was the idea that a repressor protein in the bacteria normally keeps the genes of the lactose (*lac*) operon switched off. An **operon** is a segment of DNA consisting of two or more adjacent genes whose expression is jointly regulated by regulatory sites in the DNA situated close to the **structural genes**, that is, genes that direct the synthesis of polypeptides (Figure 9-6).

The *lac* operon consists of three structural genes that code for enzymes required for the metabolism of lactose and three regulatory genes that control expression of the structural genes. The *z* gene codes for β-galactosidase, which splits lactose into glucose and galactose. The *y* gene codes for the enzyme *permease,* which is able to bind the lactose molecules that enter the cell and thereby build up the concentration of lactose inside the cell. The *a* gene codes for the enzyme *acetylase,* whose function is not yet understood.

As originally proposed by Monod and Jacob, the *lac* operon (so named because all of the genetic elements are involved in the metabolism of lactose) contains three genetic elements that regulate the expression of the three structural genes of the operon. One of these regulatory elements—the *i* gene—codes for the synthesis of repressor proteins that normally keep the *lac* operon structural genes switched off. The other two regulatory elements are called *sites* in the DNA, since these genetic elements are not

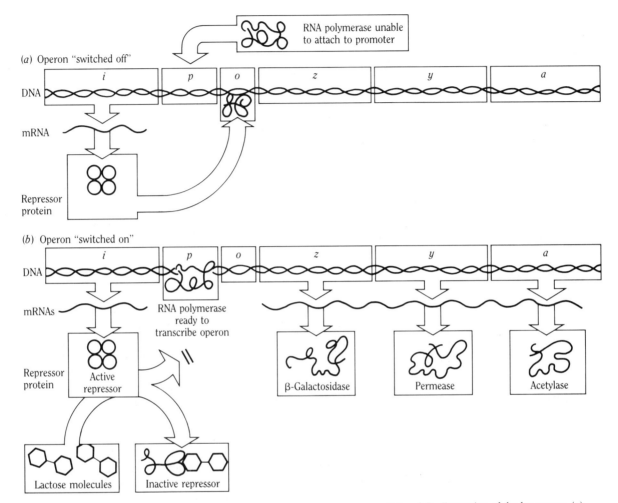

(a) Operun "switched off"

RNA polymerase unable
to attach to promoter

DNA

mRNA

Repressor
protein

(b) Operon "switched on"

DNA

mRNAs

RNA polymerase
ready to
transcribe operon

Repressor
protein

Active
repressor

β-Galactosidase

Permease

Acetylase

Lactose molecules

Inactive repressor

Figure 9-7 Expression of the *lac* operon. (*a*) The *lac* operon remains uninduced (switched off) as long as repressor protein remains attached to the operator. (*b*) In the presence of lactose, the repressor protein is inactivated and the *lac* operon is induced (switched on).

transcribed into mRNA or into proteins. The **operator**, or *o site*, consists of a stretch of about thirty base pairs in the DNA that are recognized by the repressor protein. Normally, *lac* repressor proteins are attached to the *o* site in the DNA and keep the *lac* operon switched off by preventing transcription of the *z, y,* and *a* structural genes. The **promoter**, or *p* site, consists of a slightly longer sequence of base pairs in the DNA that is recognized by the enzyme RNA polymerase, which catalyzes transcription of mRNA molecules. When the *o* site is occupied by a repressor protein, RNA polymerase cannot attach to the *p* site and thus transcription cannot occur.

Lactose molecules are like keys that unlock (switch on) expression of the *lac* operon. Lactose is referred to as an **inducer** because it induces the expression of genes that have been switched off. There are many inducers in addition to lactose in bacteria and other cells; these small molecules are able to switch on expression of genes by a variety of mechanisms. Lactose

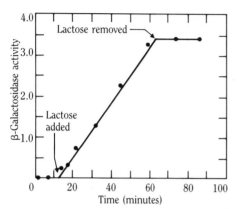

Figure 9-8.   Lactose-induced expression of the z gene, which codes for β-galactosidase. Until lactose is added to cells, no β-galactosidase is synthesized. When the inducer lactose is added, β-galactosidase is synthesized as long as lactose is present. When lactose is removed, the repressor proteins become active and β-galactosidase synthesis stops.

Table 9-1.   Regulatory Mutations Affecting the Synthesis of β-Galactosidase in *E. Coli*

Bacterial genome	Amounts of β-galactosidase (arbitrary units)	
	Excess lactose	No lactose
$i^+p^+o^+z^+$ (wild-type)	1,000	1*
$i^+p^+o^+z^-$	0	0
$i^-p^+o^+z^+$	1,000	1,000
$i^s p^+o^+z^+$	10	1
$i^+p^+o^-z^+$	1,000	1,000
$i^+p^-o^+z^+$	1	1

*Even with no lactose present, wild-type bacteria have a low level of β-galactosidase because occasionally the operon escapes from repression and an mRNA molecule is transcribed. This level of enzyme activity is called the *basal level*. With lactose present, the enzyme activity increases 1,000 times. Intermediate levels of enzyme activity are also possible as a result of mutant genes. And, of course, mutations in the z gene itself can prevent any active β-galactosidase from being synthesized.

functions as an inducer by attaching to repressor proteins, thereby preventing them from attaching to the *lac* operator (Figure 9-7). When the lactose molecules attach to the repressor, the physical shape of the repressor proteins is changed so that they cannot attach to the *o* site.

Thus, the *lac* operon exists in either of two conditions in bacteria: switched off or switched on. Both conditions can be demonstrated experimentally by measuring the synthesis of β-galactosidase in a culture of bacteria in either the presence or absence of lactose (Figure 9-8). If no lactose is present in the medium, β-galactosidase is not synthesized. When lactose is added, β-galactosidase synthesis begins and continues as long as lactose is present. If lactose is removed, β-galactosidase synthesis stops immediately.

The Monod-Jacob model for the regulation of the *lac* operon has been shown to be correct by many experiments and has provided a useful framework for formulating ideas about gene regulation in more complex organisms, including humans. Lactose is present in the milk of all mammals and must be broken down by an enzyme similar to β-galactosidase that is found in the stomachs of many (but not all) persons. The survival of human infants, of course, depends on their ability to metabolize the lactose present in mother's milk (see Box 9-1).

REGULATORY SITE MUTATIONS

Because β-galactosidase is the easiest of the three enzymes to measure, changes in its activity are used to detect mutations that affect expression of the z gene and synthesis of β-galactosidase. Table 9-1 summarizes the effects of just some of the regulatory mutations that affect synthesis of β-galactosidase in either the

Human *lac* Mutants

BOX 9-1

Mother's milk is an essential food for newborn infants. Milk is the principal food in an infant's diet for the first few months, even years, of its life. About 7–8 percent of human milk is the sugar lactose. Human babies are born with an active enzyme (lactase) in the small intestine that is essential for the digestion and metabolism of lactose, which it splits into the simpler sugars glucose and galactose. After the child reaches about four years of age, the gene that directs the synthesis of lactase is switched off, and a new gene, the adult lactase gene, is switched on. More correctly, it is switched on in *some* adults—and therein lies an interesting story about human nutrition.

Many human adults, probably a majority of the world's population, are intolerant of the sugar lactose. If these persons drink milk, they get diarrhea to varying degrees, and in the most severe cases they can dehydrate to the point of death. Only white American and European populations tend to have active adult lactase genes; most other human populations, including Africans, Chinese, and Thais, for example, are lactose-intolerant (see graph).

The ability to digest lactose is an inherited trait, although the particular human genes coding for the various lactose enzymes have

not been identified. Most of the genetic and anthropological evidence indicates that millions of years ago all human adults were lactose-intolerant. Milk was probably not generally available in the adult diet until about 10,000 years ago, when goats, cattle, sheep, and reindeer were first domesticated. If a mutation occurred in humans that permitted adults to drink large quantities of milk, those individuals with the mutant allele probably had an advantage over lactose-intolerant adults because they could adapt to a wider variety of environments and were therefore more likely to survive

until reproductive age and leave more progeny. At least this seems to explain why some populations are almost entirely lactose-intolerant while others are almost entirely lactose-tolerant.

Generally, lactose intolerance is not an all-or-nothing condition; adults who are lactose-intolerant can usually drink some milk without ill effects. And drinking even a little milk daily may serve to increase the kinds of bacteria in the gut that help break down lactose. Milk may not be the best food for everyone, but anybody can eat yogurt, cheeses, and other fermented milk products because the bacteria used in the fermenting process have digested the lactose from the milk.

Reading: Norman Kretchmer, "Lactose and Lactase." *Scientific American*, October 1972.

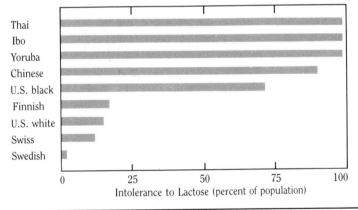

Intolerance to Lactose (percent of population)

presence or absence of lactose. In fact, it was changes in β-galactosidase activity of various bacterial regulatory mutants that led Monod and Jacob to formulate their original model for regulation of the *lac* operon.

Just as mutations in the *z* gene can destroy the activity of β-galactosidase, mutations in any one of the *lac* operon regulatory units (*i, p,* and *o*) may also affect expression of the *z, y,* and *a* structural genes. Mutations can affect the functioning of the *i* gene (which codes for the repressor protein) in a number of different ways. Some mutations destroy the repressor's ability to

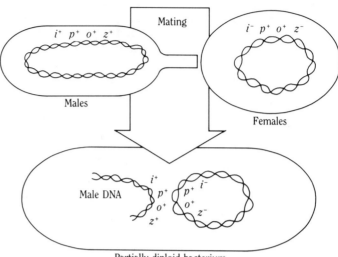

Figure 9-9 Mating of male and female bacteria carrying different mutations in genes of the *lac* operon. Bacteria that are partially diploid, that is, that carry two copies of the *lac* operon, can be used to demonstrate the existence and functions of repressor proteins.

attach to the operator site (i^- mutations), resulting in full expression of β-galactosidase activity. Other i gene mutations change the repressor's sensitivity to lactose (i^s mutations); since lactose is unable to bind to the repressor in these bacteria, the repressors remain more or less permanently attached to the operator site. In i^s mutant bacteria, lactose is unable to induce the operon, and only a small amount of β-galactosidase can be synthesized.

Mutations in the operator (o site) usually prevent the repressor from attaching to the DNA there. If the repressor is completely unable to attach, full β-galactosidase activity is observed; other operator mutations result in partial enzyme activity because of less efficient attachment of the repressor to the operator. Mutations in the promoter (p site) affect attachment of the RNA polymerase and the rate of transcription of the *lac* operon. Promoter mutations may abolish synthesis of β-galactosidase, lower synthesis partially, or even raise the level of synthesis, depending on whether the attachment of the RNA polymerase to the promoter is decreased or increased.

PROVING THAT REPRESSORS EXIST

The Monod-Jacob model predicted the existence of the repressor proteins and described how they must interact with both lactose and the operator. But at the time Monod and Jacob proposed their model, the only evidence for the existence of a repressor was inferred from genetic experiments (isolation of mutants), and it was substantiated indirectly by mating various strains of bacteria containing mutations in different genes of the *lac* operon. One such experiment is outlined in Figure 9-9. Female *E. coli* bacteria with the genotype $i^-\ p^+\ o^+\ z^-$ were mated with male bacteria with the genotype i^+ p^+ o^+ z^+. During the mating, the male bacteria transferred a part of their chromosome including the *lac* operon to the females. The female bacteria thereby became partially diploid for the *lac* operon genes; that is, they now contained two alleles of each *lac* gene, which could be the same or different. After the transfer of the *lac* operon DNA from the male to the female bacteria, the males in the medium were killed by adding T-phages that would destroy only males (the female bacteria were resistant).

Shortly after the mating, the female bacteria began to synthesize β-galactosidase, but after about 4 hours the synthesis ceased (Figure 9-10). The explanation for this pattern of enzyme synthesis is that as soon as the male's z^+ gene entered the female bacteria, β-galactosidase synthesis began because the female bacteria did not contain any repressors. The male's i gene entered a short while later, and some time had to pass before enough repressor molecules could accumulate in the female bacteria to shut off the active *lac* operon and prevent further synthesis of β-galactosidase. Finally, if lactose was added after synthesis had stopped because of repression of the *lac* operon, induction occurred and β-galactosidase synthesis resumed, as shown in Figure 9-10.

Since these mating experiments were performed over twenty years ago, all of the predictions of the Monod-Jacob *lac* operon model have been confirmed by isolation and analysis of the actual DNA and by chemical purification of the repressor, which is now known to be a protein that does indeed attach to the operator. The complete base sequence of the *lac* operon DNA and the locations of the promoter and operator sites are shown in Figure 9-11. However, what was still not explained by the model is why lactose will not induce the *lac* operon and why synthesis of β-galactosidase does not occur as long as cells are supplied with glucose (refer to Figure 9-4). The final piece of the puzzle was solved with the discovery of another protein that is required to fully switch on the *z, y,* and *a* genes.

SWITCHING ON THE *lac* OPERON

The interaction of the repressor with the operator provides an all-or-nothing switch for the *lac* operon. When the repressor is

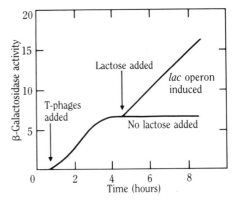

Figure 9-10   Synthesis of β-galactosidase in the mated female bacteria shown in Figure 9-9. Synthesis continues in the partial diploid until enough repressors accumulate to repress the operon. At this point addition of lactose will induce further synthesis of β-galactosidase.

Figure 9-11   The base-pair sequence in the *lac* operon DNA. The last four codons of the *i* gene are shown, along with the four amino acids at the end of the repressor. The promoter (*p* site) is recognized by two proteins: the RNA polymerase and the CAP protein (with cyclic AMP attached to it). The RNA polymerase cannot attach effectively to the promoter unless the CAP protein is also attached. The repressor attaches to the operator. Keep in mind that the DNA is not a straight rod but has a complex structure that is recognized by these regulatory proteins.

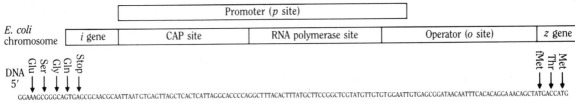

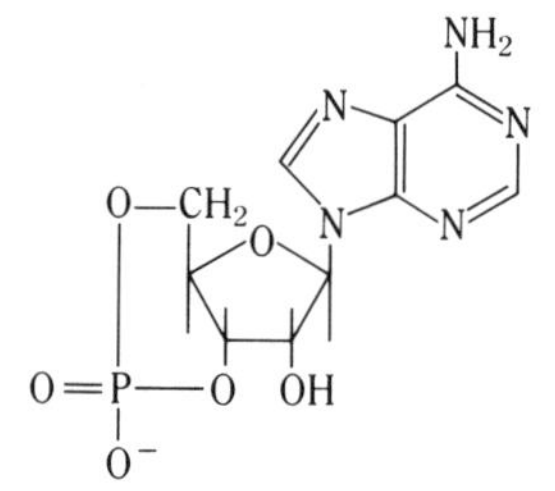

Figure 9-12 The chemical structure of an important small regulatory molecule, cyclic AMP. When cAMP attaches to the CAP in bacterial DNA, the *lac* operon is switched on. Cyclic AMP is also found in human cells, where it regulates the expression of human genes.

attached to the operator, transcription is prevented. If the repressor is removed from the DNA, the *lac* operon has the capacity for expression, but it must be positively switched on by another protein called the **catabolite activator protein (CAP)**. This protein has the potential for attaching to a section of the promoter, thereby allowing the RNA polymerase to begin transcription.

The attachment of the CAP to the promoter depends on the presence in the cell of another small molecule (about the size of lactose) called *cyclic adenosine monophosphate* (cAMP), whose chemical structure is shown in Figure 9-12. When cAMP and CAP combine, the complex attaches to the lefthand end of the promoter site and permits the RNA polymerase to begin transcription. If the CAP–cAMP complex is not attached to the promoter site, attachment of the RNA polymerase to the promoter occurs infrequently and the operon cannot be induced efficiently even in the presence of lactose.

High levels of glucose have been shown to prevent synthesis of cAMP in bacteria; this now explains why lactose cannot act as an inducer when glucose is also present. The repressor proteins function negatively to regulate the on-off switch for expression of β-galactosidase; the CAP–cAMP complex functions positively to switch on the *lac* operon when glucose, a preferred energy source, is no longer available. This combination of negative and positive gene control is a complicated business. Reread this section and the preceding one if you are having trouble keeping it all straight.

PROMOTER REGULATION OF GENE EXPRESSION

There is yet another puzzle in the mechanism of gene regulation in *E. coli* that you might already be wondering about: What regulates the regulatory proteins, and what keeps the problem of regulation from being an infinitely regressing one—regulators, regulators for the regulators, regulators for the regulators of the regulators, and so on and on? How, in other words, is the *lac* repressor protein itself regulated?

The answer is that the expression of most genes in bacteria is regulated solely by the sequence of bases in the promoter site that occurs at the beginning of each gene or operon. To begin transcription of any gene, RNA polymerase molecules must attach to the DNA at promoter sites and begin synthesizing mRNA molecules. It is now known that the *i* gene has its own promoter, which is capable of making only extremely weak attachments to RNA polymerase molecules. As a consequence, only one or two mRNAs for the *i* gene are transcribed each bacterial generation, and only about twenty or so repressor molecules are normally found in each bacterial cell, compared with the many thousands of molecules of β-galactosidase and other enzymes. Many other bacterial genes are regulated like the *i* gene; different promoters permit attachments of RNA polymerases to varying degrees. In this way, promoters determine the degree of transcription and the amounts of proteins synthesized.

It is possible to isolate mutants in the promoter for the *i* gene that greatly increase attachment of the RNA polymerase to that promoter and correspondingly increase the synthesis of repressor proteins manyfold. Mutations in the *i* gene promoter that increase the rate of synthesis of repressor proteins have also enabled researchers to isolate and chemically characterize the repressor, which normally is present in quantities too small to isolate.

Promoterlike sequences also exist in the DNA of animal cells, but regulation of gene expression in eucaryotic cells by promoters is more complex than in bacteria and is correspondingly less well understood. To date, experiments indicate that it is unlikely that genes in animal cells are organized into operons. However, it does appear that several kinds of regulatory sites in the DNA of animal cells are required to regulate the synthesis of the thousands of different proteins produced during development of animals and differentiation of their cells.

How Gene Expression Is Regulated in Animal Cells

Developing animal embryos produce differentiated cells that perform specialized functions by switching genes off or on. The different cellular activities are determined by the regulation of gene expression in the various kinds of differentiated cells (Figure 9-13). For example, in muscle cells the genes that are expressed and the proteins that are synthesized must be different from those that are expressed and synthesized in skin cells or red blood cells. Do these different animal cells contain different genes and amounts of DNA, or do the cell differences merely reflect differences in gene expression of cells with identical genomes? The answer to this important question was resolved in 1964 by John B. Gurdon's experiments with the African toad *Xenopus laevis*.

Gurdon was able to demonstrate **totipotency**: the ability of a cell to proceed through all the stages of development and ultimately produce a normal adult animal. He showed that each and every cell of an adult animal, in this case the African toad, contains all the genetic information needed for reconstructing the entire animal if that DNA is reintroduced into an unfertilized egg. Gurdon removed the nucleus from an intestinal cell of a tadpole, injected that diploid nucleus into an unfertilized egg whose own nucleus had been deliberately destroyed by a beam of UV light, and then nurtured the egg with its new nucleus through all the stages of development until an adult toad was formed (Figure 9-14). About 1–2 percent of the injected eggs subsequently developed into mature toads.

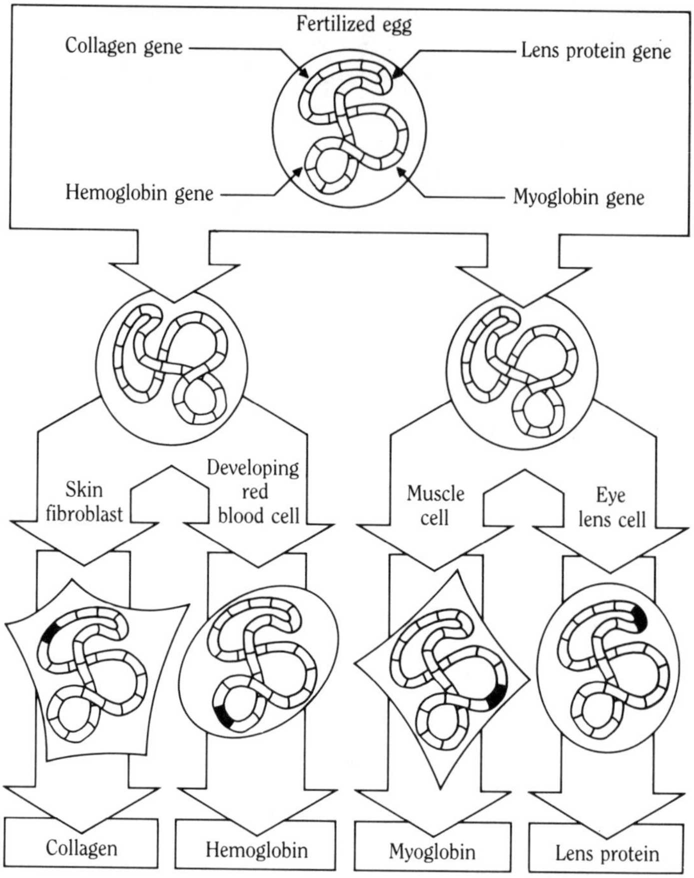

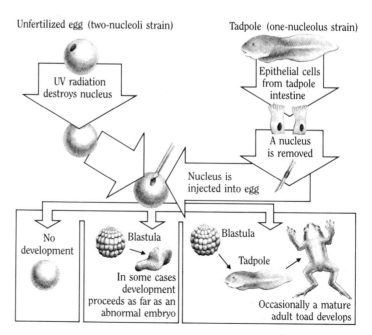

Figure 9-13 Gene regulation during the development of animals. A fertilized egg contains all of the organism's genetic information. As the embryo develops, cells differentiate as particular genes are switched on or off. All differentiated cells still contain all of the organism's genes, but only the appropriate genes are expressed.

Figure 9-14 Totipotency in the African toad *Xenopus laevis*. This was the first animal to be cloned, proving that the genetic information in the nucleus of every somatic cell is complete and potentially capable of directing development of a complete animal. All the adult toad's cells have one nucleolus, proving that its genes came from the transplanted somatic nucleus and not the egg.

192

The remarkable thing about this experiment is not that most of the eggs failed to develop properly but rather that every egg that did, did so by means of genetic information in a somatic, not a sex, cell. This proved that every somatic cell contains all the genetic information required for development to proceed normally to the adult stage.

To show that all of the genes expressed in the egg did indeed come from the transplanted nucleus and not from the egg's own nucleus due to failure of the UV light to destroy the egg nucleus, Gurdon used a genetically determined cellular marker. All intestinal cells from which nuclei were extracted came from a strain of toads having only one nucleolus (a substructure of the nucleus), whereas all of the eggs were taken from a strain of toads that had two nucleoli in all of their cells. If the cells in the resultant toads had only one nucleolus, it would prove that the transplanted nucleus had supplied all the genetic information, including that which determines the number of nucleoli. And, indeed, all of the toads that developed from the transplanted eggs did have only one nucleolus in their cells. Because such toads are genetically identical, they are called **clones** (see Box 9-2).

Cloning frogs and toads is much easier than cloning mammals. The development of toads proceeds outside the mother's body, whereas mammals develop only inside the female's uterus. In 1978 a book by David M. Rorvik, titled *In His Image: The Cloning of a Man*, was published that purported to be a true account of how a scientist, in return for a handsome payment by a millionaire, produced a son that was a clone of the millionaire by a technique similar to the one used to clone toads. A British geneticist whose name appeared in the book read the story, realized that it was a fabrication, and sued the author. In 1981 the U.S. District Court in Philadelphia ruled that the book was "a fraud and a hoax" and ordered the author and publisher to pay damages.

Cloning of mammals is technically possible, however, and in 1981 the first successful cloning of mice was reported by scientists in Switzerland. To date, however, such experiments with mammals have been successful only if the nuclei have been taken from undifferentiated embryonic cells (Figure 9-15). No mammalian cloning has yet been successful using nuclei taken from differentiated cells. Further discussion of the genetic manipulations that are possible with mammalian cells and with human embryos is deferred to later chapters.

HISTONES AND HORMONES AS REGULATORS

Although relatively little is known about the specific mechanisms that regulate gene expression in animal cells, various kinds of molecules are known to be associated with the DNA in chromo-

Plant Cloning:
Potatoes + Tomatoes =
Pomatoes

BOX 9-2

In the 1930s French plant scientists discovered that a tiny piece of tissue sliced from a carrot root would continue to grow *in vitro* in a synthetic medium quite similar to the one in which bacteria grow. Since that discovery, other researchers have found that plant cells grow quite readily in a solution of mineral salts that contains the sugar sucrose (instead of glucose) and tiny amounts of two essential plant hormones called *auxins* and *cytokinins.* In such an *in vitro* culture medium, carrot cells (and cells from many other plants) can be grown indefinitely if the tissue is transferred periodically to a fresh medium.

In the last few years researchers have found that by manipulating the relative concentrations of the auxins and cytokinins, they can cause the carrot tissues to differentiate into roots, stems, leaves, and flowers. In fact, they can regenerate complete carrot plants from single cells. Since every carrot is grown from genetically identical cells, all the carrot plants are clones.

Cloning is essential to agricultural companies that want to develop plants with the most desirable properties. (Cloning plants is nothing new; when you snip off a piece of a friend's plant to start your own, you are cloning that plant.)

The ability to grow entire plants from single cells raises some fascinating possibilities. Most plants that belong to the same family will not interbreed with each other. For example, potato, tomato, and tobacco plants are all members of the plant family *Solanaceae,* but pollen from one plant species of this family cannot fertilize eggs from a different species; you cannot cross-pollinate potato and tomato plants even by artificial pollination techniques. However, tomato-potato plant hybrids can be constructed using single cells of each species.

Plant cells are surrounded by a rigid outer layer—the cell wall—consisting mostly of cellulose. Removal of this cellulose coat with enzymes produces plant cells called *protoplasts.* Now, if protoplasts derived from potato plants are mixed with tomato cell protoplasts along with high concentrations of calcium or certain other chemicals, the different protoplasts fuse together to form a hybrid cell that contains the genetic information of both tomato and potato cells. Appropriate hormone treatment of the hybrid cells will cause the formation of new cell walls, and occasionally a complete plant can be regenerated from the hybrid cell. Such a plant could be called a "pomato."

Plants derived from the somatic hybridization of potato and tomato cells have actually been produced, but so far such plants have not had any significant agricultural value, since the hybrid plants do not bear edible tomatoes or potatoes. In the final analysis, it is the way in which the genes in such unusual hybrid plants are regulated and expressed that will determine the plant's properties and usefulness, and these properties cannot be predicted for unusual plant hybrids.

Reading: Peter Steinhart, "The Second Green Revolution." *New York Times Magazine*, October 25, 1981.

Hypothetical pomato

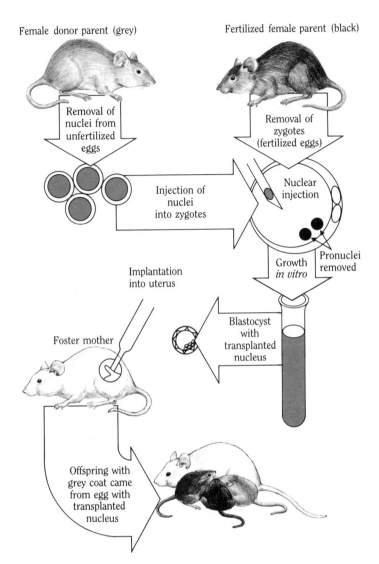

Female donor parent (grey)

Fertilized female parent (black)

Removal of nuclei from unfertilized eggs

Removal of zygotes (fertilized eggs)

Injection of nuclei into zygotes

Nuclear injection

Growth *in vitro*

Pronuclei removed

Implantation into uterus

Foster mother

Blastocyst with transplanted nucleus

Offspring with grey coat came from egg with transplanted nucleus

Figure 9-15 The first successful cloning of mice. Eggs suitable for development are obtained from a female carrying fertilized eggs. The nucleus is removed from the fertilized egg and replaced by a nucleus obtained from unfertilized eggs of the female donor. The egg with the transplanted nucleus is allowed to develop *in vitro* and is implanted in the uterus of a foster mother. Offspring mice that develop from the egg with a nuclear transplant have the coat color of the female donor parent.

somes, and these molecules have been shown to affect the transcription of DNA, which suggests that they regulate gene expression. **Histones** are a class of proteins that are attached to chromosomes in large quantities. Histones are thought to be important in determining chromosome structure, but relatively little is known about their actual functions in cells. About all that can be said with certainty is that five distinct kinds of histones can be isolated from animal cells.

Somewhat more is known about **hormones,** a large class of organic molecules produced in one tissue or organ of a plant or animal and carried to another part of the organism, where they regulate gene expression and other physiological processes (see Table 9-2). Hormones regulate gene expression in plant and

Table 9-2. Some of the Hormones That Regulate Human Gene Expression and Physiology

Hormone	Site of synthesis	Site of action	Major physiological role
Thyrotropin (TSH)	Pituitary	Thyroid gland	Stimulates synthesis and secretion of thyroid hormone
Adrenocorticotropin (ACTH)	Pituitary	Adrenal medulla	Stimulates synthesis of adrenal steroids
Growth hormone (somatotropin)	Pituitary	Many tissues	Promotes protein synthesis and skeletal growth
Luteinizing hormone (LH)	Pituitary	Gonads	Stimulates synthesis of progesterone in ovaries and testosterone in testes; causes ovulation
Follicle-stimulating hormone (FSH)	Pituitary	Gonads	Stimulates growth of ovarian follicle and of Sertoli cells of testes
Prolactin	Pituitary	Breast	Stimulates synthesis of milk proteins and growth of breast
Insulin	Pancreas	Many tissues	Promotes transport of glucose and amino acids into certain cells; synthesis of fatty acids in adipose cells and in liver; glycolysis of glucose; protein synthesis
Glucagon	Pancreas	Liver	Stimulates glycogenolysis and gluconeogenesis in liver
Antidiuretic hormone (vasopressin)	Hypothalamus	Kidney	Prevents loss of water and NaCl and controls blood pressure
Oxytocin	Hypothalamus	Milk glands	Stimulates secretion of milk and uterine contractions

animal cells by interacting with certain cellular proteins and with DNA. They are usually specific for certain tissues, and they are often found in cell nuclei. Hormones interact with receptor sites on specific cellular or chromosomal proteins and the resulting hormone-protein complex can activate a particular gene or group of genes in the chromosomal DNA.

Superficially, the interaction of a hormone with a nonhistone protein seems analogous to the interaction of lactose with a repressor protein or of cAMP with the CAP protein. However, little is known about the actual mechanism of hormone-protein interaction, and even less is known about the regulation of gene expression by these hormone complexes. One important application of recombinant DNA techniques (discussed in Chapter 10) is the capability of inserting specific animal or plant regulatory genes into bacteria. Because genes from animal cells can be integrated into bacterial DNA, the regulation of these genes can be studied by researchers using techniques that have been successful in understanding the regulation of the *lac* operon and other genes in bacteria.

Hormone	Site of synthesis	Site of action	Major physiological role
Parathyroid hormone	Parathyroid gland	Bone, kidney	Acts to increase Ca^{2+} in blood
Calcitonin	Thyroid C cells	Bone, kidney	Acts to decrease Ca^{2+} in blood
Cortisol	Adrenal glands	Liver and peripheral tissue	Stimulates glycogenolysis and synthesis of certain liver proteins; promotes protein breakdown in peripheral tissues
Aldosterone	Adrenal glands	Kidney	Aids in NaCl retention
Estradiol	Ovary	Uterus, breast	Stimulates development of secondary female sex characteristics
Testosterone	Testis	Spermatogonia	Promotes sperm synthesis; stimulates development of secondary male sex characteristics
Progesterone	Ovary, placenta	Uterus, breast	Helps preserve pregnancy
Vitamin D_3	Skin	Bone, kidney	Stimulates Ca^{2+} transport by small intestine; acts on bone to increase Ca^{2+} in blood
Epinephrine	Adrenal glands	Liver, heart, adipose tissue	Stimulates glycogenolysis and breakdown of fats; increases cardiac output
Norepinephrine	Adrenal glands	Heart, adipose tissue	Increases blood pressure; acts as neurotransmitter
Thyroid hormones	Thyroid gland	Many tissues	Increases respiration; required for nervous system growth in fetus and young animal; stimulates synthesis of certain enzymes

SPLIT GENES AND RNA SPLICING

In procaryotic cells, the information in a gene is contained in a continuous sequence of bases in DNA that is transcribed into a continuous sequence of bases in mRNA. Until 1977, there was no reason to think that the situation is any different in eucaryotic cells. In that year, however, researchers made the surprising discovery that in eucaryotic cells (such as human cells) the information in many genes is contained in discontinuous segments of DNA that are separated from one another by sequences of bases whose information is not expressed (Figure 9-16). This organization of genetic information in eucaryotic cells is referred to as **split genes**.

An easy way to grasp the concept of split genes is to imagine a human gene that codes for a hypothetical protein called "information." If the gene (word) is organized so that "information" is written as inxxxxforxxmaxxxxtion, the gene (word) is split. The segments of DNA containing information that codes for amino acids are called **exons** (*expressed regions*); the pieces of noncoding DNA separating the information-containing segments are

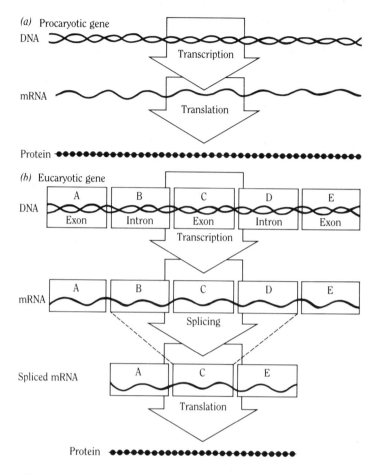

(a) Procaryotic gene

DNA

Transcription

mRNA

Translation

Protein

(b) Eucaryotic gene

DNA

| A | B | C | D | E |
| Exon | Intron | Exon | Intron | Exon |

Transcription

mRNA

| A | B | C | D | E |

Splicing

Spliced mRNA

| A | C | E |

Translation

Protein

Figure 9-16 The "split" genes of eucaryotic cells. (a) The genetic information in a procaryotic gene is contained in a continuous sequence of bases in DNA that is transcribed into a continuous sequence of bases in messenger RNA. (b) The genetic information in many eucaryotic genes is contained in discontinuous segments of DNA called exons. The intron segments are removed from the mRNA by splicing enzymes before the mRNA is translated.

called **introns.** In our imaginary protein inxxxxforxxmaxxxxtion, the x's are introns.

When mRNAs are transcribed from split genes, the extra bases must be removed before the mRNA is read; otherwise, proteins with extra and incorrect amino acids will be synthesized. For example, the gene that codes for the β-polypeptide of human hemoglobin consists of three exons and two introns (Figure 9-16b). The β-hemoglobin mRNA is transcribed with all of the information—exons and introns—that is in the gene. But before the mRNA is translated, **RNA splicing enzymes** remove the bases corresponding to the introns from the mRNA and rejoin the remaining RNA fragments correctly. The reconstructed mRNA can now be translated into the correct amino acid sequence. Of the few eucaryotic genes that have been analyzed so far, most contain both introns and exons in the DNA.

Why are genes split in eucaryotic cells and why are genes continuous in procaryotic cells? How many different RNA splicing enzymes are used in eucaryotic cells to ensure accurate splicing of all mRNAs? Have introns and exons played important roles in the evolution of eucaryotic organisms? Are split genes or continuous genes the more primitive form of gene organization? These are but a few of the as yet unanswered questions that have been raised by the discovery of split genes. A few preliminary experiments suggest that each exon might have originally directed the synthesis of short chains of amino acids that functioned as simple enzymes. During the course of millions of years of evolution, chance recombination of segments of DNA may have joined various exons and introns together and placed the entire sequence of bases under the regulation of other genes.

Until more is known about the functions of introns and exons and about the mechanisms that regulate gene expression in eucaryotic organisms, little can be said that is not speculative. The techniques of recombinant DNA now provide powerful tools for studying the organization and regulation of eucaryotic genes, since segments of DNA from any organism, including humans, can be cloned in bacteria and other microorganisms where it can be conveniently studied.

differentiation   the process by which cells become increasingly (and irreversibly) specialized in their functions in tissues and organisms during development.

regulation of transcription   a process that prevents or alters synthesis of RNA generally and mRNA in particular.

regulation of translation   a process that prevents or alters synthesis of proteins from mRNA.

feedback inhibition   a mechanism by which the final product of a biosynthetic pathway (say an amino acid) inhibits the activity of the first enzyme in that pathway.

β-galactosidase   an enzyme that breaks apart lactose into the sugars glucose and galactose.

operon   a segment of DNA consisting of two or more adjacent genes capable of synthesizing polypeptides, along with the regulatory sites that govern their expression. The genes in an operon are transcribed together into continuous mRNA molecules.

structural gene   a gene that codes for the synthesis of a polypeptide or protein.

operator (o site)   a sequence of base pairs in DNA to which repressor proteins attach, preventing transcription of structural genes of the operon.

promoter (p site)   a sequence of base pairs in DNA to which RNA polymerase enzymes attach to initiate transcription of structural genes.

inducer   any small molecule (for example, lactose) that is able to switch on the expression of one or more genes.

catabolite activator protein   the positive controlling element for glucose-sensitive operons in bacteria.

totipotency   the ability of a cell to proceed through all the stages of development, producing a normal adult organism. The nucleus from a single cell, in principle, contains all of the information for reconstructing the complete organism.

clone   individuals with identical genetic information. Identical human twins are clones since both developed from a single fertilized egg that split into two early in development.

histones   a small group of proteins that are thought to regulate gene expression and chromosome structure in animal cells.

hormones   a large class of different molecules in plants and animals that regulate gene expression and other physiological processes.

split genes   eucaryotic genes in which genetic information is encoded in the DNA in discontinuous segments.

exons   the discontinuous segments of DNA in eucaryotic genes that carry information that is translated into the sequence of amino acids in proteins.

introns   the segments of DNA in eukaryotic genes that separate exons and that are untranslated.

RNA splicing enzyme   an enzyme that removes bases from eucaryotic mRNA corresponding to introns and that splices the exons together in mRNA before translation occurs.

Additional Reading

Beermann, W. and V. Clever. "Chromosome Puffs." *Scientific American,* June 1964.

Check, W. "The Regulation of Gene Expression." *Mosaic*, November/December 1982.

Davidson, E. H. "Hormones and Genes." *Scientific American*, June 1965.

Gurdon, J. B. "Transplanted Nuclei and Cell Differentiation." *Scientific American*, December 1968.

Hilts, P. "Mark Ptashne's Molecular Mission." *Science 82*, December 1982.

Ptashne, M., Johnson, A. D. and C. O. Pabo. "A Genetic Switch in a Bacterial Virus." *Scientific American,* November 1982.

Treichel, J. A. "Blood Diseases Partially Corrected by Gene Alteration." *Science News*, December 18–25, 1982.

Review Questions

1. How many different molecules are involved in regulating the *lac* operon?

2. What molecule binds to promoter sites in DNA?

3. What kinds of cells carry split genes?

4. Why is it not safe for some adults to drink milk?

5. Do human clones exist?

6. What regulates the synthesis of various kinds of repressor proteins in bacteria?

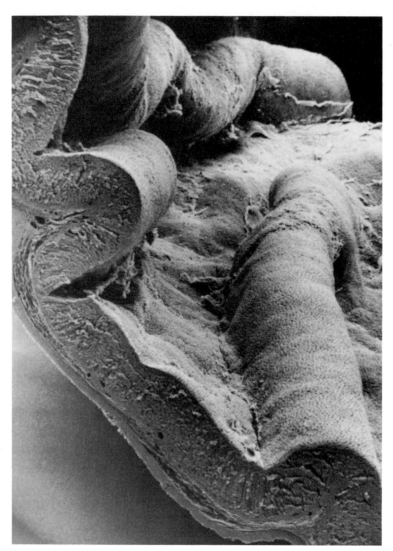

THE FOLDS IN THE WALL OF AN EMPTY STOMACH. (28 ×)

10

RECOMBINANT DNA TECHNOLOGY

*Opening
Pandora's
Box*

*It becomes increasingly evident that the
prime, urgent need of our times is not for
more science and improved technology,
medical, agricultural, or otherwise, but for
some new ethical policies and moral
guidelines to live and govern by that will
work against overpopulation, pollution,
depletion of resources, and so on.*

ROGER SPERRY

10
RECOMBINANT DNA TECHNOLOGY: OPENING PANDORA'S BOX

Of what use is basic scientific research to society? The answer to this frequently asked question is that basic research has immense practical value. Countless examples can be offered of how we have benefited from basic scientific discoveries that at first had no practical significance or application. To cite just a few instances:

- Faraday's experimental investigations of electricity and magnetism made possible the dynamo and the electric motor.
- Research in organic chemistry laid the basic scientific groundwork for the petrochemical industries: petroleum-derived fuels, rubber, plastics, synthetic fibers, and so on.
- Ongoing basic research in physics and chemistry underlies the development of the ever smaller, more capable, and more affordable computers, hand calculators, and other electronic devices that we depend on.
- Mendel's experiments with pea plants (Chapter 11) revolutionized our understanding of heredity, which in turn has advanced our understanding of (among other things) hereditary diseases, some of which can now be treated.

In the realm of biology and medicine, recent advances using recombinant DNA techniques provide the most dramatic and promising answer to the question, "Of what value is basic biological research to society?" Articles in magazines and newspapers daily announce new discoveries emerging from recombinant DNA research. Dozens of genetic engineering companies employ scientists and technicians who are applying basic genetic discoveries to produce useful microorganisms, drugs, plants, and chemicals. Valuable human proteins such as insulin, somatostatin, interferon, and growth hormone are already being synthesized in genetically engineered bacteria and are being clinically tested preparatory to their routine commercial use.

Yet few people, even among those who argue the pros and cons of recombinant DNA technology, realize that the techniques for splicing together genes from different organisms had their origins in experiments involving the growth of phages in bacteria—experiments that were of interest to only a handful of scientists for many years until the potential of their findings suddenly became apparent. Today, recombinant DNA technology, the outgrowth of basic microbiological and genetic research, is revolutionizing the chemical and pharmaceutical industries by producing new kinds of microorganisms that have been genetically engineered to produce useful and profitable substances. Recombinant DNA research has provided insights into gene regulation and gene expression that were unexpected and unpredicted just a few years ago.

A whole new industry, already consisting of hundreds of

Table 10-1. Breakthroughs Expected in Recombinant DNA research by the Year 2000

Area	Expected breakthrough
Basic knowledge	Understanding of gene regulation in humans
	Increased understanding of human development and aging
	Manipulation of DNA repair mechanisms
	Development of bacteria capable of producing desired antibodies
	Understanding of mechanisms of cancer cell formation
	Understanding of mechanisms for evolutionary changes in DNA
Medicine	Synthesis of human biological substances (proteins, hormones) in microorganisms
	Transplanting of genes into human cells and into people
	Isolation of genes responsible for genetic diseases
	Replacement of missing enzymes in cells
	Improved genetic screening for and detection of hereditary diseases
Agriculture	Development of plants capable of fixing nitrogen from air
	Development of unusual hybrid plants
	Development of plants resistant to diseases, drought, salt, and so on
	Development of plants with improved nutritional characteristics
	Production of proteins from single-celled organisms
Industry	Development of bacteria that can destroy dangerous chemicals
	Use of microorganisms for pollution control
	Use of microorganisms for extraction of ores
	Use of microorganisms to generate fuels and energy

companies, large and small, has sprung up world-wide to explore and exploit the potentials of the new gene splicing techniques. The pace of current recombinant DNA research and the rate of new discoveries is truly staggering. It is safe to predict that our lives will be changed by the discoveries and products of recombinant DNA research. A few of the expected developments are listed in Table 10-1. Besides realistic projections of discoveries such as those listed in Table 10-1, ridiculous forecasts have also appeared, such as those shown in Figure 10-1 from an article in the *National Enquirer*. Millions of Americans read this article, and many of them probably believe such pseudoscientific hokum.

A Cross Between Human Beings and Plants...

SCIENTISTS ON VERGE OF CREATING PLANT PEOPLE

...Bizarre Creatures Could Do Anything You Want

by **RICHARD BAKER** and **CHERIE HART**

In a world-shaking breakthrough that will forever change life as we know it, scientists reveal they're on the verge of creating an astonishing race of plant people.

These bizarre creatures, a cross between humans and plants, will become our pets and slaves, doing anything we want—including think!

And as incredible as it sounds, one scientist in the U.S. has already made it a partial reality — by successfully fusing the cells from a woman with those of a tobacco plant.

Excited researchers say the plant people will:

• Work in factories and on farms, where they'll pick their own fruit and walk for miles to find the best soil and climate.

• Serve as soldiers — capable of firing bullet-like seeds with deadly force.

• Operate as human spare part kits — sprouting arms and legs for use in transplant operations.

• Produce their own gasoline-like fuel, then fill up our vehicles at service stations run by plant people.

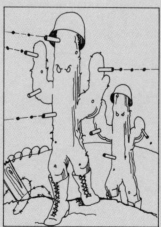

KILLER CACTUS: Plant people could serve as guards or soldiers using very special weapons — deadly seeds.

PUMPING GAS will be revolutionized by a tree recently discovered in Brazil that produces a kind of diesel fuel. It could be trained to fill up autos by itself.

Figure 10-1 Articles in newspapers and magazines often sensationalize the potentials of recombinant DNA technology. Such ideas may capture readers' imaginations but are often fictitious. (Used by permission of National Enquirer, July 1, 1980.)

In this chapter we describe the original basic research that led to today's recombinant DNA technology. We also outline the techniques of the basic process that is used to move genes from one kind of cell into another, a process called **DNA cloning** because billions of identical copies can be produced. And finally, we discuss the social problems and ethical questions raised by recombinant DNA research and technology.

What Is Genetic Engineering?

Recombinant DNA, genetic engineering, and *gene splicing* are more or less equivalent terms that refer to the same procedures for manipulating pieces of DNA. Basically, these techniques all entail the extraction of DNA from cells of one kind of organism and the insertion of fragments of that DNA into cells of a different kind of organism, usually a microorganism. In order for a piece of "foreign" DNA (DNA from a different species) to be cloned, it must be extracted from cells and cleaved into small fragments by any one of a number of recently discovered proteins known as **restriction enzymes** (Figure 10-2). The foreign DNA fragments are then inserted into circular plasmid DNAs previously isolated from donor microorganisms, such as bacteria or yeast. Next, the plasmids into which the foreign DNA has been spliced are introduced into the recipient microorganisms, where they are replicated. Finally, the foreign fragments of DNA are propagated by the growing microorganism, and the genes carried in the foreign DNA can be expressed.

As you can see, recombinant DNA techniques are, in principle, quite simple. However, cloning particular DNA pieces that are interesting or useful requires both scientific expertise and that elusive quality—which even scientists value highly—called luck.

HOW GENETIC ENGINEERING STARTED

Bacteria that belong to the same species and share many characteristics but that differ from one another in some regard are called **bacterial strains.** For example, some strains of *E. coli* are lysogenic for λ-phages while others are not (λ-phages and bacterial lysogens were discussed in Chapter 8). Special tests can be applied to distinguish lysogenic bacteria from cells of the nonlysogenic type. Exposing bacteria of each type to small doses of UV light, for instance, will destroy the lysogenic bacteria while scarcely harming the nonlysogenic bacteria because phages are produced that destroy the cell in one case but not the other.

Lambda phages can infect most strains of *E. coli* (except for those that already carry the λ-prophage in the DNA), and most of the infected bacteria will lyse, each cell producing hundreds of new λ-phages. However, a puzzling fact emerges when λ-phages are used to infect two different strains of bacteria, one called *E. coli* K and the other called *E. coli* R. Lambda phages synthesized in the K strain are labeled λ(K)-phages, and λ-phages produced in the R strain are labeled λ(R)-phages. After the λ(K)- and λ(R)-phages have been produced, they are again used to infect each different bacterial strain separately.

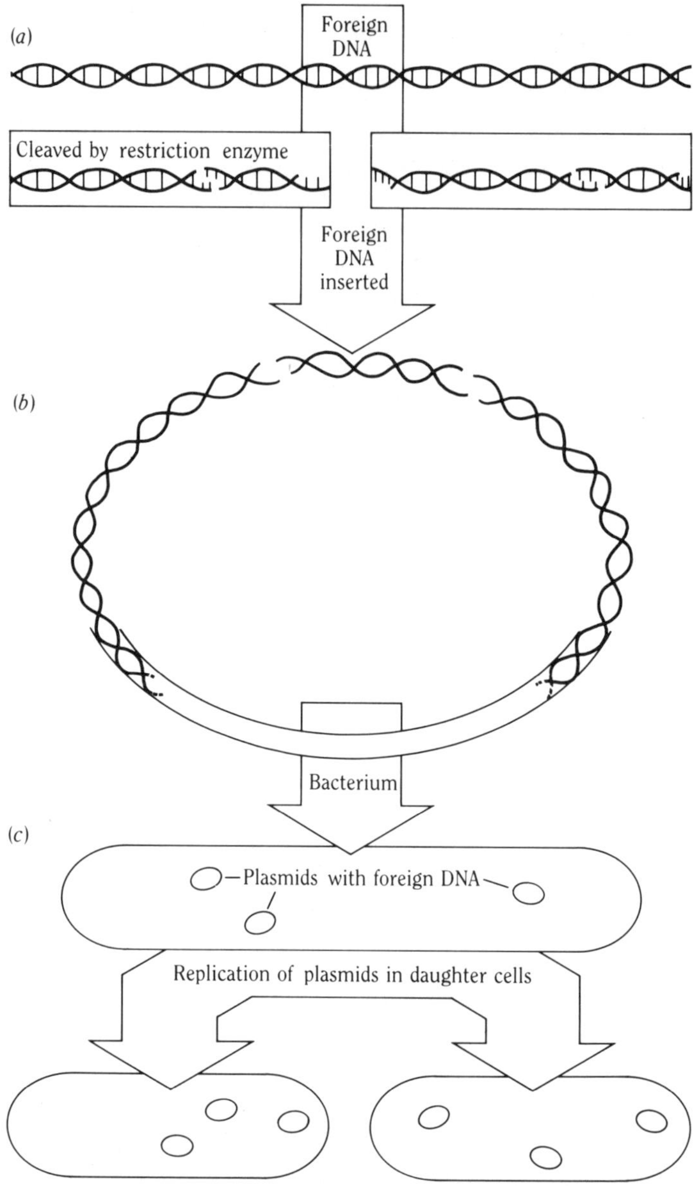

Figure 10-2 The basic steps in genetic engineering. (a) DNA from the cells of any organism (called the *foreign DNA*) is cleaved into pieces by restriction enzymes. (b) These pieces of DNA are randomly inserted into bacterial plasmid DNA in a series of enzyme-mediated reactions. (c) Plasmids carrying pieces of foreign DNA (foreign genes) are reintroduced in bacteria. Plasmids are capable of self-replication in the bacteria and replicate their own genes along with the foreign genes. Plasmids may exist in as few as one copy per cell to as many as several hundred. Each time the bacteria divide, daughter cells receive one or more plasmids, which then replicate until the characteristic number of plasmids in the cell is restored.

What is observed in these infection experiments is not easily explained. On the one hand, λ(R)-phages are able to grow in either bacterial strain and to synthesize new λ-phages. On the other hand, λ(K)-phages are able to grow *only* in the *E. coli* K strain and are unable to grow in the R strain (Figure 10-3). The initial observations of this phenomenon were made by researchers in the early 1950s, but the explanation for this puzzling difference was to require almost twenty years of further research to uncover. Indeed, only a few scientists in the 1950s thought the λ-phage growth differences were of sufficient interest to warrant

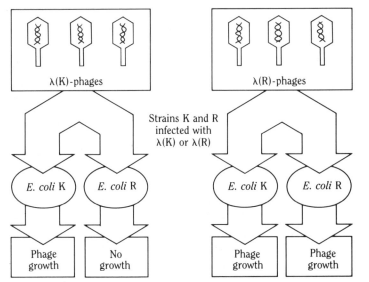

Figure 10-3 Pattern of growth of λ-phages in bacterial strains. Phages grown in one bacterial strain are sometimes unable to grow in another strain. Lambda phages grown in *E. coli* strain K are labeled λ(K); phages grown in *E. coli* strain R are labeled λ(R). When these phages infect the same two bacterial strains, the λ(R)-phages grow on either strain but λ(K)-phages grow only on strain K.

further research. Yet once researchers had discovered why λ-phages grew differently in various bacterial strains, the possibility of gene splicing became obvious to many.

By the middle of the 1960s, researchers who had pursued the question had discovered that *E. coli* strain R contains an R plasmid in all of its cells (the R plasmid was discussed in Chapter 4). Moreover, it is the expression of one particular gene carried in the DNA of the R plasmid that prevents the growth of the λ(K)-phages. As the story unfolded in many laboratories, it became clear that all DNA present in bacteria containing plasmids is modified as the bacterial DNA is synthesized. Subsequently, it was discovered that the modified DNA is protected against destruction by an enzyme produced by a different plasmid gene.

When the λ(K)-phage DNA enters bacteria that carry the R plasmid, the λ-DNA is recognized as foreign and the enzymes—now known to be restriction enzymes—attack the DNA of the λ(K)-phages, thereby preventing the growth of λ-phages. How DNA molecules are modified, recognized, and destroyed by plasmid-encoded enzymes is now understood and provides the basis for all recombinant DNA experiments.

Restriction and Modification of DNA

When DNA is replicated in bacteria, methyl (CH_3) groups are attached to bases in the newly synthesized DNA at specific sites defined by a sequence of six to ten nucleotides in both strands. The sites are recognized by **modification enzymes** that attach the methyl groups. These enzymes can be coded for either by genes in the bacterial chromosome or by genes in plasmids, but the plasmid-encoded enzymes are the ones responsible for the differ-

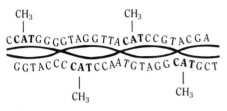

Figure 10-4 Modification of DNA by methyl groups (CH_3). DNA of different bacterial strains is modified by enzymes coded for by either bacterial or plasmid genes or both. In this example, the modification enzymes recognize the CAT sequence of bases in either strand and attach a methyl group to adenine.

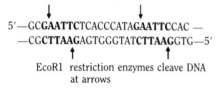

EcoR1 restriction enzymes cleave DNA at arrows

Fragments of DNA produced

Figure 10-5 Cleavage of DNA by restriction enzymes. Restriction enzymes recognize specific sequences of bases in DNA. If the DNA is not modified (protected) by the appropriate pattern of methylation, the restriction enzymes cleave the DNA into fragments. The EcoR1 restriction enzyme coded for by the R plasmid recognizes the sequence GAATTC and makes staggered cuts in the DNA strands, as shown by the arrows. Each DNA fragment has single-stranded AATT tails.

ences in λ-phage growth. It is the attachment of methyl groups to particular bases that modifies the DNA and protects it from being destroyed by restriction enzymes in the cell (Figure 10-4).

The restriction enzymes recognize the specific patterns of DNA methylation produced by corresponding modification enzymes. The specific methylation pattern of a bacterium's DNA provides it with a means of distinguishing its own DNA from that of any other DNA that might happen to enter the cell. For example, bacteria may be infected by different kinds of phages, but all of the phage DNA will itself be methylated according to the pattern characteristic of the particular strain of bacteria in which the phages were grown. Phages will be restricted from growing if the methylation pattern of the phage DNA tells the infected bacterium that the DNA is not the same as that of the bacterial strain. If bacteria of different strains or species mate, any DNA that is transferred and is recognized as having a different pattern of methylation will also be destroyed.

Any DNA molecule whose pattern of methylation is different from that cell's own DNA is broken down by the restriction enzymes (Figure 10-5). Each kind of microorganism contains restriction enzymes that recognize a different sequence of bases and pattern of methylation. Therefore, a restriction enzyme, which breaks DNA molecules only at certain sites, will attack any unmethylated DNA or any DNA whose methylation pattern is different from that of its own DNA. The restriction enzyme named EcoR1, for example, is coded for by a gene on the R plasmid of *E. coli* (hence the name EcoR1).

As you may have already guessed, the pattern of λ-phage growth shown in Figure 10-3 is explained in terms of modification and restriction enzymes. Each *E. coli* K strain produces a modification enzyme that methylates DNA in a pattern specific to that strain and also produces restriction enzymes that recognize that pattern and break down DNA that is unmethylated or methylated differently. *E. coli* R strains possess two different methylation enzymes and two different restriction enzymes. One pair of the modification-restriction enzymes is coded for by genes on the bacterial chromosome; the other pair is coded for by genes on the R plasmid.

The modification-restriction enzyme system can be thought of as a means of coding the DNA that is analogous to color coding. Imagine that the enzymes in the *E. coli* K strain are blue; this means that the methyl groups on the DNA are blue also, and the blue restriction enzymes will not attack that DNA. The modification-restriction enzymes in *E. coli* R can be thought of as both blue and red. The DNA in *E. coli* R strains, therefore, is colored both red and blue and will not be attacked by either red or blue restriction enzymes. Similarly, λ-phage DNA that is blue—λ(K)-phage DNA—can infect and replicate in bacteria with blue re-

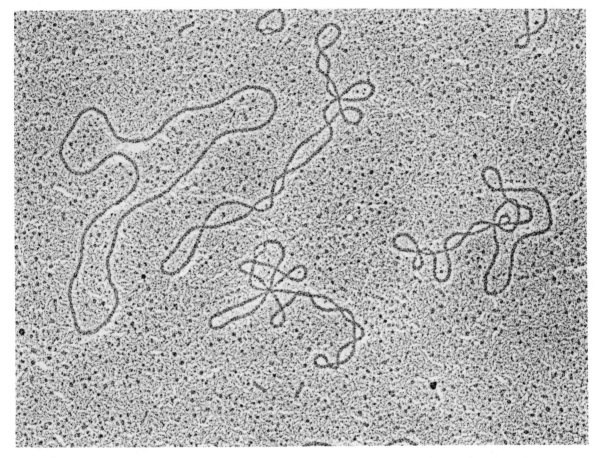

Figure 10-6 An electron micrograph of several small circular cloning vehicles—plasmids that replicate their own DNA and any foreign DNA inserted into them. Plasmids exist in coiled forms or uncoiled circles.

striction enzymes (strain K) but will be destroyed in bacteria having red restriction enzymes (strain R). However, λ-phage DNA that is colored blue and red—λ(R)-phage DNA—will survive and replicate in bacteria that have red, blue, or both red and blue restriction enzymes (both K and R strains).

As soon as the enzymatic mechanisms of DNA restriction and modification became understood, many scientists realized that it should be possible to join together pieces of DNA from completely different organisms. The next step was to construct cloning vehicles: small self-replicating plasmids that would carry along and replicate any piece of foreign DNA that had been inserted into their own circular DNA molecules.

PLASMIDS AS CLONING VEHICLES

A vehicle is something used to transport an object from one place to another. **Cloning vehicles** (also referred to as *cloning vectors*) are plasmids or viruses whose DNA has been constructed in such a way that foreign DNA fragments produced by restriction enzymes can easily be inserted into them (Figure 10-6). When these

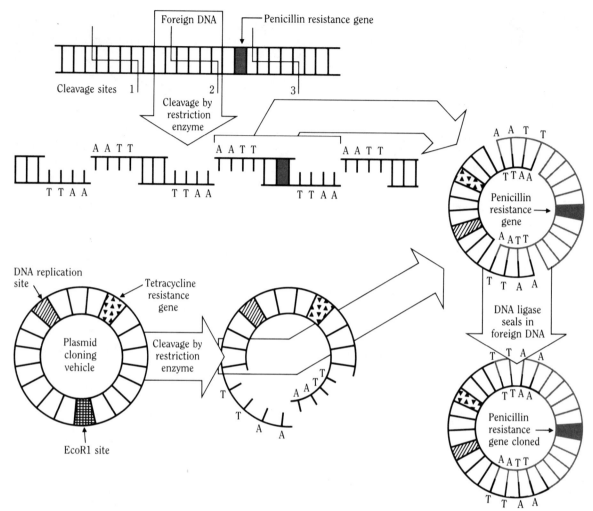

Figure 10-7 Cloning the penicillin resistance gene by recombinant DNA techniques. Foreign DNA containing the penicillin resistance gene is cleaved with EcoR1 restriction enzymes, as is the plasmid cloning vehicle. The foreign DNA fragments are inserted into the plasmids and sealed in by the enzyme DNA ligase. Only a few plasmids will receive the DNA fragment carrying the penicillin resistance gene.

plasmid or virus cloning vehicles are reintroduced into microorganisms, they replicate both their own DNA and the foreign DNA fragment they carry. The foreign DNA is thus cloned, since identical copies are made each time the cloning vehicle replicates itself. Of course, the microorganisms used for cloning must be genetically altered so that they no longer produce restriction or modification enzymes.

The first functional cloning vehicle was constructed from an R plasmid bearing genes that made the bacteria containing it resistant to the antibiotic tetracycline. The antibiotic resistance genes of the plasmid provide researchers with a simple means of identifying bacteria that contain the cloning vehicle, thereby facilitating detection of foreign DNA carried in the plasmid.

Suppose that we want to clone the genes that are responsible for making microorganisms resistant to destruction by the antibiotic penicillin. First we isolate DNA from a naturally occurring penicillin-resistant microorganism (called the "foreign DNA" in Figure 10-7). Then we cleave the foreign DNA into pieces with a restriction enzyme such as EcoR1. Next we open up the circular plasmid DNA by exposing plasmids to the same EcoR1 enzyme. Then we randomly insert and seal the foreign DNA fragments into the plasmid DNA using the enzyme ligase. Some plasmids, by chance, will contain the gene or genes that code for penicillin resistance.

How do we identify and select, from all the plasmids, only those that carry the penicillin resistance gene? We begin by reintroducing all the plasmids containing foreign DNA pieces into bacteria that no longer synthesize either restriction or modification enzymes. Treating *E. coli* bacteria with calcium chloride facilitates the entry of the plasmid DNA into the bacteria (Figure 10-8). Next, to select from the mixture of plasmid-containing bacteria only those relatively few that contain the plasmids with the penicillin resistance genes, we spread the entire mixture of bacteria onto Petri plates containing both tetracycline and penicillin in addition to the nutrients necessary for bacterial growth. Of the millions of bacteria spread on the plates, virtually all will be destroyed by the antibiotics, except for the one in a million that is resistant to both antibiotics. We then allow the antibiotic-resistant bacteria to grow into colonies. Each one of the millions of bacteria in these colonies must contain the penicillin resistance gene. In this way, the gene for penicillin resistance has been cloned and can be studied in detail.

Similar cloning techniques are used to create **gene libraries**: millions of bacteria that contain all of the genetic information from any other organism whose DNA has been cloned in the bacteria (see Box 10-1). In principle, any gene from any organism can be cloned and isolated in microorganisms if a selection procedure to identify the gene can be devised.

OTHER TECHNIQUES FOR CLONING GENES

The techniques we have described permit the construction of gene libraries and the cloning of specific genes such as the penicillin resistance gene. However, most genes or pieces of DNA cannot be selected for, as was possible for the penicillin resistance gene, so cloning foreign DNA is not of much use if the particular bacterial clone that contains the foreign DNA of interest cannot be identified and isolated from all of the rest. For example, no convenient method exists for selecting bacteria that contain the genes coding for human growth hormone or genes coding for the human antiviral substance interferon. Thus, other methods for cloning desired pieces of foreign DNA had to be devised.

The two other methods of cloning genes that have been developed are based on the principles of transcription and translation described in Chapter 5. The first method involves isolating

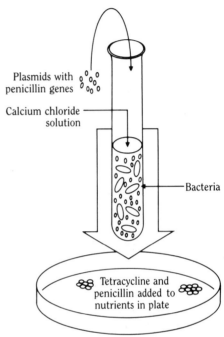

Figure 10-8 Treating bacteria with calcium chloride to facilitate reentry of the plasmid cloning vehicles into the bacteria. The few bacteria that carry and express the penicillin resistance gene can be selected in the laboratory on a medium containing tetracycline (which ensures the presence of the plasmid) and penicillin (which ensures the presence of the desired piece of foreign DNA).

213

Constructing Gene Libraries

BOX 10-1

Gene splicing techniques are now being used to store all of the genetic information for a specific organism in a "library" of bacteria. In this way, the genes of species that are in danger of extinction can be preserved. Suppose, for example, we want to preserve all of the genetic information carried in the cells of a condor—a species of bird that is on the verge of extinction in California because so few individuals remain. Here's how it can be done.

DNA is extracted from condor cells and is cleaved into small pieces by one or more restriction enzymes. (To get pieces of DNA that are small enough to clone, cleavage by several enzymes is usually required.) The small condor DNA fragments are then inserted into plasmid DNA in a random fashion. The mixture of plasmids carrying different fragments of condor DNA is reintroduced into bacteria, so that each bacterium contains a plasmid with some of the condor's genes. Among the billions of bacteria that carry the plasmid, all of the condor's genetic information is preserved. And when a single bacterium grows into a colony on a Petri plate, millions of copies of certain condor genes will be cloned as the bacteria grow and divide.

Because billions of bacteria can be frozen in a few drops of liquid and stored almost indefinitely, bacterial libraries provide a means of preserving the genetic information of any species. Less than 100,000 bacteria clones are needed to preserve the entire human *genome* (all of the genetic information in a sperm or egg), and such libraries have already been constructed.

In a conventional library the information is catalogued and physically organized so that it is readily accessible: the books and periodicals are not just randomly piled together. But in a bacterial library the genes are randomly inserted into the bacteria, which are indistinguishable one from another. Thus, in order to find a particular gene in a bacterial gene library, we must first devise a procedure that will enable us to identify and to isolate the few bacteria containing the gene of interest from all the rest. Clever selection procedures have indeed been devised for a number of genes, but often a large number of bacterial clones must be examined to find the desired piece of DNA—an arduous task.

Libraries of human genes have been used to isolate genes for hemoglobin and other human proteins. Plant and animal breeders may have use for gene libraries containing the genetic information from rare or extinct species. Once a species becomes extinct, its genetic information is lost forever; gene libraries provide a way of preserving the genetic information for the future when it may be needed.

Reading: C. Grobstein, "The Recombinant DNA Debate." *Scientific American*, July 1977.

the mRNA transcribed from a specific gene. If sufficient amounts of the specific mRNA can be purified, the information in the sequence of bases in the mRNA can be converted into a DNA copy by using the enzyme reverse transcriptase (Figure 10-9). Once the DNA has been synthesized, it can be inserted into a plasmid, which is then cloned in the usual manner. Because all of the pieces of DNA are identical (because they were all copied from the same mRNA), the bacterial clones will all be the same, and no laboratory selection procedure is necessary.

Sometimes it is not even possible to isolate the particular mRNA of interest from the millions of other mRNAs in cells. Often, however, the polypeptide or protein product that is desired can be isolated. If sufficient quantities of the protein can be purified, the complete amino acid sequence of the protein can be determined by chemical techniques. For example, the amino acid

Figure 10-9 *(opposite page)* Synthesis of genes from mRNAs. If a particular mRNA can be isolated and purified in sufficient amounts, the enzyme reverse transcriptase can synthesize a piece of DNA whose base sequence is complementary to the base sequence of the mRNA. This enzymatically constructed gene can then be cloned in a plasmid in the usual manner.

sequences for the human hormones somatostatin and insulin have been determined in this fashion. Using the genetic code, it is possible to determine the bases in the DNA that code for the particular amino acid sequence in either of these protein hormones. Of course, since the genetic code is degenerate, more than one codon could be chosen to specify any particular amino acid. Fortunately, one of the several possible codons for an amino acid is usually preferred in *E. coli*, so the most frequently used bacterial codon is the one chosen for each amino acid. Once the sequence of amino acids is known and the corresponding sequence of bases has been determined from the genetic code, the gene can be chemically synthesized and cloned in the usual manner. In this way the DNA corresponding to genes coding for human hormones whose amino acid sequences are known has been constructed and cloned.

To summarize, three basic methods are used in recombinant DNA technology for cloning genes of interest: (1) isolating and cloning the piece of DNA itself, (2) cloning DNA that has been synthesized from specific mRNAs that have been isolated and purified, and (3) cloning DNA that has been chemically synthesized by knowing the amino acid sequence of the gene product.

As you might have already guessed, in all of these recombinant DNA techniques there are many problems and stumbling blocks that must be overcome before the product of a foreign gene can be obtained. First, the correct regulatory sites must be situated at the beginning of the cloned DNA fragment to ensure correct transcription and translation of the gene. However, even if the gene is correctly transcribed and translated, the products of the foreign gene may be recognized as foreign proteins in the microorganisms and destroyed before they can be isolated. And the recently discovered fact that eucaryotic DNA is organized as split genes with introns and exons means that genes cloned directly from animal cell DNA cannot be translated successfully in microorganisms because microorganisms do not have the necessary RNA splicing enzymes that are present in animal cells. Despite these difficulties, genetic engineering technology has advanced at a phenomenal pace—and some genetically engineered microorganisms and products have already been patented (Box 10-2).

Genetic Engineering Applications

Genetic engineering techniques have already found a wide range of useful applications, and some spectacular successes have been reported. Genes coding for the expression of human hormones have been inserted into *E. coli*, and these bacteria are being used to manufacture human insulin, growth hormone, and somatostatin. These hormones have been shown to be effective and safe

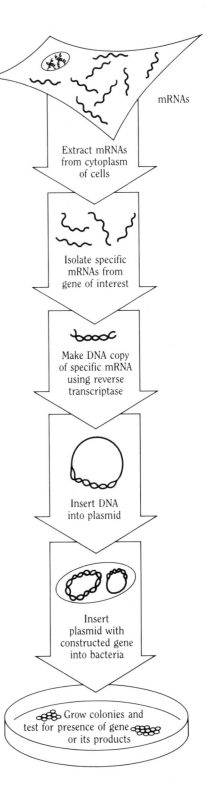

mRNAs

Extract mRNAs from cytoplasm of cells

Isolate specific mRNAs from gene of interest

Make DNA copy of specific mRNA using reverse transcriptase

Insert DNA into plasmid

Insert plasmid with constructed gene into bacteria

Grow colonies and test for presence of gene or its products

In June 1980 the U.S. Supreme Court ruled, five votes to four, that "a live, human-made microorganism is patentable." This history-making legal decision involved the case of *Diamond v. Chakrabarty*. Sidney Diamond, the commissioner of patents and trademarks, took the position that patents should not be awarded for living organisms. Chakrabarty, a scientist who worked for the General Electric Company, had constructed a microorganism that was able to use petroleum as its nutrient source. General Electric, which believed that Chakrabarty's organism might have commercial use in cleaning up oil spills, wanted a patent on the microorganism so that it could profit by controlling the microorganism's commercial use.

As a result of the Supreme Court's decision, new organisms that are constructed by genetic engineers can now be patented, and the patent owners have exclusive right to profit from them. While officials at genetic engineering companies have been pleased with the ruling and view the decision as necessary to protect their interests and investments, some critics worry about the long-term consequences of allowing life forms to be patented. Although no one is too concerned about patenting bacteria, the issue becomes more controversial when we consider the implications of patenting human cells, or even genetically engineered mice.

For example, Sheldon Krimsky, a sociologist at Tufts University, points out that "the right to patent microorganisms, including human cells, will foster manipulation of human genes that could breach an ethical threshold for our society." And Jonathan King, an MIT biologist, believes that "forms of life, be they microbial, plant or animal, are too important to be allowed to be in private ownership."

What do you think?

Reading: S. Krimsky, "Social Responsibility in an Age of Synthetic Biology." *Environment*, July/August 1982.

in clinical trials with humans, and they will probably be prescribed by physicians in the near future, just like other drugs. In fact, human insulin (marketed as *humulin*) is now routinely manufactured in bacteria and was the first drug produced by recombinant DNA technology to be approved for human use.

Microorganisms can be genetically engineered to manufacture a wide range of drugs more efficiently and, it is hoped, more inexpensively than they can be obtained in other ways. Table 10-2 lists some of the dollar values of substances whose synthesis can be vastly improved by recombinant DNA techniques, and the profit potential shows why there is such great interest in recombinant DNA technology.

The applications of genetic engineering are not limited to improving the manufacture of drugs and other chemicals. Recombinant DNA research is also being used in a promising effort to develop microorganisms that can economically convert agricultural waste into fuels such as hydrogen, ethanol, and methanol. And because microorganisms play vital roles in the manufacture of beverages and foods, including beer, wine, bread, pickles, cheese, yogurt, tofu, and soy sauce, it is likely that the fermentations carried out by these microorganisms can

Table 10-2. Approximate Value of Some Drugs Currently or Potentially Produced by Microorganisms

Type of drug	Approximate sales (1979)
Penicillins	$221,000,000
Other antibiotics	$638,000,000
Sulfanilamides	$ 47,000,000
Vaccines	$ 90,000,000
Antifungal drugs	$104,000,000
Digestive enzymes	$ 17,000,000
Prescription vitamins	$134,000,000

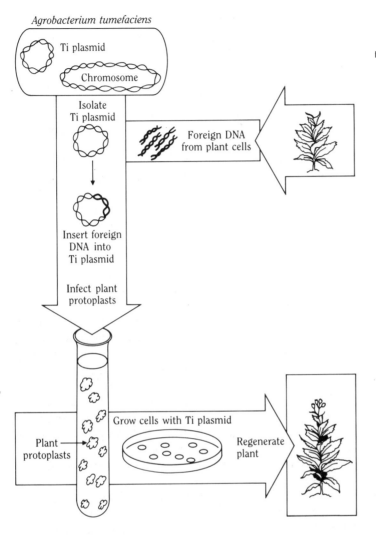

Figure 10-10 Construction of new kinds of plants using recombinant DNA technology. The Ti plasmid from *Agrobacterium tumefaciens* is used as a cloning vehicle for plant cells. Foreign DNA containing genes from other plants can be inserted into Ti plasmids, which are subsequently introduced into plant protoplast cells (cells from which the cell walls have been removed). Cells containing the Ti plasmid carrying foreign DNA are grown, and in some cases the foreign DNA will integrate into the cell's chromosomes. These cells can then be used to regenerate the entire plant.

be made more efficient and more productive by recombinant DNA technology.

Many scientists believe that the greatest practical benefits of genetic engineering will be in agriculture. Genes will eventually be moved from one kind of plant into another in order to improve virtually any desired crop characteristic—disease resistance, for example, or flavor, or nutritional value. The cloning vehicle that seems to have the greatest potential for engineering new plant varieties is the Ti plasmid. This plasmid was originally isolated from *Agrobacterium tumefaciens,* a bacterium that causes tumors to form on many varieties of plants, as we noted in Chapter 8. The Ti plasmid has been genetically altered so that pieces of foreign DNA can be inserted in it at specific restriction enzyme recognition sites. The Ti plasmid (Figure 10-10) can then be used to transfer desirable genes into the chromosomes of plant cells. As described in Box 9-2, entire plants can, in some instances, be regenerated from individual plant cells. This capability holds out the promise of our eventually being able to construct entirely new species of plants from single cells.

To date only a few plants have been successfully regenerated from single cells. Tobacco was the first and is still the easiest plant to grow from single cells. When tobacco plants have been regenerated from single cells containing Ti plasmid vehicles, the foreign DNA has been found to be present in all of the tissues in some plants. Thus, it seems likely that as plant cell culture techniques improve, useful genes can be inserted into cells and new kinds of plants produced. Genetic engineering using the Ti plasmid may give rise to a great variety of new plants that could never be produced by traditional methods of plant breeding and selection.

Hazards of Recombinant DNA Technology

In February of 1975 scientists from all over the world gathered at Asilomar, California for a meeting that was without precedent in the history of science. The scientists met to discuss and adopt guidelines under which all further recombinant DNA research should proceed. For almost a year before the Asilomar meeting, these scientists had observed a self-imposed moratorium on recombinant DNA research until safety procedures could be evaluated and agreed upon. Most of them were anxious to get on with the experiments they all knew would yield exciting information on how genetic information is stored in DNA and how genes are regulated and expressed in different kinds of cells.

After much discussion, the scientific and nonscientific participants at Asilomar agreed upon guidelines for recombinant DNA research, and these guidelines were formally adopted by the National Institutes of Health in Washington, which provides much of the funds used to support basic biological research in the United States.

The guidelines deal with the danger of potentially hazardous microorganisms escaping from research laboratories by a combination of physical and biological containment rules. The degree of health or environmental hazard is to be assessed for each general type of recombinant DNA experiment. For example, it was generally agreed that the potentially most hazardous type of experiment is the insertion of DNA from oncogenic (tumor) viruses into *E. coli* bacteria, a species that inhabits the human gut. Although such experiments would be invaluable for studying how viruses cause cancer, researchers worried about what might happen if bacteria carrying DNA capable of producing cancer were to multiply in the gut of humans.

Paul Berg, a Nobel laureate and one of the originators of recombinant DNA techniques, summed up the scientists' concern at that time: "We are placed in an area of biology with many unknowns; indeed, the greatest risk may well be our ignorance. And it is this ignorance which compels us to pause, reflect, and assess the magnitude of the potential risks associated with this line of research."

According to the recombinant DNA guidelines, physical containment of microorganisms carrying foreign DNA would be accomplished by working in specially constructed laboratories and by sterilizing all equipment used in such experiments. Biological containment would be ensured by inserting the foreign DNA into enfeebled strains of *E. coli*—bacteria that have been genetically altered so that they can grow only under laboratory conditions. In other natural environments or in an animal's intestines, the enfeebled bacteria cannot survive, and the plasmids containing foreign DNA are destroyed along with the microorganisms.

Since 1975, an enormous amount has been learned about recombinant DNA molecules and about microorganisms that carry plasmids with foreign DNA. Furthermore, public and scientific concern over the hazards has abated considerably. Even though vast numbers of recombinant DNA experiments have been performed in the past few years, no adverse effects on the environment or on peoples' health have been reported. Some of the guidelines for recombinant DNA research are still in effect, but the rules have been greatly simplified, and most of the original restrictions have been eased or eliminated.

In 1981 Vincent W. Franco of the National Institutes of Health presented the prevailing view of most scientists and experts:

It is clear that the wealth of evidence and expert testimony gathered to date support the conclusion that recombinant DNA research poses no real hazard to public health, laboratory workers, or the environment. The benefits reaped and projected for humanity surpass all unsubstantiated risks, and without question justify continuation of the research.

This is currently the consensus regarding recombinant DNA research, but there are still many ethical, legal, and social questions that are not easily answered (see Box 10-3). The consequences of the burgeoning recombinant DNA technology on people and society will be evaluated by the next generation. We cannot predict the outcome in any detail. But we can confidently say that the impact of recombinant DNA technology on agriculture, medicine, and industry—as well as on the biological sciences themselves—will be nothing short of revolutionary.

Gene Splicing—
The Real Concern?

BOX 10-3

In recent years, God's monopoly (or nature's monopoly, if you wish) has been broken, and man, making use of refined methods of cross-breeding and of the newly discovered technique of gene splicing, in which a gene belonging to one organism is fitted into the genetic substance of another, has acquired for his own use the power to create the "actual natures" of living things. Some scientists as well as laymen have feared that artificially produced organisms, not envisioned by God or evolution, could prove harmful, or even fatal, to the existing biosphere, and in 1976 the National Institutes of Health announced regulations to govern continued research. To be sure, scientists are far from being able to engineer living things out of dead matter—out of the "dust," as God reportedly did—and, up to now, can do little more than tamper with existing forms of life. God appeared

to have no particular purpose in bringing forth His creation. The Bible provides no record that He rushed off to the patent office on the seventh day, or sought to exploit the creation in any other manner. Instead, He restricted himself to observing that it was all "very good," and took a rest. The new creation, on the other hand, promises to be a utilitarian affair—one that is wholly subservient to the needs of men, and in which every creature comes with its price tag attached.

The ancient Greeks, who believed that man reached for godlike powers only at his extreme peril, would no doubt have warned us that by trying to turn the power of birth and life to our dubious human ends we were courting a fall. The Biblical story

of the tower of Babel points a similar moral: God confounded its builders because He foresaw that "this is only the beginning of what they will do; and nothing that they propose to do will now be impossible for them." However that may be, man's intervention in evolution's workshop certainly has its ironic aspect, for even as we set about the manufacture of new organisms we go ahead with no less vigorous and ingenious preparations for our own extinction, in a nuclear holocaust. In this, we are like a builder who keeps adding floors to a building whose foundation is cracking, or a general who campaigns in ever more far-flung territories while behind him his native city falls to the enemy. With the marvels of our skill and imagination multiplying around us, we ourselves are left undefended and afraid.

Source: *The New Yorker*, April 7, 1980. Used by permission.

Key Terms

recombinant DNA (genetic engineering; gene splicing) DNA created by the joining together of segments of DNA from different organisms *in vitro*.

DNA cloning the production of billions of copies of a piece of DNA as part of a plasmid introduced into a microorganism.

restriction enzymes enzymes that recognize particular sequences of bases in DNA and cleave the DNA at or near that site. If the DNA is modified by a particular pattern of methyl groups, the DNA is protected from cleavage by particular restriction enzymes.

bacterial strains bacteria of the same species that share many phenotypic characteristics but differ from one another in some respect.

modification enzymes enzymes that recognize particular sequences of bases in DNA and attach methyl (CH_3) groups to one of the bases in the recognized site. The pattern of methylation (modification) of the DNA protects it from attack by the restriction enzymes of that strain.

cloning vehicles (vectors) small DNA molecules—usually plasmids or viruses—into which foreign DNA (genes) is inserted. When the cloning vectors are reintroduced into microorganisms, the plasmids or viruses replicate their own DNA as well as the foreign DNA. Thus, billions of identical copies of the foreign genes are produced (cloned).

gene libraries millions of bacteria containing all the genetic information from an organism whose DNA has been cloned in the bacteria.

Additional Reading

Cherfas, J. *Man-made Life*. New York: Pantheon, 1983.

Cohen, S. N. "The Manipulation of Genes." *Scientific American*, July 1975.

Davis, B. D. "The Recombinant DNA Scenarios: Andromeda Strain, Chimera and Golem." *American Scientist*, September/October 1977.

Gilbert, W. and L. Villa-Komaroff. "Useful Proteins from Recombinant Bacteria." *Scientific American*, April 1980.

King, J. "Patenting Modified Life Forms: The Case Against." *Environment*, July/August 1982.

Lipsey, C. E. and C. P. Einaudi. "Patenting Modified Life Forms: The Case For." *Environment*, July/August 1982.

Rensberger, B. "Tinkering with Life." *Science 81*, November, 1981.

Stockton, W. "On the Brink of Altering Life." *The New York Times Magazine*, February 17, 1980.

Valentine, R. C. "Genetic Blueprint for New Plants." *The Sciences*, February 1978.

Wetzel, R. "Applications of Recombinant DNA Technology," *American Scientist*, November/December 1980.

Review Questions

1. What class of enzymes is crucial for all genetic engineering?

2. What is the function of modification enzymes in bacteria?

3. What bacterial structures are most frequently employed as cloning vehicles?

4. How many basic methods are available for isolating and cloning specific genes of interest?

5. What sequence of bases is recognized by EcoR1 restriction enzymes?

6. What molecule is synthesized by gene machines?

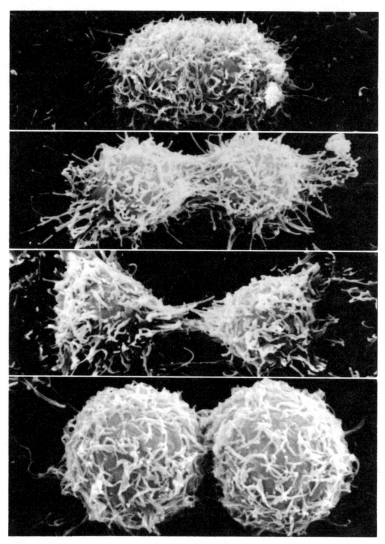

GROWTH AND DIVISION (MITOSIS) OF AN ANIMAL CELL. (2,400 ×)

11

Mendel's Experiments

The Foundation of Genetics

The most significant, the most profound, the most disturbing (and to many the most frightening) consequence of the development of Science lies not in the industrial and technical revolution, but in the agonizing reappraisal, which Science forces upon Man, of his deepest rooted concepts of himself and of his relationship to the universe.

JACQUES MONOD

Figure 11-1 Gregor Mendel, founder of the science of genetics. (Courtesy of National Library of Medicine.)

The science of genetics formally began in 1865 when an Augustinian monk reported the results of his experiments with various strains of peas. The discoveries of Gregor Mendel and his successors concerning the patterns of inheritance of specific traits among sexually reproducing plants and animals constitute the period of *classical genetics*—a period extending from 1865 to about 1940. Genetic experiments during this period were conducted without any knowledge of the chemical nature of the genetic information that was being transmitted from one generation to the next. Modern molecular genetics is generally regarded as having begun with the discovery of the structure of DNA in 1953 (although many experiments bearing on the chemical properties of genes and chromosomes were performed much earlier).

By analyzing the results of crosses between pea plants with different traits, Mendel was able to discover the universal rules of inheritance. Mendel was the first person to correctly explain the phenomenon of **hybrid plants**—progeny plants with combinations of traits that are different from those in the parent plants. Mendel was also the first to propose that traits are determined by discrete hereditary factors, which have since come to be called genes. And many years before it became commonplace to use numerical analysis in biological experiments, Mendel showed that he could predict the proportion of progeny plants that would inherit a particular trait or combination of traits.

To appreciate the significance of Mendel's accomplishments, one must realize that for thousands of years prior to his efforts, the very different concept of **pangenesis** was accepted as the explanation for inheritance of traits. Even Charles Darwin, who was unacquainted with Mendel's discoveries or ideas, believed in this ancient theory. The basic idea of pangenesis, originally proposed by the Greek physician Hippocrates, was that each part of an adult organism produces a tiny replica of itself and that all of the individual replicas are collected in the "seed" of that organism and are transmitted to individuals in the next generation. It was Mendel's bold and brilliant insights that helped disprove the incorrect notions of pangenesis and pangenes that had persisted for thousands of years (see Box 11-1).

Gregor Mendel entered the Augustinian monastery in Brünn, Austria (now Brno, Czechoslovakia) in 1843. These Augustinian monks belonged to a teaching order, and so Mendel was sent by his monastery to study zoology at the University of Vienna. After several years of study, Mendel twice failed to pass his exams and finally had to leave the university without receiving a degree. Sometime after he returned to the monastery in 1853, Mendel began his breeding experiments with the common garden pea, *Pisum sativum.*

Although we will probably never know precisely what motivated Mendel to carry out his pea experiments, he may have been

From Pangenes to Genes: A Journey of 2,000 Years

Since the beginning of recorded history, people have wondered how organisms reproduce and how traits are inherited. Why do the phenotypes of each species remain recognizable generation after generation—why do cats always beget cats and not mice or dogs? When plants and animals reproduce, the progeny are never exactly like their parents in all respects, but they are invariably similar. Some traits appear to be identical with parental traits; others are distinctly different.

One theory that attempted to explain heredity was *pangenesis* (*pan* = all, *genesis* = origination). Among scientists and philosophers who believed in this theory were such famous figures as Aristotle, Hippocrates, and Darwin. The basic idea of pangenesis, particularly as it related to human reproduction, was that each part of the body produced tiny particles that, taken all together, carried all of the individual's traits. The invisible particles, variously referred to as pangenes, gemmules, or plastidules, were carried by the blood to the organs of reproduction. As the different particles from each of the body's parts accumulated in the female womb, they became capable of reproducing all of the parts and traits of a new individual.

It was thought that the pangenes actually contained miniature replicas of each organ, tissue, or other body part. When the male and female pangenes came together in the womb, the particles grew and were modified, and a new individual with a mixture of both parents' traits was produced. Such were the ideas that persisted for thousands of years and that were used to explain why children resemble their parents.

Pangenesis also explained how parents' defects or diseases could be passed on to their sons and daughters. If the father was mentally defective, the particular pangenes copied from his diseased brain would also be defective, and the child would inherit the defects of the father. Or if the mother had contracted cancer or tuberculosis, the pangenes made from her cancerous organ or diseased lungs would also be diseased (or at least prone to disease).

Of course, people had noticed that children did not always turn out like their parents. This was supposedly because defective male particles could be modified in the mother's womb during pregnancy and because the mother's contribution was not considered to be nearly as great as the father's. (Male chauvinism has a long history, particularly in matters of inheritance; even today most couples express a preference for having sons.) The ideas of pangenesis persisted over the centuries and supported prejudices toward male lineage and maintenance of "royal blood lines" (because the pangenes were thought to be carried in the blood).

Leonardo da Vinci was one of the few creative geniuses of the past who believed that men and women contribute equally to the traits of their offspring. In the late fifteenth century he wrote: "The blacks of Ethiopia are not the products of the sun; for when black gets black with child in Scythia, the offspring is black; but if a black gets a white woman with child, the offspring is grey in color. Now this proves that, with regard to the embryo, the seed of the mother has the same vigor as that of the father."

Gregor Mendel appears to have been the first scientist to realize that the theory of pangenesis was all wrong and to correctly describe how traits are transmitted from generation to generation. It was Mendel who conceived of and documented the existence of discrete hereditary factors, which later came to be called genes. The word *gene* was derived by geneticist W. Johannesen from the term *pangenesis* many years after Mendel's discoveries. In 1909 Johannesen wrote, "We will simply speak of the 'gene' and the 'genes' instead of 'pangene' and the 'pangenes.' The word 'gene' is completely free from any hypothesis; it expresses only the evident fact that, in any case, many characteristics of the organism are specified in the germ cells." The discrete hereditary factors of Mendel eventually became the genes on chromosomes of classical genetics and, almost a century later, the sequences of bases in DNA.

Reading: F. H. Portugal and J. S. Cohen, *A Century of DNA*. Cambridge, Mass.: MIT Press, 1979.

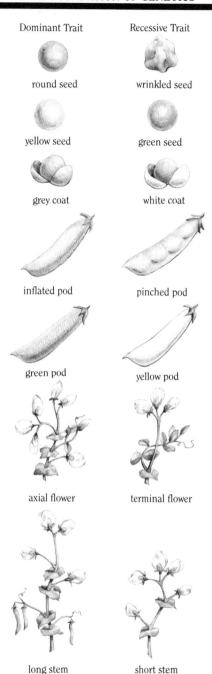

Dominant Trait	Recessive Trait
round seed	wrinkled seed
yellow seed	green seed
grey coat	white coat
inflated pod	pinched pod
green pod	yellow pod
axial flower	terminal flower
long stem	short stem

Figure 11-2 Seven different pairs of traits Mendel crossed in pea plants. From the results of his crosses, he conceived the idea of dominant and recessive traits.

stimulated by two things. First were the ideas of evolution and natural selection that were generating so much interest and controversy at the time. (Although Darwin was not informed of Mendel's work, Mendel had read Darwin's book *Origin of the Species* by the time he published his own results.) Second was the problem of plant hybrids that had puzzled breeders for centuries. How could new combinations of traits appear in later generations of plants when the same combination of traits was never observed in either parent? Like other plant breeders before him, Mendel may have been puzzled over this phenomenon.

After ten years of careful breeding experiments, Mendel wrote up his results and conclusions, which were published in 1865. It is this report that placed the study of inheritance on a sound scientific basis and that is regarded as the beginning of the science of genetics. However, Mendel's results were totally ignored by the scientific community for thirty-five years. In the opinion of plant geneticist Verne Grant, "The originality of Mendel's paper, not to mention its revolutionary implications for biological thought, was almost enough in itself to guarantee its early neglect and tardy acceptance, in view of the well-known phenomenon of resistance to new ideas in science."

Mendel did not live long enough to receive any recognition for his genetic discoveries, but he understood the significance of his undertaking even if no one else did. In the introduction to his now famous paper entitled "Experiments on Plant Hybrids," he wrote:

> . . . among the numerous experiments not one has been carried out to an extent or in a manner that would make it possible to determine the number of different forms in which hybrid progeny appear, permit classification of these forms in each generation with certainty, and ascertain their numerical interrelationships. It requires a good deal of courage, indeed, to undertake such a far-reaching task; however, this seems to be the one correct way of finally reaching the solution to a question whose significance for the evolutionary history of organic forms must not be underestimated.

It is not often appreciated that Mendel, a monk in a strict religious order, was aware of the evolutionary significance of his experiments. This chapter describes Mendel's classic experiments with peas, explains his conclusions, and discusses other important findings of classical genetics.

The Law of Segregation

Mendel observed that among the natural varieties of pea plants in his garden, some plants differed with respect to certain traits,

Figure 11-3 Artificial pollination of plants has been performed for thousands of years. This Assyrian relief, dating from the ninth century BC, shows masked priests pollinating date palms. (Courtesy of Hans Stubbe, Akademie der Wissenschaften der DDR.)

such as color or seed shape. That is, some plants produced round seeds, while others produced wrinkled seeds. Thus, the trait of seed shape could occur in either of two forms. This fact ultimately led Mendel to the idea of dominant and recessive traits that we now understand are caused by alternate alleles of the gene that determines a particular characteristic. (Dominant and recessive alleles were explained in Chapter 7.) Eventually, Mendel was able to identify and use seven pairs of traits in his hybrid crosses (Figure 11-2).

Peas normally inbreed (eggs in a pea flower are fertilized by pollen from the same flower), and so true-breeding pea plants for each trait were probably available in Mendel's garden. Mendel carefully crossed plants by dusting flowers of one true breeding type—such as one with round seeds—with pollen obtained from plants of another true-breeding type—one with wrinkled seeds. He also prevented self-pollination by "emasculating" the recipient flowers.

Techniques of artificial pollination had been known since antiquity (Figure 11-3), but they had been applied without Mendel's purpose and without numerical analysis of the proportion of

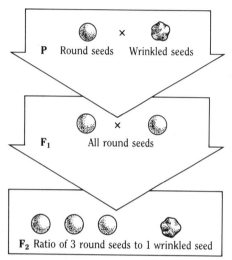

Figure 11-4 Mendel's cross between true-breeding parent plants (P) having either round or wrinkled seeds. In the first (F₁) generation all plants produced round seeds. When F₁ generation plants were crossed, the plants in the F₂ generation produced round and wrinkled seeds in the ratio of three round-seed plants to one wrinkled-seed plant.

the two traits that appeared in the progeny. Mendel collected seeds produced from these crosses, planted them, and counted the number of plants in the first (called the F₁) generation with each form of the trait—that is, with round or wrinkled seeds. The seeds produced by the F₁ generation plants were all round; no wrinkled seeds were observed (Figure 11-4). After the F₁ plants had grown into adults, Mendel allowed them to self-fertilize, producing the second, or F₂, generation. Then he counted the number of F₂ plants with each trait.

After counting hundreds of plants in the F₂ generation, Mendel realized that for every plant bearing wrinkled seeds, there were three plants bearing round seeds: on the average, seed shape traits appeared in the F₂ generation in the proportion of three round to one wrinkled. Over the years Mendel performed many crosses with other pairs of the seven traits and always observed this same pattern of inheritance. In a cross of true-breeding plants, only one of the parental traits would appear in plants of the F₁ generation. Then, in the F₂ generation, both traits would appear, always in the ratio of 3 to 1.

To understand in greater depth how Mendel performed his experiments and reached his conclusions, we will reconstruct Mendel's experiments using an imaginary plant we shall call *Pakalolo aromatica*. Let's suppose one variety of *Pakalolo* plants breeds true for the trait of narrow leaves and that another variety breeds true for wide leaves. We will suppose, as Mendel might have, that leaf shape is determined by a pair of hereditary factors and that the phenotype of each plant is determined by the particular combination of the dominant and recessive factors. In true-breeding plants the two factors are identical; in our cross, factors *N* and *N* produce narrow leaves and factors *n* and *n* produce broad leaves.

A cross involving narrow-leaf and broad-leaf plants is outlined in Figure 11-5. This particular method of analyzing a genetic cross was devised by geneticist R. C. Punnett many years after Mendel's death and is known as the *Punnett square*. What we observe in this cross is that all plants in the F₁ generation have narrow leaves. However, when these F₁ plants are interbred, we find that in the F₂ generation one-fourth of the plants have broad leaves and three-fourths have narrow leaves. From this reproducible pattern of inheritance we conclude, as Mendel did after running many trials, that traits segregate in a 3:1 ratio in the F₂ generation. This fact is now known as Mendel's first law: the law of segregation.

To explain the 3:1 segregation of traits in the F₂ generation, Mendel postulated the existence of two particulate hereditary factors and he introduced the concepts of *dominant* (able to mask) and *recessive* (able to be masked) traits to characterize these factors. For the first time in history, as far as we know, the correct explanation for the transmission of hereditary traits from generation to generation had been described.

The Law of Independent Assortment

Mendel's second discovery, the law of independent assortment, is merely a logical extension of the law of segregation, but it is an important one. Suppose a cross is performed involving two pairs of traits simultaneously, such as a cross between tall *Pakalolo* plants with narrow leaves (*TTNN*) and short *Pakalolo* plants with broad leaves (*ttnn*). If the pattern of inheritance of both traits is studied in a series of crosses, it can be shown that both traits segregate independently from each other; that is, the 3:1 ratio of one pair of traits does not affect the 3:1 ratio of the other pair (Figure 11-6).

z

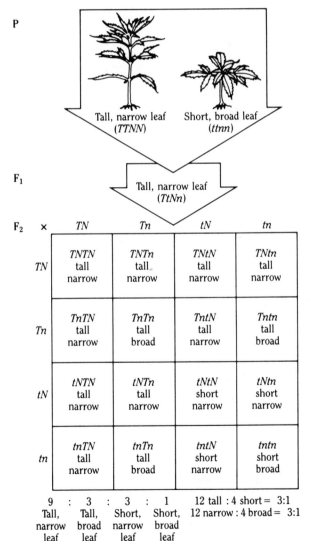

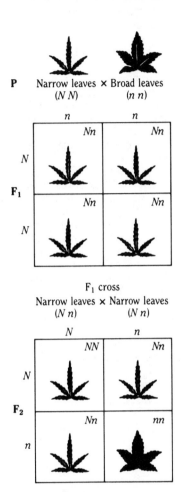

Figure 11-5 (*above*) Law of segregation. True-breeding *Pakalolo* plants that differ in a single trait (narrow leaves versus broad leaves) are crossed. In this example *N* is the hereditary factor for narrow leaves. Since *N* is dominant and *n* is recessive, F₁ generation plants all have narrow leaves. In the F₂ generation, plants with the factors *NN* or *Nn* will have narrow leaves, but one-fourth of all plants will receive the recessive factors *nn* and will have broad leaves.

Figure 11-6 (*left*) Law of independent assortment. True-breeding tall, narrow-leaved plants are crossed with true-breeding short, broad-leaved plants. Because the hereditary factors for different traits assort independently from each other at meiosis, they will appear in the F₂ generation with a 9:3:3:1 ratio of the four phenotypes.

The Mystery of the Source of Mendel's Ideas

Box 11-2

No one questions the remarkable insights Mendel had into the mechanisms of inheritance. His discoveries are all the more astonishing because he worked alone in a remote monastery in what is now Czechoslovakia, with no collaborators or support for his ideas or his experiments. Even when he communicated his ideas to other scientists in letters, his research was either ignored or regarded as having no merit.

The importance of Mendel's discoveries has been discussed in this chapter—his view of a discrete particle, or factor, of inheritance; the bold concept of dominant and recessive factors; and the realization that precise statistical analysis of phenotypes can correctly indicate genotypes. There is no mystery regarding these accomplishments themselves. However, there is considerable mystery surrounding the source of his ideas and the remarkable precision of the data he obtained.

The first mystery concerns his selection of traits to study, not to mention how he chose the pea plant over all other plants. Keep in mind that at the time Mendel began his experiments, nothing was known about the existence of chromosomes, genes, or chromosomal recombination. Yet Mendel chose to study seven traits—and it just so happens that peas have seven haploid chromosomes that segregate independently during meiosis! Even more fortunate for Mendel, four of these traits are determined by genes that lie on different chromosomes, and the other three genes are so loosely linked on the chromosomes so that they behave as if they assort independently because the genes become randomly separated by recombination during meiosis.

How did Mendel manage to choose seven traits that would all behave as if they were unlinked and segregate randomly to produce 3:1 ratios? Was it luck, or did he do preliminary experiments to determine which traits would behave "correctly"? If the latter, then Mendel must have known the answer to his experiments even before he began them. As far as we know, no other person in the world at that time knew what genes were, much less that different chromosomes assort independently from one another.

Then there is the mystery of how Mendel chose seven traits that unambiguously expressed dominant or recessive characteristics. Was this also luck, or did he deliberately exclude from his experiments any traits that did not behave in the desired way? And how did he conceive of the idea of heredity being determined by discrete factors whose expression is always dominant or recessive?

The logic of the experiment is quite simple. If both traits from each parent were inherited together—that is, if traits that appeared together in the parent plants always appeared together in the offspring plants—then the pattern of inheritance would be identical to the pattern observed for inheritance of a single trait. Thus, in the F_2 generation the expected 3:1 ratio should be found for both traits inherited as a single unit. In other words, three-fourths of the plants would be tall and have narrow leaves, and one-fourth would be short and have broad leaves. But this pattern is not observed. Instead, four different classes of plants appear in the ratio of 9:3:3:1, and each of the two traits (tallness of plant, narrowness of leaf) is found in the 3:1 ratio—twelve tall plants to four short plants (3:1), and twelve narrow-leafed plants to four broad-leafed plants (3:1). This pattern of two independently inherited pairs of factors defines Mendel's law of independent assortment.

To appreciate Mendel's insight, one must be aware of the fact

While this question cannot ever be answered with any assurance, one speculation is that Mendel got his brilliant ideas by thinking about sex. He may have noticed that individuals are always either male or female, never a mixture of the two. Because male and female characteristics are not blended in progeny but remain unchanged generation after generation, Mendel might have reasoned that this should be true for other traits as well.

Finally, how did Mendel obtain such good data? The numbers of plants of each phenotype that Mendel reported in 1865 are extremely close to the expected 3:1 ratio (see table)—too close, according to recent statistical analysis of the data. If the same experiments are repeated today and comparable numbers of plants of each phenotype scored, the ratios rarely approximate the 3:1 ratio Mendel reported for all of his experiments. Did Mendel deliberately manipulate his data—an inexcusable sin among scientists— until they fit the 3:1 ratio? If so, it would again imply that he knew the results of the experiments beforehand or that he changed the actual data to make them conform more closely to the 3:1 ratio.

We will never know how Mendel came upon his genetic insights. The monk who became head abbot of the monastery after Mendel's death destroyed all his notebooks and papers. There are no other clues to help solve the mystery of Mendel's discoveries.

Reading: H. Stubbe, *A History of Genetics*, Cambridge, Mass.: MIT Press, 1968.

Mendel's Data

Parent characteristics	F_1	F_2	F_2 ratio
Round × wrinkled seeds	All round	5,474 round: 1,850 wrinkled	2.96:1
Yellow × green seeds	All yellow	6,022 yellow: 2,001 green	3.01:1
Gray × white seedcoats	All gray	705 gray: 224 white	3.15:1
Inflated × pinched pods	All inflated	882 inflated: 299 pinched	2.95:1
Green × yellow pods	All green	428 green: 152 yellow	2.82:1
Axial × terminal flowers	All axial	651 axial: 207 terminal	3.14:1
Long × short stems	All long	787 long: 277 short	2.84:1

that the 9:3:3:1 ratio is obtained *only* if the two different factors (genes) are inherited independently from each other and *only* if both traits are expressed completely in a dominant and recessive manner. We now know that if the pairs of genes for the different traits Mendel studied had been located close to each other on the same chromosome or if they had affected each other's expression, Mendel would not have observed the 9:3:3:1 pattern in the F_2 generation.

Mendel knew nothing about chromosomes or genes, yet his reasoning and conclusions regarding the inheritance of traits were correct. Several mysteries surrounding Mendel's discoveries can never be solved because all of his notes were destroyed after his death (see Box 11-2). Yet Gregor Mendel, an academic failure whose work was ignored during his lifetime and for many years after his death, had discovered the basic rules of inheritance.

Mendel's important conclusions can be briefly summarized:

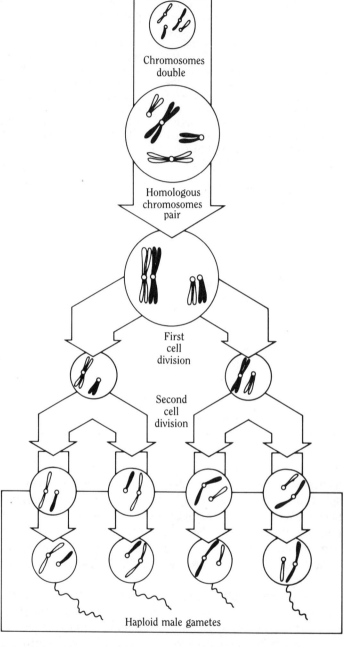

Figure 11-7 The process of meiosis, which produces the haploid number of chromosomes in gametes. Each chromosome doubles in a sex cell, followed by two cell divisions that reduce the chromosomes to the haploid number.

1. Traits are determined by particulate factors that are transmitted unchanged from one generation to the next.
2. These factors can be expressed in either a dominant or recessive manner.
3. The factors assort independently from each other.
4. The traits appear in the F_2 generation in predictable ratios.

Table 11-1. Differences in Chromosome Number Among Plants and Animals

Animal	Diploid chromosome number	Plant	Diploid chromosome number
Turkey, *Meleagris galiopavo*	82	Upland cotton, *Gossypium hirsutum*	52
Chicken, *Gallus domesticus*	78	Potato, *Solanum tuberosum*	48
Dog, *Canis familiaris*	78	Tobacco, *Nicotiana tabacum*	48
Horse, *Equus caballus*	64	Bread wheat, *Triticum aestivum*	42
Donkey, *Equus asinus*	62	Cherry, *Prunus cerasus*	32
Chimpanzee, *Pan troglodytes*	48	Tomato, *Solanum lycopersicum*	24
Rhesus monkey, *Macaca mulatta*	48	White oak, *Quercus alba*	24
Man, *Homo sapiens*	46	Yellow pine, *Pinus ponderosa*	24
Rat, *Rattus norvegicus*	42	Rice, *Oryza sativa*	24
House mouse, *Mus musculus*	40	Bean, *Phaseolus vulgaris*	22
Cat, *Felis domesticus*	38	Radish, *Raphanus sativus*	18
Frog, *Rana pipiens*	26	Garden pea, *Pisum sativum*	14
Housefly, *Musca domestica*	12	Barley, *Hordeum vulgare*	14
Fruit fly, *Drosophila melanogaster*	8	Rye, *Secale cereale*	14
Mosquito, *Culex pipiens*	6	Cucumber, *Cucumis sativus*	14

Meiosis: Segregation of Chromosomes

The reason that certain traits segregate and assort in predictable ratios is that many genes, certainly all of the ones Mendel analyzed, are located on chromosomes. The inheritance of various combinations of chromosomes is due to the process of *meiosis* (defined in Chapter 1) that occurs in the reproductive cells of plants and animals. Meiosis ensures that only one copy of each chromosome pair—called homologous chromosomes—is put into each gamete (pollen or sperm in males and eggs in females). Gametes carry an organism's genetic information from one generation to the next. Unlike mitosis, which doubles the chromosome number prior to each cell division, meiosis first doubles each chromosome pair but then follows with two successive cell divisions that result in cells with only one chromosome member of each original pair (Figure 11-7).

Meiosis occurs only in reproductive cells and is responsible for maintaining the diploid number of chromosomes in plants and animals from one generation to the next. If it were not for meiosis, each succeeding generation would have *twice* as many copies of each chromosome—clearly a situation that would soon get out of hand! Most species of plants and animals are diploid, although the number of different chromosomes varies greatly from one species to the next (Table 11-1).

Meiosis provides the structural basis for the law of segrega-

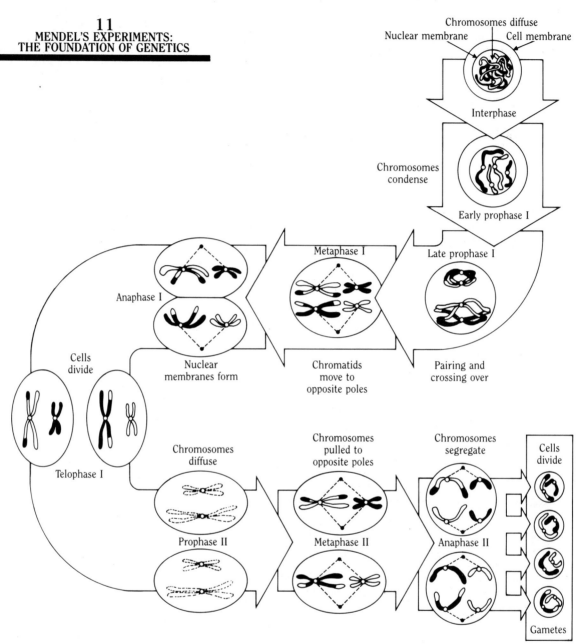

Figure 11-8 The various stages of meiosis I (the first cell division) and meiosis II (the second cell division). The pairs of chromosomes have a characteristic appearance at each stage of meiosis.

tion, and the random inclusion of one member of each pair of chromosomes into gametes provides the structural basis for the law of independent assortment. If Mendel had known about meiosis, his discoveries would not have been quite so remarkable, but he had no knowledge of chromosomes or meiotic production of gametes.

In the Punnett squares in Figures 11-5 and 11-6, which show how traits segregate in a genetic cross, you will see that the genes for each trait listed at the top and the left-hand edge of each square actually reflect the different alleles in the male and female gametes that are segregated during meiosis. Meiosis produces the characteristic phenotypic ratios Mendel observed.

STAGES OF MEIOSIS

Meiosis involves eight distinguishable stages that occur in all the reproductive cells of sexually reproducing plants and animals. The various stages of duplication, pairing, recombination, and segregation of chromosomes into gametes are outlined in Figure 11-8. One crucial difference between mitosis and meiosis is the pairing and joining together of the homologous chromosomes in an early stage of meiosis (prophase I). Each chromosome previously received from the male parent pairs with the corresponding (homologous) chromosome received from the female parent in a process known as **synapsis**. It is shortly after this pairing that exchange of information between chromosomes—the process of recombination—takes place. During recombination, different segments of DNA in the paired chromosomes are physically exchanged, which produces new gene combinations that are passed on to progeny.

CROSSING OVER BETWEEN CHROMOSOMES

By rearranging the combinations of genes inherited from each parent, recombination serves as an important mechanism for generating genetic diversity among organisms. The biological significance of recombination is that these new gene combinations may produce an individual that is better adapted to its environment or that reproduces more efficiently.

Recombination during prophase I of meiosis involves four chromosomes (pair of duplicated homologous chromosomes) that are fully aligned along their entire lengths. These four chromatids taken together are called a tetrad (Figure 11-9a). Within each tetrad a chromosome and its exact replica are called *sister chromatids* since their DNAs have identical base sequences (assuming a mutation has not occurred during replication). When recombination occurs, it happens in the tetrads and then only between nonsister chromatids. This physical exchange between genetically dissimilar DNA molecules, called **crossing over**, results in new arrangements of the various alleles carried on each chromosome (Figure 11-9b). As a result of segregation, progeny may inherit chromosomes that are identical to ones transmitted to them from their parents. However, since crossing over is a frequent event during the pairing of homologous chromosomes at meiosis, progeny also may receive recombinant chromosomes that contain new gene combinations (Figure 11-9c). Crossing over also results in a cytologically (observable under the microscope) visible structure termed a **chiasma** (plu-

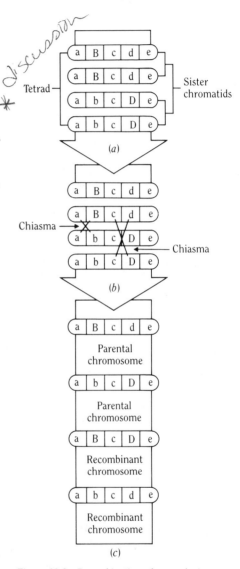

Figure 11-9 Recombination of genes during meiosis. (*a*) Each pair of homologous chromosomes is duplicated and synapse (pair up) along their entire length. (*b*) Crossing over occurs between nonsister chromatids and produces chiasmata that are visible under the microscope. (*c*) In this example the products of meiosis and crossing over are two parental chromosomes and two recombinant chromosomes.

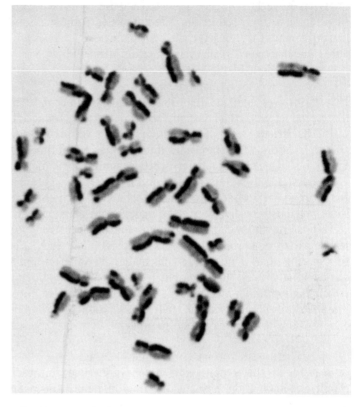

Figure 11-10 Photograph of chromosomes from a human lymphocyte cell. The paired chromosomes have been stained to show the exchange of chromosomal segments that results from the formation of chiasmata and crossing over. (Courtesy of Sheldon Wolff; Samuel Latt.)

ral: chaismata) that is the physical structure in the tetrad formed during crossing over of nonsister chromatids (Figure 11-10).

To appreciate the variety of gametes that can be generated by meiosis, consider the following calculation. Human cells contain forty-six chromosomes (diploid number) and thus twenty-three pairs of chromosomes (the haploid number). Any sperm or egg may receive either the father's or mother's chromosome for any one of the twenty-three homologous pairs. Therefore, the number of different gametes that can be produced by segregation and independent assortment alone is 2^{23} or more than 8 million. And if we consider the number of different gene combinations produced by crossing over, probably no two gametes in an organism are ever genetically identical.

Meiosis accomplishes three essential functions in the reproduction and transfer of genetic information from one generation to the next:

1. The diploid number of chromosomes is maintained by reducing the chromosomes to the haploid number in the gametes. When the sperm and egg combine to produce a new individual, the diploid number is reestablished.

2. Genetic diversity is ensured by the random segregation and independent assortment of chromosomes into gametes.

Figure 11-11 The kernels on ears of corn demonstrate Mendel's law of segregation. The top ear gives a ratio of three colorless to one colored kernel. The bottom ear shows three colored to one colorless kernel, reflecting a different allele segregation. (Courtesy of M. G. Neuffer, University of Missouri. From *The Mutants of Maize*, 1968, Crop Science Society of America.)

3. Recombination creates new gene combinations that result in organisms that may be better adapted to their environments.

Gene Linkage

Genes may be located either close together or far apart on the chromosomes. Most chromosomes contain thousands of genes, so it is useful to know where each gene is located relative to the others along the chromosome. By measuring the degree of linkage between genes, biologists can infer the relative distance between any two genes on a chromosome. More specifically, **gene linkage** is a measure of the degree to which a pair of genes assorts independently at meiosis or when crosses are performed between genetically different individuals. Genes that are close together on a chromosome are tightly linked because they are invariably transmitted jointly to gametes following meiosis. Genes that are far apart on a chromosome are loosely linked because crossing over during meiosis often separates the pair of genes in the resulting chromosomes.

Corn has long been a favorite organism of geneticists and plant breeders for a variety of reasons. Because of its importance as a food, breeders have long sought to develop improved varieties. And for geneticists corn is an instructive organism because they can study gene interactions and gene linkages by observing the phenotypes of the hundreds of kernels (seeds) that appear on each ear of corn. Just as genes affected the shape and color of Mendel's pea seeds, genes also affect the shape and color of the corn kernels. For example, the 3:1 segregation of alleles that affects the color of kernels can be observed on a single ear of corn (Figure 11-11).

Corn has ten pairs of chromosomes, and genes have been mapped on each of the ten different chromosomes. A **genetic map** shows the relative positions of genes (alleles) along a

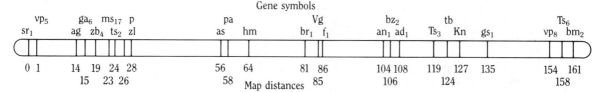

Gene symbols

vp₅	ga₆ ms₁₇ p	pa	Vg	bz₂	tb	Ts₆
sr₁	ag zb₄ ts₂ zl	as hm	br₁ f₁	an₁ ad₁ Ts₃ Kn gs₁	vp₈ bm₂

0 1   14 19 24 28   56 64   81 86   104 108   119   127   135   154 161
              15   23 26   58   85   106   124   158

Map distances

Figure 11-12 Gene map of a corn chromosome. The relative distances between genes on the genetic map reflect the frequency with which recombination takes place between alleles in genetic crosses.

chromosome. To construct a genetic map, geneticists perform genetic crosses and measure the degree to which recombination occurs between pairs of genes. The relative distances along the genetic map of a chromosome reflect the differences in the frequencies of recombination between genes when homologous chromosomes synapse and recombine during meiosis (Figure 11-12). If only a few chiasmata occur on a chromosome pair and they are dispersed randomly along the chromosome, the probability of recombination between any pair of genes depends on their distance apart along the chromosome.

Because of recombination between genes on homologous chromosomes, corn plants will be produced that show new combinations of traits that were not present in the parent plants. Geneticists look for these new combinations by performing crosses between plants differing in traits, and they count the number of progeny with the parental traits and with new combinations. From these measurements, map distances such as those shown in Figure 11-12 can be calculated. The distance between genes on a genetic map is directly related to the frequency of recombination that occurs between them.

A genetic map of any chromosome is much like a road map. Genes are the "towns" along the "highway," and the numbers tell you the distances between genes. In the case of corn, knowing which chromosome carries which gene and knowing the gene's relative position are necessary in order to perform genetic analyses. Breeders may use the information in genetic maps to estimate their chances of breeding plants with particular desirable characteristics. As discussed in Box 11-2, Mendel was fortunate in that he chose to analyze traits that were governed by genes that were either loosely linked or located on different chromosomes. If any pair of genes that Mendel worked with had been tightly linked on a chromosome, he would have observed recombinant phenotypes and would not have observed independent assortment for that pair of genes. This might have prevented his reaching the conclusions that he did.

How Genes Act and Interact

Most phenotypic traits, particularly in animals, are not determined by single dominant or recessive genes that are inherited in a simple Mendelian fashion. Interactions among genes and among the products of genes often complicate the genetic analy-

Three Human Traits Governed by Single Dominant-Recessive Genes

Box 11-3

Numerous traits in plants—including height at maturity; the color of flowers, leaves, and seeds; and the shape of leaves and fruits—are determined by single genes that are often expressed in a dominant or recessive manner. Certain analogous traits in animals—height; color of skin, hair, or eyes; size and shape of parts of the body—are also governed by simple dominant-recessive genes. However, in humans at least, these traits and most others are governed by complex interactions of several genes with each other and with the environment. Thus, they do not obey the simple rules of Mendelian inheritance, but rather obey the more complex rules of *polygenic* inheritance (discussed in Chapter 13). Although the rules of inheritance for people *are* the same as for plants, human traits often appear to be inherited in a more complex manner because of the ways genes are expressed and the ways they interact with other genes and with the environment. Mendel's rules are not the only rules of inheritance.

However, three easily recognized human traits *are* deter-

mined by single genes that are expressed in a dominant or recessive fashion. The ability to roll one's

tongue is determined by a dominant autosomal allele. If you possess one or both copies of the dominant allele, and about 85 percent of people do, you can roll your tongue without any effort (see photo). The product of this gene controls transverse muscles in the tongue. If you possess only recessive genes for this trait, no amount of effort will enable you to roll your tongue.

Another normal human trait is pattern baldness, which also appears to be determined by a single autosomal gene. This trait,

however, is said to be *sex-limited* because the expression of the gene and the degree of baldness is affected by the presence or absence of particular male and female sex hormones. Men who carry either one or both of the dominant pattern baldness genes lose their hair (regardless of what magazine and newspaper advertisements say about preventing baldness and restoring hair). In women, on the other hand, even if both dominant genes are present, there may be some thinning of the hair but rarely complete baldness. This example demonstrates how cellular chemistry may affect gene expression.

A third human trait that is the result of a single dominant-recessive gene is the ability to taste the bitterness of the chemical phenylthiocarbamide (PTC). About 70 percent of white Americans and 90 percent of black Americans are tasters—that is, they experience a bitter taste if a speck of PTC is placed on their tongue. Because the gene for PTC taste behaves in a Mendelian fashion, a taster has at least one dominant allele of this gene; nontasters have two recessive alleles.

sis of how genes segregate and how they are expressed in progeny. Relatively few human traits show clear-cut Mendelian dominant or recessive patterns of inheritance. Three common traits that do are described in Box 11-3.

Most genes code for proteins, and these proteins must interact with other cellular components to produce the observed phenotype. The expression of any gene, and the way it interacts with other genes, is usually influenced by the organism's environment, which may also affect cellular chemistry. When genes are not expressed in a well-defined dominant or recessive manner, they are said to have **variable expressivity**. This simply means that the effect of the gene on the phenotype depends on unknown factors. Variable expressivity is a vague term that geneticists use to account for the fact that the phenotype determined by a particular gene is not always predictable.

A more specific kind of gene–gene interaction is **epistasis**, which refers to the complete masking of a gene's expression by the presence of another gene located elsewhere in the genome. In this situation the dominant or recessive nature of the first gene cannot be determined because the second gene (the epistatic one) interferes with the expression of the first gene.

Another property of genes—one that applies to a population of individuals rather than to a single individual—is **penetrance**. In some instances, identical genes carried by individuals are not expressed in all members of the population. Penetrance refers to the proportion of individuals in the population possessing the allele who actually express the trait. Penetrance is often important in the analysis of genetic diseases (see Chapter 13). For example, members of the same family who are known to carry an identical mutant (disease-causing) gene because of their genetic relatedness may manifest mild symptoms, severe symptoms, or no symptoms of the disease at all. Many genetic and environmental factors, most of which cannot be identified, affect the degree to which genes are expressed. And sometimes people with a disease-causing gene may experience its effects differently at different periods of their life.

The number of copies of genes carried by an organism—the **gene dosage**—also affects its phenotype. Many abnormal phenotypic effects can result from having more or less than the normal number of genes. For most organisms two copies of each gene (except for genes on the X chromosome) is the normal number, since most organisms are diploid. In **aneuploidy**, the organism has one chromosome too many or one too few. Aneuploidy usually results in serious phenotypic abnormalities in animals. The best-known example of aneuploidy in humans is **Down's syndrome**, a serious disorder caused by a complete extra chromosome 21 or a sizable part of it. Generally, the phenotype of plants is less adversely affected by abnormal numbers of chromosomes, and plant breeders often deliberately construct polyploid plant varieties that have desirable characteristics.

Most genes, unlike the ones Mendel studied or the ones described in Box 11-3, affect more than one trait. Any gene that affects several unrelated traits of an organism is said to show **pleiotropy**. Mutant genes that cause diseases in humans, especially those that affect basic metabolic cellular processes, generally produce pleiotropic effects. For example, **phenylketonuria (PKU)** is a serious human hereditary disease that is caused by a defect in a single gene that codes for the enzyme that converts the amino acid phenylalanine into the amino acid tyrosine. The pleiotropic effects of this single gene defect include progressive deterioration of the central nervous system, accumulation of toxic levels of phenylalanine in blood and urine, and mental retardation. (The inheritance of PKU is discussed in Chapter 13.)

Sex Linkage

So far we have assumed that genes are situated on chromosomes and that genes segregate or recombine in predictable and reproducible patterns. Today, of course, we know that these assumptions are correct. For many years, however, they were untested. Proof of their correctness emerged from genetic studies, using the fruit fly *Drosophila melanogaster,* that were performed at the beginning of this century by T. H. Morgan, C. B. Bridges, and other "fruit fly" geneticists. Morgan discovered that genes located on the X chromosome (one of the sex chromosomes) of these flies affected inheritance patterns in predictably different ways than genes located on the *autosomes* (chromosomes other than the sex chromosomes).

This group of *Drosophila* geneticists studied how eye color segregated in crosses between male and female flies having either red or white eyes. By these genetic crosses they were able to show conclusively that (1) hereditary information *is* contained in chromosomes and (2) certain genes that determine eye color are located on the X chromosome. These fundamental insights were reached many years prior to any knowledge of DNA structure or of the chemical nature of genes.

The genetic cross in the classic fruit-fly experiments was performed between true-breeding white-eyed female flies and red-eyed male flies. Alleles that determine red eye color are dominant; mutant alleles that determine white eyes are recessive and were discovered when rare white-eyed mutants were found in a population of red-eyed flies. (If you want to examine fruit flies for yourself, put a banana peel in a jar whose mouth is blocked, as shown in Figure 11-13, so that flies can fly in but cannot fly out; then place the jar outdoors for several days.)

We now know that male flies (and male humans) have an X and a Y chromosome while females have two Xs. Thus, the results of a cross between white-eyed female flies and red-eyed male flies can be understood if we assume that the alleles for eye color are located on the X chromosome. In the cross shown in Figure 11-14, all female flies of the F_1 generation have red eyes and all males have white eyes. In the F_2 generation produced by mating F_1 generation males and females, red and white eyes appear in a 1:1 ratio in both males and females instead of the expected 3:1 ratio. The Punnett squares show the genetic analysis of the crosses assuming that the alleles for eye color reside on the X

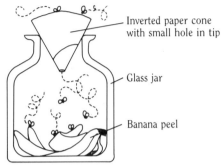

Inverted paper cone with small hole in tip

Glass jar

Banana peel

Figure 11-13 How to construct a device for collecting fruit flies.

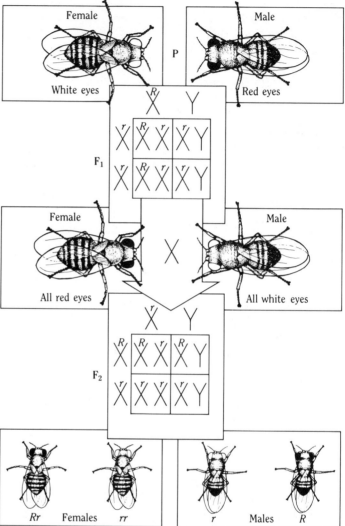

Figure 11-14 A genetic cross between white-eyed female flies and red-eyed males. In the F₁ generation all females are red-eyed and all males are white-eyed. If the F₁ generation males and females are crossed, the expected 3:1 ratio is not observed in the F₂ generation. Rather, red-eyed flies and white-eyed flies occur in a ratio of 1:1 among males and females. This deviation from the expected Mendelian pattern of inheritance led to the discovery that eye color genes are located on sex chromosomes and are inherited in different patterns from genes on autosomes.

chromosome. If a female fly has either *RR* or *Rr* alleles, her eyes will be red; because males have only one X chromosome, a single allele determines their eye color.

The reciprocal cross involves mating red-eyed females with white-eyed males; this genetic cross gives a very different pattern of inheritance for eye color in male and female flies (Figure 11-15). In the F₁ generation both male and female flies have red eyes; in the F₂ generation all females have red eyes while half of the males have red and half have white eyes. A sex-linked pattern of inheritance accounts for the difference in the results between the crosses shown in Figures 11-14 and 11-15. When genes are located on autosomes, the inheritance pattern does not change regardless of the arrangement of alleles in males or females.

Occasionally, fruit flies were discovered that had abnormally

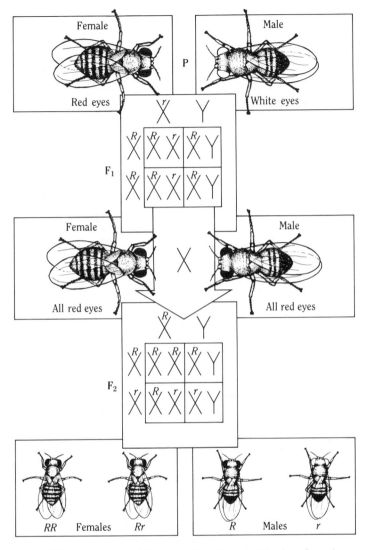

Figure 11-15 A genetic cross between red-eyed female flies and white-eyed males. In the F₁ generation both male and female flies have red eyes. In the F₂ generation all of the female flies have red eyes while half of the males have red eyes and half have white eyes.

shaped X chromosomes that could be distinguished in the microscope from normal X chromosomes. With this discovery it was possible to prove that the gene governing eye color was always associated with a particular chromosome. This observation proved the chromosomal basis of inheritance and established the association of genes with particular chromosomes.

Nondisjunction of Chromosomes

In performing matings between red-eyed and white-eyed flies, the *Drosophila* researchers had to examine thousands of fruit flies. On rare occasions they observed an unusual pattern of inheritance in a cross that otherwise gave normal results. For example, on the basis of patterns of inheritance as they were understood at that time, in a cross between white-eyed females and red-eyed

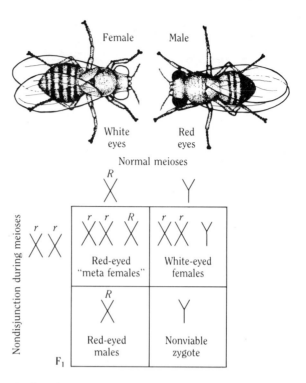

Figure 11-16 Nondisjunction of homologous X chromosomes during meiosis in formation of fly eggs. When such eggs are fertilized by normal sperm, abnormal flies result that have more or less than the normal number of X chromosomes.

males, F_1 female progeny should *always* have red eyes and F_1 males should *always* have white eyes, as indicated in Figure 11-14. However, among many thousands of flies produced in such a cross, an occasional white-eyed female or red-eyed male was observed. In genetics the rare exceptions to the predictable patterns of inheritance often provide the keys to some new discovery.

C. B. Bridges studied these exceptional flies and eventually determined that the white-eyed daughters must have inherited, in some abnormal fashion, both of their mother's X chromosomes. Correspondingly, the red-eyed males must have received their single X chromosome from their father instead of from their mother. But how could the X chromosome be inherited in this unconventional manner?

If any two homologous chromosomes (or chromatids) fail to separate from each other during meiosis—a process called **nondisjunction**—gametes will be produced that have more or less than the normal number of chromosomes. For example, if, in female flies, the two X chromosomes fail to separate at meiosis, genetically abnormal eggs will result. Consequently, when female flies carrying the abnormal eggs mate with normal males, four abnormal combinations of sex chromosomes can be expected to occur in the progeny flies (Figure 11-16). The white-eyed female flies are genetically XXY (also showing that the Y chromosome does not always determine maleness in fruit flies), while the red-eyed males have only a single X chromosome (suggesting that it may be the number of X chromosomes that determines sex in fruit flies). When Bridges examined the chromosomes from cells of the unusual male and female flies, chromosomal abnormalities were found. Female flies with three X chromosomes—"metafemales"—were identified, and eventually nonviable embryos containing the single Y chromosome were also discovered.

Nondisjunction of the X chromosome—an aberration of meiosis—occurs in humans of both sexes as well as in fruit flies. In humans nondisjunction of the X chromosome causes serious phenotypic changes in the individual's sexual characteristics (as we will see in Chapter 12). The most frequent observable occurrence of nondisjunction of autosomes in humans is with chromosome 21 and results in Down's syndrome.

All higher plants and animals reproduce by sexual fusion of gametes. Meiosis is the cellular process that produces haploid gametes by segregating chromosomes in a random manner and that produces new gene combinations in the process of crossing over. Thus, the variety of life is both maintained and changed as a result of matings between male and female individuals. How sex types are determined and how abnormalities in sex phenotypes occur are the subjects of the next chapter.

hybrid plant a progeny plant with a different combination of traits than those observable in either parent.

pangenesis the concept that each part of an adult organism produces a tiny replica of itself that is collected in the "seed" of the organism and then transmitted to offspring.

law of segregation Mendel's law explaining the separation of homologous chromosomes or genes into separate gametes during meiosis.

law of independent assortment Mendel's explanation of the failure of two pairs of genes to segregate together.

homologous chromosomes the two members of a chromosome pair that were inherited from each parent and that pair up (synapse) at meiosis.

synapsis the side-by-side pairing of homologous chromosomes during the prophase stage of meiosis.

tetrad the two pairs of sister chromatids formed from a pair of homologous chromosomes.

crossing over the exchange of segments between pairs of homologous chromosomes during meiosis.

chromatid one of the two sisters formed after a chromosome divides.

chiasma (pl. chiasmata) the visible physical crossing over between the nonsister chromatids of homologous chromosomes at meiosis.

gene linkage a measure of the degree to which a pair of genes assort independently at meiosis or when crosses are performed between genetically different individuals. Also, a measure of the relative distance between genes on chromosomes.

genetic map the imputed arrangement of genes (alleles) on a chromosome according to the observed frequency of recombination between them in genetic crosses.

variable expressivity a characteristic of genes that are expressed in a dominant or recessive manner but whose expression is affected by other genes or by the environment.

epistasis the masking of the expression of a gene by a gene located elsewhere in the genome.

penetrance the proportion of individuals carrying a particular allele that actually express the trait.

gene dosage having more or less than the normal number of genes.

aneuploidy the characteristic of having one chromosome more or less than the set normally found in an organism.

Down's syndrome an inherited human disease caused by an extra chromosome number 21 in all of the individual's cells.

pleiotropy the ability of a gene to affect several unrelated traits in an individual.

phenylketonuria (PKU) an inherited human disease caused by a defect in the gene that converts phenylalanine to tyrosine.

nondisjunction failure of homologous chromosomes (or chromatids) to separate properly at meiosis, resulting in gametes with two few or too many chromosomes.

Additional Reading

Davis, B. D. "Frontiers of the Biological Sciences." *Science,* July 4, 1980.

Keller, E. "McClintock's Maize." *Science 81,* August 1981.

Mendel, G. "Experiments in Plant Hybridization." In C. I. Davern, ed., *Genetics.* San Francisco: W. H. Freeman, 1981.

Milunsky, A. *Know Your Genes.* New York: Houghton Mifflin, 1977.

Myers, R. H. "The Chromosome Connection." *The Sciences,* May/June 1980.

1. What can you infer about the genetic basis of a trait that segregates in a 3:1 ratio in the F_2 generation?

2. How many different traits did Mendel study in peas?

3. What biological process produced the phenomena Mendel observed?

4. How many different gametes can be produced by segregation alone in *Drosophila?*

5. Are identical genes expressed to the same degree in all individuals?

6. How many genes determine eye color in *Drosophila?*

SPERM ARE PRODUCED IN SPECIAL CELLS OF THE TESTIS. (1,370 ×)

12

Human Reproduction

*How
Sex Is
Determined*

*The living world is a continuum in each
and every one of its aspects. The sooner we
learn this concerning human sexual
behavior, the sooner we shall reach a
sound understanding of the realities of sex.*
ALFRED C. KINSEY

12
HUMAN REPRODUCTION: HOW SEX IS DETERMINED

Over the centuries philosophers and natural historians have observed that the principal goal of all life is the preservation and continuance of life. What this means is that the act of reproduction is the most important behavioral and biological function carried out by any organism. All organisms, from the simplest bacteria to the most complex animals, propagate themselves and reproduce their likeness. Microorganisms and some plants can reproduce asexually (without mating) but virtually all animals rely on sexual matings in order to reproduce. Sexuality is the most essential biological behavior for the continuance of life. Thus, it is not surprising that sex is one of the most fundamental of all biological processes.

In infancy we automatically begin to explore the sexual parts of our bodies. In early childhood we begin to notice that boys and girls are physically different and behave differently. During adolescence, we pass through the psychological stress of establishing our sexual identities and of learning how to relate to persons of the opposite sex. During adolescence, too, our bodies complete the process of sexual differentiation, and we mature into men and women who are capable of carrying on the endless cycle of human reproduction.

For many species of animals, especially birds, the observable physical differences between males and females are dramatic; generally, the male is the larger and more colorful of the pair. **Sexual dimorphism** is the term for this distinctly different physical appearance of males and females of a species. Darwin was very aware of sexual dimorphism among the different species of animals and devoted an entire book to a discussion of the various mechanisms of sexual selection and their relevance to evolution. It is clear from Darwin's work that he viewed sexual selection as being only slightly less important than natural selection in creating new types of individuals and species of organisms.

Despite the enormous effort that has been devoted to studying *sexual differentiation*—how males become males and females become females—only the general scheme of the biological processes that produce males and females is understood even today. Sexual differences, sexual preferences, and mating mechanisms are thought to play important roles in the survival and evolution of organisms, but the evidence supporting these quite reasonable ideas is still inconclusive. Even the question of whether males choose the females they mate with or whether it is females who choose males is a difficult one to answer from studies of animal species. Human sexual behaviors are even more complex than those of other animals because much sexual behavior—and the attitudes underlying it—is learned, not instinctive.

Some plants and a few animals have developed strategies for

Parthenogenesis:
Could It Happen in Humans?

Box 12-1

Some species of plants and animals are able to reproduce without fertilization of the egg by sperm—a process known as parthenogenesis. Reproduction by parthenogenesis occurs only in female plants and animals; therefore, all parthenogenetic progeny are also female. And since all the daughters are genetically identical to their mother, they are clones.

Parthenogenesis is rare in animal species. It has been observed among some species of insects, fishes, salamanders, and lizards, but it has never been reliably demonstrated for any species of mammal. Nevertheless, numerous claims of human parthenogenesis—"virgin birth"—have appeared over the years.

Within the past few years it has become possible to make a careful study of parthenogenesis in a species of whiptail lizards that live in New Mexico (see figure). These lizards can now be raised in the laboratory, and occasionally a female that is isolated from male lizards will give rise to daughters who, in turn, grow up and produce more daughters without any male contact. In nature, the whiptail lizards normally reproduce sexually and the female's eggs are fertilized by male sperm in the usual manner. However, in whiptail lizards, and in certain other species of lizards as well, the ability for certain females to reproduce by parthenogenesis arose as a consequence of one or more mutations in some individuals. The mutations alter genes that affect the meiotic process in such a way that the egg retains the diploid number of chromosomes instead of having the haploid number. These diploid eggs are able to develop without being fertilized by sperm because they have received all of the necessary chromosomes from their mother.

In environments where male lizards are scarce, parthenogenetic females have a distinct reproductive advantage over sexually reproducing lizards. When males are scarce, parthenogenesis allows females to reproduce, maintaining the survival of the population. Although parthenogenesis has not been observed in mammals, the chance of its happening as a result of mutations cannot be ruled out, given the degree of similarities of biological processes among different animals—particularly similarities in the processes of meiosis and egg formation.

These simple facts about lizard reproduction might be blown up into science fictional stories in which parthenogenetic women repopulate and take over planet earth after the depletion of males by war. Perhaps women in the future will evolve to produce only sterile males, as happens among the wasps and honeybees. Or maybe small harems of sperm-bearing males will be maintained solely for female pleasures and for maintaining a certain amount of genetic diversity in the populations.

Even the "battle of the sexes" may become a thing of the past. If you read one day that scientists have found a way to construct a parthenogenetic woman, it may be time to prepare for the best of all possible worlds—or the worst, depending on your sex. What do you think?

Reading: Charles J. Cole, "The Value of Virgin Birth." *Natural History*, January 1978.

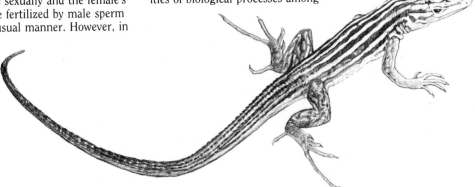

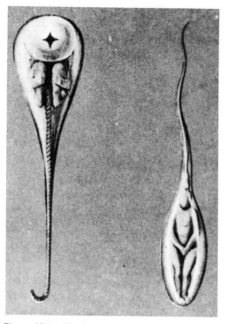

Figure 12-1 The homunculus theory. For over two millennia, it was believed that little persons called homunculi (singular: homunculus) were carried in sperm. This drawing is a seventeenth-century interpretation. (Courtesy of Dorsey Stuart, University of Hawaii.)

reproduction that do not involve matings between individuals of the opposite sex. About half of all flowering plants are **hermaphrodites**; that is, an individual plant (such as the pea plants in Mendel's experiments) has both male and female reproductive organs. Hermaphrodites are much less common among animal species; they are not a normal developmental pattern in humans, although, on rare occasions, human hermaphrodites have been observed. Many female plants and some female animals can reproduce without being fertilized by male gametes. The process by which this is accomplished is called **parthenogenesis**; such females produce only female offspring (see Box 12-1).

The biological mechanisms of sex determination and sexual reproduction vary from species to species, but for most organisms male and female characteristics are genetically determined by a special pair of chromosomes, the *sex chromosomes*. The sex chromosomes, however, only start the overall process of sexual differentiation resulting in males and females. This developmental process is also affected by specific genes on other chromosomes and by the production of sex hormones at the appropriate time in each organism's development. For humans, the psychological and social environment in which the newborn infant is raised also affects its sexual development.

This chapter discusses the genetic basis for sexual differentiation and the functions of the sex chromosomes in sexual development. Most of us hold strong beliefs about what we regard as "normal" sexual behaviors, sexual appearance, and reproduction. Yet, as you will discover in this chapter, external sex characteristics can be misleading indicators of what sex chromosomes people carry and even of their reproductive capability as males or females. A thorough understanding of the biological basis of sexuality is essential if we hope to comprehend the variety of sexual behaviors exhibited in human societies.

Discovery of Sex Chromosomes

Until the beginning of this century, nothing was known about how plants and animals produce either male or female offspring. Since the time of the early Greeks, people had held to the notion that a miniature being, called a **homunculus** ("little man"), existed inside the head of each sperm. It was thought that after a sperm was deposited in the woman's womb, the homunculus in the sperm head increased in size until it was finally born (Figure 12-1).

Sexual characteristics are among the most easily recognized of all traits, yet before Mendel conceived of the idea of particulate hereditary factors (genes), no one had ever been able to formulate a genetic basis for sex determination because all views of heredity were based on the incorrect ideas of pangenesis (discussed in Box 11-1). Following the rediscovery of Mendel's ideas at the beginning of this century, scientists began to look for hereditary factors that would explain sex determination and that would be analogous to the ones Mendel had shown to exist for other traits.

Around 1900 it was discovered that male and female characteristics are determined not by single genes but rather by a pair of sex chromosomes. Microscopic examination of chromosomes in insect reproductive cells revealed that the cells of some individuals contained a single unusual chromosome; it was named "X" because its function at that time was unknown. Other reproductive cells that did not have the X chromosome had a much smaller chromosome that was called "Y." Eventually, it became apparent that the reproductive cells of all animals contain either a large X chromosome or a small Y chromosome in addition to the normal number of autosomes. Examinations of sperm from various species of animals ultimately produced the hypothesis that cells contain specific sex chromosomes that determine the sexual phenotype of the individual.

In most species of plants and animals, male somatic cells contain both X and Y chromosomes, and male reproductive cells produce sperm bearing either an X or a Y chromosome. Female somatic cells contain two X chromosomes and female reproductive cells produce eggs that all bear one X chromosome (Figure 12-2). Individuals that produce gametes of only one type with respect to the sex chromosomes are called the **homogametic sex;** those that produce gametes with two types of sex chromosomes are called the **heterogametic sex.** In humans, females are the homogametic sex and males are heterogametic.

Variation in Sex Determination

Although in many species of plants and in all species of mammals (including humans) females have two X chromosomes and males have an X and a Y, other forms of sex determination are also found in nature (Figure 12-3). In many species of birds, the males and females are determined by X and Y chromosomes, but the chromosomal constitutions of the sexes are reversed when compared to mammals. In chickens, for example, roosters are the homogametic sex (two X chromosomes), while hens are the heterogametic sex (an X and a Y chromosome).

In some insect species, including grasshoppers, the cells have no Y chromosome and gender is determined by the number of X chromosomes. Thus, female grasshoppers have two X chromosomes whereas males have only one. Honeybees streamline the process of sex determination even further; their cells do not contain any specific sex chromosome. Instead, all fertilized eggs of honeybees are diploid and automatically become females; any eggs that remain unfertilized are haploid and automatically develop into males.

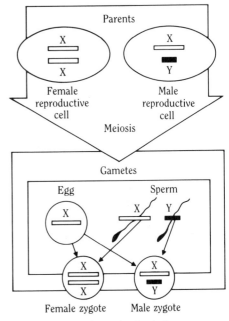

Figure 12-2 Sex determination in humans.

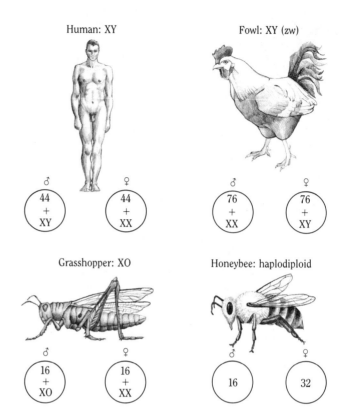

Human: XY

Fowl: XY (zw)

♂ 44 + XY   ♀ 44 + XX

♂ 76 + XX   ♀ 76 + XY

Grasshopper: XO

Honeybee: haplodiploid

♂ 16 + XO   ♀ 16 + XX

♂ 16   ♀ 32

Figure 12-3 Variation in sex determination among species of animals. In humans, males are the heterogametic sex (XY) and females are the homogametic sex (XX). Chickens are reversed; hens are XY (usually the letters ZW are used instead of XY) and males are XX. Grasshoppers have no Y chromosomes; males have one X chromosome and females have two. The sex of honeybees is determined by whether the individual has the haploid or the diploid number of chromosomes.

Table 12-1. Sex Determination in *Drosophila*

Sex phenotype	Number of X chromosomes	Sets of autosomes	Ratio of X to autosomes
Meta female (sterile)	3	2	1.5
Normal female (tetraploid)	4	4	1.0
Normal female (triploid)	3	3	1.0
Normal female (diploid)	2	2	1.0
Intersex (sterile)	2	3	0.67
Normal male	1	2	0.5
Meta male (sterile)	1	3	0.33

Sex determination in *Drosophila* depends not only on X and Y chromosomes but more importantly on the particular ratio of the autosomes to the X chromosomes. Normal female fruit flies are produced whenever the ratio of sets of autosomes to X chromosomes is 1:1 and normal male flies are produced whenever the ratio is 2:1 (Table 12-1).

Among animals whose sex is determined by the usual complement of XX and XY chromosomes, the sex and fertility of individuals is often affected by the presence of more or less than

Table 12-2. Effects of Abnormal Numbers of Sex Chromosomes in Different Species

Sex chromosome constitution	Human	Mouse	*Drosophila*
XX	Normal female	Normal female	Normal female
XY	Normal male	Normal male	Normal male
XXY	Sterile male	Sterile male	Fertile female
XYY	Fertile male	Semisterile male	Fertile male
XO	Sterile female	Fertile female	Sterile male

the normal number of sex chromosomes. For example, a single X chromosome (XO) results in a sterile human female, whereas fruit flies with only one X chromosome become males. And in mice a single X chromosome results in the development of fertile females (Table 12-2). The presence of an extra X chromosome in cells of human males and in male mice (XXY) results in sterility, but an extra X chromosome in fruit flies (XXX or XXY) produces fertile females. This brief survey of the diverse effects of the number of sex chromosomes in different animal species shows that male and female characteristics and the organism's fertility are determined in quite different ways by the complement of X and Y chromosomes.

Abnormal Human Sex Determination

Determining a person's sex usually poses no special problem: we do not need to know anything about the numbers of sex chromosomes to decide whether a person is male or female. Whenever we find ourselves in a new group of people, we unconsciously make note of the sex of the individuals around us on the basis of their appearance or behaviors. Indeed, the first piece of information that our minds instinctively seek about another person is whether that person is male or female. Generally, we infer this information from the person's dress, hairstyle, presence or absence of facial hair, and body build.

SEX CHROMOSOME ABNORMALITIES

At birth the first determination of a newborn's status is whether it is male or female. Sex assignment is based on the type of external genitals: scrotum and penis for boys, vagina and clitoris for girls. However, ambiguous sexual phenotypes that are not detectable at birth occur rather frequently in humans as well as in other animals. If nondisjunction of the X chromosome occurs during formation of the mother's egg or the father's sperm (discussed in Chapter 11), offspring having either an XO or an XXY sex-chromosome constitution can be conceived and will usually survive. At birth, females with XO and males with XXY genotypes are

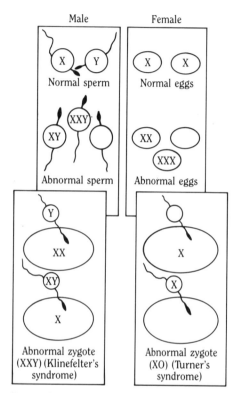

Male Female

Normal sperm | Normal eggs

Abnormal sperm | Abnormal eggs

Abnormal zygote (XXY) (Klinefelter's syndrome) | Abnormal zygote (XO) (Turner's syndrome)

Figure 12-4 Nondisjunction at meiosis in human males and females. Nondisjunction produces gametes with abnormal sex-chromosome constitutions. Abnormal zygotes are produced by various combinations of normal and abnormal gametes.

usually undistinguishable from normal infants. Other sex chromosome abnormalities, such as males with an XYY complement or females with three or more X chromosomes, are also found and result from abnormal partitioning of the sex chromosomes into male or female gametes at meiosis.

In some individuals the abnormal sex-chromosome constitution has little noticeable effect; however, in other cases development does not proceed normally, and the phenotypic effects of the sex-chromosome abnormalities become apparent later in life. For example, some women with **Turner's syndrome** (XO) do not become aware of their condition until they try to become pregnant and cannot. One of the major characteristics of Turner's syndrome is sterility. A variety of other characteristics are often associated with this syndrome. (A *syndrome* is a group of symptoms or traits that tend to occur together and that are used to characterize particular diseases.) Many such women may be shorter than normal, may have immature development of breasts and external genitals, and may age prematurely.

Turner's syndrome is thought to result from fertilization of an abnormal egg that lacks an X chromosome as a consequence of meiotic nondisjunction (Figure 12-4). If such an egg is fertilized by a sperm carrying the father's X chromosome, a female will be produced whose cells contain only a single X chromosome. Such a woman is sterile because the female reproductive organs do not develop in the embryo. Approximately one out of every 5,000 infants classified as female at birth has an XO karyotype, or chromosome complement (Figure 12-5a), and has Turner's syndrome.

In 1942 H. F. Klinefelter described a group of phenotypically abnormal human males who had unusually small testes, were sterile, had certain female physical characteristics (including enlarged breasts and wide hips), and in some instances had subnormal mental capabilities. About one in every 1,000 males exhibits these symptoms, which together are known as **Klinefelter's syndrome**. Genetic studies of the chromosomes in these Klinefelter's males show that all of them have an XXY karyotype (Figure 12-5b).

These two examples of abnormal sex determination show that many phenotypic characteristics, including fertility in men and women, are determined by their sex-chromosome constitution. Infertility results from the absence of functional reproductive organs (ovaries and testes) in people whose cells contain one X chromosome too many or too few.

As mentioned earlier, hermaphroditism (a term derived from the Greek god Hermes and the goddess Aphrodite) is an exceedingly rare sexual abnormality in humans. In the last seventy years only about 350 examples of true human hermaphrodites have been reported. These individuals have both ovarian and

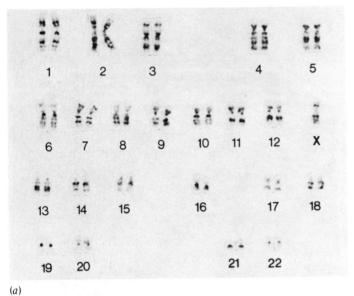

(a)

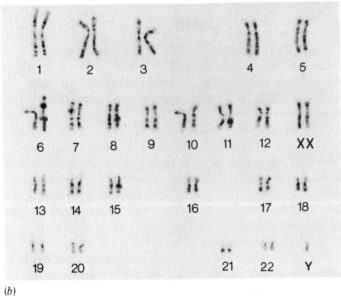

(b)

Figure 12-5 Karyotypes for (a) Turner's syndrome and (b) Klinefelter's syndrome. (Courtesy of Patricia Jacobs, University of Hawaii.)

testicular tissue, but it is nonfunctional. The tissues of true hermaphroditic individuals may consist of cells with either an XX or XY karyotype.

HORMONE ABNORMALITIES

One of the most remarkable of the many kinds of human sexual abnormalities is **androgen insensitivity syndrome** (also called *testicular feminizing syndrome*), in which a person's outward physical sexual appearance is completely female despite the fact that she is genetically male (XY) in all of her somatic cells. Such

Figure 12-6 Because of the biblical description, artists traditionally show Eve being formed from Adam's rib. Could it have been the other way around? (Courtesy of Archive Alinari, Florence, Italy.)

individuals generally become aware of their unusual condition during adolescence, when they fail to begin to menstruate, or even later when they discover that they are sterile. They have internal testes, which quite often produce functional sperm. However, the sperm cannot be transported to any external sexual organ because the sperm duct has not developed and the external sex organs are female.

It is now known that androgen insensitivity syndrome is caused by a mutant allele of a gene located on the X chromosome. The product of this gene, which is normally present on the surface of male cells, allows sexually undifferentiated cells to respond to male hormones (called **androgens**) during development of the embryo. If the cells are unable to respond to male hormones, embryo development continues as female.

It should be emphasized, particularly in view of the widespread misguided notions of male dominance in our society, that all human embryos will develop as females, irrespective of whether the sex-chromosome constitution is XX or XY, unless the appropriate male hormones are produced and function properly at early stages of development (Figure 12-6).

Another unusual abnormality in human sex determination can be caused by another kind of hormone imbalance that occurs early in development. Some individuals begin with a normal female karyotype (XX) but develop as males and are classified as boys at birth. It turns out that the developing female embryo can be masculinized by a genetic defect that blocks production of the hormone cortisol in the embryo's adrenal glands, causing them to release male hormones (androgens) instead. Thus, we can see from these several examples of abnormal sexual phenotypes that human sex determination results from a series of complex interactions involving sex chromosomes, autosomal genes, and hormones.

MALE-FEMALE DIFFERENCES

All babies are sexed at birth, and except for the most obvious abnormalities, are classified as either male or female. But can any individual be regarded as 100 percent male or 100 percent female? It seems reasonable to suppose that within these two absolute sex categories, small differences in sex determination also occur. All normal males may possess testes and male genitals and produce viable sperm, but individuals may differ in hormone levels, sperm production, and other aspects of masculinity. Corresponding small differences in female sexual development may produce individuals with different female hormone levels and other feminine characteristics.

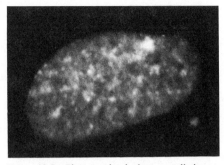

Figure 12-7 Photograph of a human cell showing a Barr body (the bright area on the upper right). An inactive X chromosome is observable as a Barr body in female cells. (Courtesy of A. J. R. de Jonge, Erasmus University, Rotterdam.)

In addition to human biological sex-determination differences, a person's sex-role identity is influenced by the environment in which he or she is raised. Our sexual development and preferences depend on the particular sexual behaviors and acts that are encouraged or discouraged by the society we live in.

Over the centuries women have been victimized by **sexism**, the prejudice held by many men (and women also) that women are biologically, emotionally, and mentally the inferior sex (see Box 12-2). In recent years this unfounded belief in female inferiority and weakness has been actively opposed by organized groups of both men and women. While it is true that there are biological differences between men and women, it certainly has not been shown that sex-determined differences affect any person's overall capacities any more than other genetic differences among individuals can be shown to be primarily responsible for their mental capabilities or social behaviors (discussed in Chapter 19).

Barr Bodies

In the late 1940s two Canadian scientists, Murray L. Barr and Ewart G. Bertram, were staining the chromosomes in the nuclei of animal cells with dyes so that they could study them under the microscope. They repeatedly noticed a particularly dark-staining spot in the nuclei of cells from female animals that did not appear in the nuclei of male cells. Such dark-stained nuclear spots, which are characteristic of female cells, became known as **Barr bodies** (Figure 12-7).

Sexism:
Prejudice Against Women

Box 12-2

Throughout history, women have been regarded as fundamentally different from men—and "different" has almost invariably meant inferior—in both physical and mental capabilities. Women have been defined by men as "the weaker sex": physically weaker, psychologically more unstable, and in general biologically inferior. In addition to being physically smaller (hence, weaker), women have been regarded as having more "irritable" nervous systems, which accounts for their more frequent "nervous breakdowns" and higher incidence of insanity. In the nineteenth century, brain size was assumed to measure intelligence, and the mental inferiority of women was "proven" by their smaller brain size, as well as by the overall lack of intellectual accomplishments by women. After all, men argued,

how many great artists, musicians, scientists, or philosophers have been women?

Even the astute reasoning of Charles Darwin (who is regarded by many as exemplifying a male of particular brilliance) led him to biased and unfounded conclusions on the issue of female intelligence. In his book *The Descent of Man and Selection in Relation to Sex,* published in 1871, Darwin wrote: "The chief distinction in the intellectual powers of the two sexes is shown by man's attaining to a higher eminence, in whatever he takes up than can woman— whether requiring deep thought, reason, or imagination, or merely the use of the senses and hands." Even Darwin was unable (or unwilling) to consider that social and cultural prejudices and condi-

tions may have prevented women from "attaining to a higher eminence" throughout history.

Controversy over female intellectual ability is still very much in evidence today. In 1980 the psychologists C. P. Benbow and J. C. Stanley published a study on the mathematical abilities of seventh and eighth grade boys and girls. Both groups had taken the same number of math courses, yet the boys achieved better grades on the math sections of scholastic aptitude tests. From this evidence, Benbow and Stanley concluded that boys are biologically (genetically) more talented in math than girls. The conclusions of these psychologists are not generally accepted for many reasons, not the least being that the social and cultural experiences that boys and girls are exposed to are very different. For example, how many

---

Some years later, it was noticed that somatic cells taken from men with Klinefelter's syndrome (XXY) also have a Barr body, whereas somatic cells from normal males do not. This fact, and the detection of several Barr bodies in abnormal cells with multiple X chromosomes, prompted a British geneticist, Mary Lyon, to suggest that each Barr body is, in reality, a condensed, inactive X chromosome. She reasoned that since women have two X chromosomes and men have only one X chromosome, one of the X chromosomes in every female cell must be inactivated in order to compensate for gene dosage effects. If both X chromosomes were active in female cells, she argued, twice as much gene expression would occur for the hundreds of genes on the X chromosome in women than in men. Because it was known that too little or too much of even a single gene product could have serious consequences during development of animals, it seemed reasonable to her that X chromosome inactivation is nature's way of equilibrating the expression of genes on the X chromosome.

The **Lyon hypothesis,** as these ideas came to be called, pro-

girls receive a calculator or a computer as a present? (The effects of genetic and environmental influences on complex human behaviors—an issue referred to as "nature versus nurture"—is discussed in more detail in Chapter 19.)

As emphasized in this chapter, the biological basis of sex determination involves complex interactions of chromosomes, genes, and hormones. It is likely that hormonal and developmental differences cause male and female brains to develop differently in cellular or biochemical ways that are measurable. Biological differences between any two groups of individuals can always be demonstrated, as can differences between any two individuals. Superimposed on biologically determined sex characteristics are the social and psychological differences in the ways boys and girls, men and women are treated. Defining a "normal" man or a "normal" woman is quite impossible unless we restrict the definition to reproductive capability. But even that

definition would imply that an infertile male or female is "abnormal." It remains a fact, even though many people (including some scientists) choose to ignore it, that the genetic and environmental contributions to complex human behaviors, such as intellectual ability, femininity, and masculinity, are so complex that they cannot be separated by any experiment yet devised. This being the case, it is unjust to define a person's abilities in terms of his or her sex.

Lest any reader think that sexism is much overblown or that women's groups make much ado about nothing, it is worth quoting what a respected male scientist, Gustave Le Bon, a founder of social psychology and a contemporary of Darwin, wrote in 1879:

"All psychologists who have studied the intelligence of women . . . recognize today that they represent the most inferior forms of human evolution and that they are closer to children and savages than to an adult civilized man. They excel in fickleness,

inconstancy, absence of thought and logic and incapacity to reason. Without doubt there exist some distinguished women, very superior to the average man, but they are as exceptional as the birth of any monstrosity, as for example, of a gorilla with two heads.

Even though this obviously sexist statement was made a hundred years ago, less blatant sexist views are often expressed by contemporary psychologists and scientists in language that is seemingly more rational. Sexism is only one example of the use of biological and genetic facts to justify prejudice against a particular group of people.

Readings: Stephen Jay Gould, "The Brain Appraisers." *Science Digest,* September 1981.
Jon Beckwith and John Durkin, "Girls, Boys and Math." *Science for the People,* September/October 1981.
D. Gelman et al., "Just How the Sexes Differ." *Newsweek,* May 18, 1981.
C. Benbow and J. C. Stanley, "Sex Differences in Math Reasoning." *Science News,* 119 (1981) 147.

poses that only one X chromosome is allowed to be active in any cell; additional X chromosomes must be inactivated so that none of the genes will be expressed. The Lyon hypothesis also predicts that inactivation is a random event in cells and that early in development either one of the two X chromosomes in each cell can be inactivated. Once the choice is made, however, all progeny cells produced by cell division will have the same X chromosome inactivated. The X chromosome inactivation only affects gene expression; its movement during mitosis and cell division is normal. The Lyon hypothesis has been shown to be essentially correct, and each Barr body does represent an inactivated X chromosome. (The Y chromosome has very few active genes and is not inactivated even when present in more than one copy, such as in XYY males.)

X Chromosome Mosaicism

One of the testable predictions of the Lyon hypothesis is that, if the inactivation of X chromosomes is random in cells, females should be *mosaic* (show different allele activities in different

Table 12-3. The Number of Barr Bodies in Human Cells as a Function of the Number of X Chromosomes

Sex chromosome constitution	Number of Barr bodies
XX	1
XY	0
XO	0
XXY	1
XXX	2
XYY	0
XXXY	2
XXXXY	3

cells) for traits governed by genes on the X chromosome. For example, a woman who is heterozygous for genes on the X chromosomes that code for the synthesis of the enzyme glucose-6-phosphate dehydrogenase (G6PD) could have two forms of the enzyme in every cell, since she will have two alleles for G6PD. However, if one X chromosome is randomly inactivated in each cell as her tissues develop, her cells should express either one allele or the other, but both forms of the enzyme should not be present in a single cell.

When individual cells from a woman heterozygous for G6PD are analyzed, one enzyme activity or the other is indeed found, and both forms of the enzyme are never found in the same cell. This experimental evidence provides strong support for the idea that most, if not all, genes are inactivated on one of the two X chromosomes in female cells, although it is not known precisely when in female development the inactivation occurs. In animals that have an abnormal number of X chromosomes, the number of Barr bodies in the cell nuclei increases in proportion to the number of X chromosomes, indicating that only a single X chromosome remains active in any given cell (Table 12-3). At this point you might be wondering why all women do not have Turner's syndrome if only one X chromosome is active in a normal woman's cells. The explanation is that X chromosome inactivation occurs after sex determination has been initiated. Exactly how or when the X chromosomes are randomly inactivated in cells after female sex development has been initiated is not understood.

A practical application of the correspondence between Barr bodies and X chromosome inactivation is its use in determining the sex of athletes in international competitions. Women athletes are required to have a Barr body in their cells; male athletes should have none. If a female athlete's cells do not show a Barr body, she is not allowed to compete as a woman. Olympic athletes are now routinely examined for Barr bodies using a test known as a *buccal smear,* in which cells scraped from the inner lining of the athlete's cheek are stained and examined. This test is deemed necessary because on various occasions sports officials have had good reason to believe that some athletes in women's events were actually males. Indeed, several Olympic athletes competing as women have been disqualified because their cells failed to show any Barr bodies, demonstrating that they had a male genotype.

The H-Y Antigen and Male Development

While studying the acceptance and rejection of various skin grafts in highly inbred strains of mice, a group of scientists in the 1930s noticed that the grafts were always successful between any pair of inbred strains of mice *except* when the donor mouse was male and the acceptor mouse was female. From this observation, it was eventually discovered that male mice have a protein on the surface of their cells that is recognized by antibodies circulating in the blood of female mice and that causes the females to reject the male tissue (tissue rejection and the immune system are discussed in Chapter 16).

This male-specific protein is now called the **H-Y antigen** (H-Y stands for *H*istocompatibility–*Y* chromosome) and is thought to be the protein that controls the formation of testes in male embryos and the subsequent development of male sexual organs. The H-Y antigen is thought to be coded for by a gene or genes on the Y chromosome, but its definite assignment to the Y chromosome is still uncertain. If the H-Y antigen is eventually located on the Y chromosome, it will be the first definite assignment of a gene to that chromosome. (A gene for hairy ears in men is often said to reside on the Y chromosome, but that assignment has been questioned.)

If no H-Y antigen is produced early in the embryonic development of a genetic male, female sexual development continues and ovaries are formed. Thus, the H-Y antigen seems to be the key element in determining whether or not male development occurs after an egg is fertilized by Y-bearing sperm. Although a few cases have been reported of apparently normal human males that lack a Y chromosome, these males do have H-Y antigen on the surface of their cells. Presumably, the H-Y antigen genes in these individuals have, in some way, become transferred to one of their other chromosomes. It is not yet known whether proteins or hormones other than the H-Y antigen are necessary for the formation of the male reproductive organs during early embryonic development.

Hormones and Sexual Development

The complete development of a male or female human is conveniently described as a three-stage process. The first step in sex determination begins with the chromosome constitution of the zygote. As we have noted, two X chromosomes usually lead to the development of a female, and X and Y chromosomes usually lead to the development of a male. Chromosomes determine **primary sex differentiation**: the formation of female ovaries or male testes, which in the adult will produce eggs or sperm, respectively. However, once the embryonic gonads have formed (usually around the sixth week after fertilization), the internal sex organs themselves produce hormones that influence further sexual differentiation during development of the embryo and that continue to affect sexual development after birth. The hormones produced by the gonads affect other tissues and direct what is called **secondary and tertiary sex differentiation**: the formation over time of a penis, facial hair, deep voice, and male body structure in men, and of a clitoris, breasts, and a female body build in women (Figure 12-8).

Regardless of a person's chromosome constitution or even

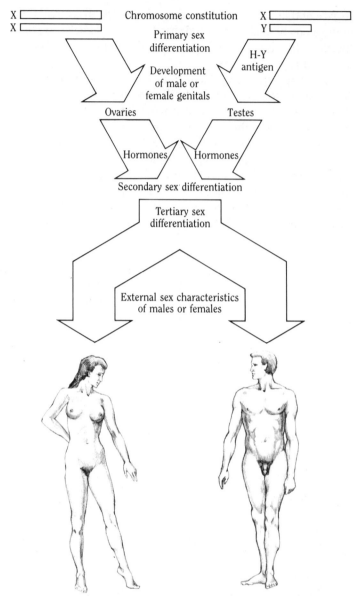

Figure 12-8 Stages in the sexual development of males and females. In humans, XX chromosome constitution directs female development; XY constitution directs male development. Primary sex differentiation in males is determined by the H-Y antigen. In the absence of this protein, ovaries develop. Hormones determine the secondary male and female sex characteristics.

the presence of ovaries or testes, secondary and tertiary sex characteristics can still be modified or reversed by appropriate surgery and hormone therapy. Some individuals are born with abnormal secondary sex characteristics that may have been caused by abnormal chromosome constitution, by gene defects, or by hormone imbalances. If such persons are identified soon enough after birth, modern medical therapies and psychological counseling can help them cope with their aberrant sexual phenotypes (Figure 12-9).

It is important to remember that nature does not construct perfect "all male" or "all female" individuals. Each person is a

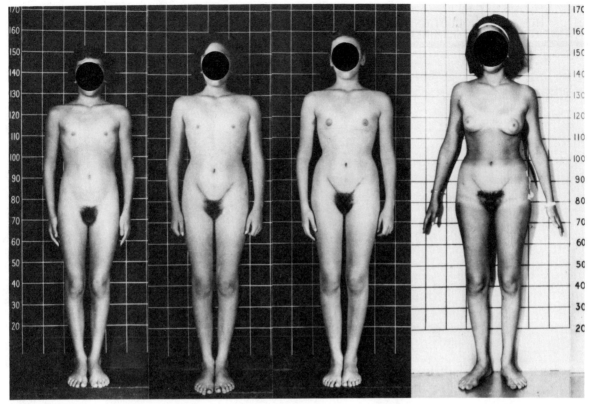

Figure 12-9 Medical transformation of a male (XY) individual into a female. The individual, shown at ages eleven, twelve, thirteen, and nineteen, was raised as a male until age eleven, when it was decided to change her secondary sexual characteristics. At ages eleven and nineteen she received surgical feminization; in the intervening years she was given hormone therapy. (Courtesy of Johns Hopkins School of Medicine and Johns Hopkins University Press.)

mixture of male and female characteristics that are both biologically and socially determined. Thus, considerable variation in sexual development is possible, and these differences may be manifested in adult sexual behaviors. In recent years, our society has become more tolerant of people whose sexual characteristics and preferences fall outside previously acceptable biological and social norms. However, many individuals still adhere to narrow definitions of what is acceptable sexual behavior for men and for women as well as what constitutes normal sexual appearance.

Human Gametes

Plants and animals normally produce new individuals by the fusion of male and female gametes that form a **zygote**, a fertilized egg that is the first cell of a new individual. The time scale of development in different animal species varies enormously: A new fruit fly larva develops from a fertilized egg in about 13 hours, whereas a new human requires about nine months of development (Figure 12-10).

In humans each male gamete contains the haploid number of chromosomes in a sperm cell that is only about 0.06 mm in length. Each sperm carries either an X or a Y chromosome in addition to the twenty-two autosomes. About 300 million sperm,

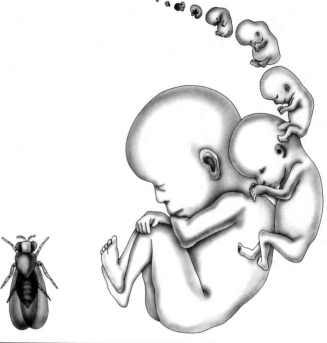

Event	Time for fruit flies	Time for humans
Fertilization of egg	zero	zero
First cleavage division	23 minutes	30 hours
Eighth cleavage division	90 minutes	3 days
Early formation of nervous system	4 hours	3½ weeks
Hatching/birth	24 hours	9 months

Figure 12-10 The development times for fruit flies and for humans.

synthesized in the male testes, are released in each ejaculation. Many sperm will attach to an egg; however, one and only one sperm will actually penetrate and fertilize the egg.

Male chromosomes are contained in the nucleus of the sperm. The nucleus is located beneath the **acrosomal cap**, which contains proteins that recognize the egg's surface and help the sperm attach to and penetrate it (Figure 12-11). Mitochondria, located in the tail near where it joins the head, provide the energy that the sperm needs to swim vigorously through the vagina and uterus en route to an egg traveling toward it down the Fallopian tube. Depending on conditions in the vagina and uterus, the journey of the millions of sperm may take hours or even days before they reach the egg and one fertilizes it.

Human eggs (*ova*), which are the largest of all human cells (about 0.1 mm in diameter), are produced in the ovaries. At birth, a woman's ovaries have already produced all of the eggs she will have during her lifetime, and the eggs will have all undergone the first meiotic division. At puberty, when a woman begins to menstruate, the first eggs are released in a process that will be repeated each month for thirty to forty years.

The human egg is much more complex in structure than the sperm. In addition to the haploid number of chromosomes consisting of an X chromosome and 22 autosomes in the nucleus, eggs have a complex internal structure that provides all of the cellular components necessary for the early stages of development. The outer surface of the egg is surrounded by a gelatinous substance called the **zona pellucida**, which acts as a barrier against the egg's being fertilized by sperm from other species. The zona pellucida is also thought to provide a mechanism that allows only a single sperm to penetrate the egg.

When a sperm finally becomes attached to the surface and penetrates the egg, the sperm's tail is left behind on the egg's inner surface, where it disintegrates. The head of the sperm is absorbed into the egg and begins to swell prior to releasing its nucleus. A human sperm can attach to the egg and be swallowed in less than a minute. Within 20 minutes the sperm and egg nuclei have fused, the diploid chromosome number has been restored, and the process of **fertilization** has been accomplished (Figure 12-12). Once the egg has been fertilized, development proceeds very rapidly. DNA synthesis begins almost immediately, and within a few hours the zygote undergoes its first mitosis and division into two cells (Figure 12-13).

Perhaps what is most remarkable about fertilization and embryonic development is that the process functions correctly so much of the time. However, there are many ways in which reproduction can go awry. During the production of hundreds of millions of sperm (many billions over a man's lifetime), the synthesis of abnormal sperm is not uncommon. Sperm with two tails or two heads are often observed. If the proportion of abnormal sperm in a man's ejaculate is too high, he may be infertile. Recently, sophisticated tests have been developed that assess male infertility by measuring the ability of sperm to fertilize an egg *in vitro* (see Box 12-3).

Reproductive abnormalities in the woman can also prevent fertilization and cause infertility. After the sperm have been deposited in the woman's vagina (either by intercourse or by artificial insemination), they undergo chemical changes called **capacitation** as they swim through the secretions in the vagina and uterus and into the Fallopian tubes, where fertilization occurs. The secretions in a woman's reproductive system may fail to capacitate sperm, or they may even inactivate them. Another possible cause of female infertility is defective Fallopian tubes. After it is released from the ovary, the egg must travel down the Fallopian tubes in order to become fertilized; then it must attach to the wall of the uterus. If these tubes are blocked as a consequence of disease, injury, or malformation, fertilization may be prevented. Another reason for infertility is that the uterus may not have been adequately prepared by hormones for the egg's attachment.

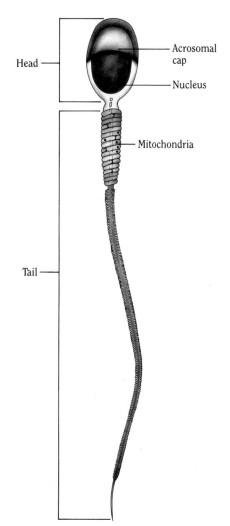

Figure 12-11 Diagram of a human sperm. The acrosomal cap consists of proteins that help the sperm attach to an egg. The nucleus contains the haploid set of chromosomes. The mitochondria provide the energy that allows the sperm to swim.

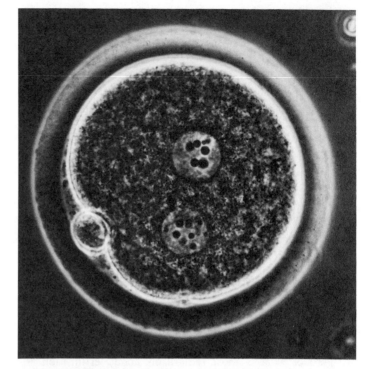

Figure 12-12   A fertilized egg showing both the egg's nucleus and the sperm's nucleus. The structure on the lower left is a polar body, one of the meiotic products that does not become an egg. (Courtesy of Jane Rogers, University of Hawaii Medical School.)

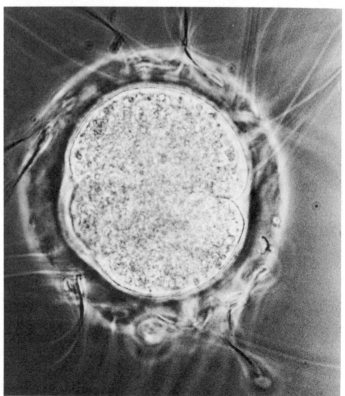

Figure 12-13   A fertilized egg about to undergo the first cell division. The nuclei from the egg and sperm have fused and the chromosomes are no longer visible. (Courtesy of Jane Rogers, University of Hawaii Medical School.)

"Why Can't We Have a Baby?"

Box 12-3

Millions of couples who want children have not been able to conceive any. Such human infertility has many causes. The first step in helping an infertile couple is to determine whether the reproductive problem exists in the man or in the woman.

In testing for male fertility, sperm are examined for the number produced, for normal appearance, and for the capacity to swim vigorously. If any of these properties are markedly abnormal, it may mean that the sperm are not sufficiently active or are not present in sufficient numbers to fertilize an egg.

Until quite recently, it was not possible to actually measure the capacity of human male sperm for fertilization, but now tests that measure this capacity are available. Using eggs obtained from hamsters or guinea pigs, researchers can observe whether or not a sample of human sperm is able to fertilize eggs from another mammalian species and, by inference, to fertilize human eggs, too.

Normally, sperm from one animal species are unable to fertilize eggs of another species. However, this interspecies barrier can be removed in the laboratory by stripping the zona pellucida—the outer layer—from the hamster or guinea pig eggs. Without this gelatinous barrier, normal human sperm can attach to the egg's surface, penetrate the egg, and actually stimulate a few cell divisions before the egg aborts, indicating that the sperm have the capacity for fertilization. Defective sperm—even those that appear normal by other tests—are unable to fertilize the hamster or guinea pig egg. This test, then, either assigns a couple's infertility problem to a defect in the sperm or indicates that the woman may have a reproductive problem. If the sperm prove normal, the woman may elect to undergo tests to determine whether her reproductive process is blocked at a certain stage and whether the problem can be corrected.

Reading: Laurence, E. Karp, "Genetic Crossroads," *Natural History*, October 1978.

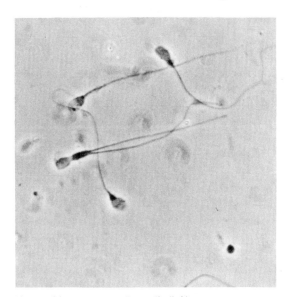
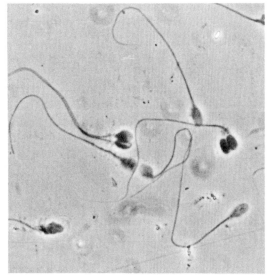

Abnormal human sperm: Two tails (left) or two heads (right). (Courtesy of Jane Rogers, University of Hawaii Medical School.)

Genetic abnormalities may also cause abnormal development and miscarriage. If either the egg or the sperm has an abnormal number of chromosomes, the embryo may abort spontaneously early in development. During embryonic development, cells must undergo hundreds of cell divisions, and each time the cell's chromosomes must undergo mitosis and the cells must divide with precision. Thus, genetic errors may arise in chromosomes during development of the embryo, causing it to abort. Finally, during the many months of embryonic development, genes must be switched off and on and expressed in a precise sequence if abnormal development is not to result. That so many of us are born biologically and genetically normal testifies to the many failsafe mechanisms that nature has developed.

Key Terms

sexual dimorphism   the distinctly different physical forms of males and females in a species; for example, male birds are generally larger and more colorful than their female counterparts.

hermaphrodites   individuals in which both male and female reproductive organs are present.

parthenogenesis   reproduction by females without fertilization by male gametes. This form of reproduction produces only female offspring.

homunculus   a little individual that early Greeks believed was contained in the head of sperm.

homogametic sex   the sex of a species that produces gametes containing only one type of sex chromosome.

heterogametic sex   the sex of a species that produces gametes with two types of sex chromosomes.

Turner's syndrome   a syndrome characteristic of women who have an XO chromosome constitution.

Klinefelter's syndrome   a syndrome characteristic of men who have an XXY chromosome constitution.

androgen insensitivity syndrome   a syndrome characteristic of women who have an XY chromosome constitution.

androgens   a group of male hormones (the major one is testosterone) that are produced early in development and result in formation of testes.

sexism   prejudice based on a person's sex; discrimination against women.

Barr bodies   dark staining spots in cell nuclei that contain a condensed, inactive X chromosome.

Lyon hypothesis   the suggestion that only one X chromosome is expressed in each female cell and that the other X chromosome in that cell is inactive.

H-Y antigen   a male-specific protein that initiates development of male sexual organs.

primary sex differentiation   the constitution of the sex chromosomes in the zygote.

secondary and tertiary sex differentiation   the development of male or female sex organs and masculine or feminine physical characteristics.

zygote   a fertilized egg formed by the fusion of male and female gametes; the first cell of a new individual.

acrosomal cap   proteins surrounding the head of sperm that facilitate attachment to the egg.

zona pellucida   the gelatinous outer covering of an egg that allows sperm of the same species to attach and that also facilitates fertilization by a single sperm.

fertilization   the fusion of a sperm and an egg to form a zygote.

capacitation   chemical changes produced in sperm as they swim through the vagina and uterus.

Additional Reading

Hogan, B. "Biological Signals in Early Embryos." *New Scientist,* July 29, 1976.

Leff, D. "The Way of a Sperm with an Egg." *Mosaic,* July/August 1981.

Money, J. and P. Tucker. *Sexual Signature: On Being a Man or a Woman.* Boston: Little, Brown, 1975.

Nelkin, D. and C. A. Raymond. "Tempest in a Test Tube." *The Sciences,* November 1980.

Warner, R. "Metamorphosis." *Science 82,* December 1982.

Review Questions

1. What kinds of cells are essential for the creation of a new person?

2. What evolutionary process is affected by exotic coloration and behaviors in birds?

3. What kinds of chemicals determine secondary sexual characteristics?

4. What are the most common human defects or diseases caused by nondisjunction of chromosomes?

5. Which chromosomes are contained in Barr bodies?

6. What is the primary function of the Y chromosome in humans?

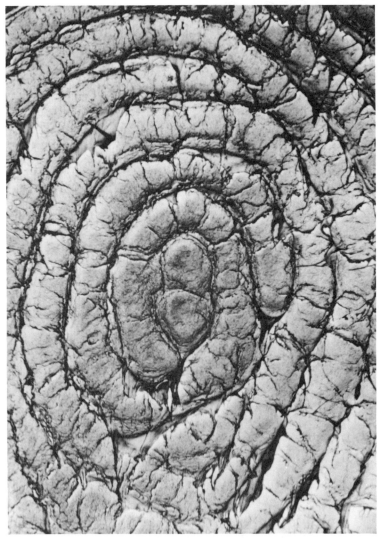

THE SKIN ON THE TIP OF EACH FINGER HAS A DISTINCTIVE FINGERPRINT. (27 ×)

13

HUMAN HEREDITARY DISEASES

How Mutations Are Transmitted

One of the things that each of us has to do in our lives is to discover, as far as possible, the grounds for believing what we are asked to believe.

STEPHAN L. CHOROVER

13
HUMAN HEREDITARY DISEASES: HOW MUTATIONS ARE TRANSMITTED

Enormous strides have been made in the past century in preventing and curing many serious human diseases. Epidemics that formerly decimated human populations have been eliminated in many areas of the world. Bacteria that caused outbreaks of plague, cholera, pneumonia, and typhoid fever have been virtually eliminated in the United States as a result of improved sanitary conditions, better nutrition, and modern medical care. Many significant viral diseases have also been combatted successfully: smallpox is thought to be totally eradicated throughout the world, and polio has nearly disappeared as a result of vaccination. In short, modern medicine has been extremely successful in eliminating or reducing many infectious diseases. However, it has been much less successful in coping with hereditary diseases, either by prevention or by treatment.

Hereditary diseases—more commonly called **genetic diseases**—result from abnormal chromosomes or mutant genes that are passed from one or both parents to their offspring. Although these inherited defects exist in the egg or sperm at the moment of fertilization, the phenotypic effects are not usually observable until birth or, in some instances, until much later in life.

Many people are confused about what causes hereditary diseases and what can be done about them. The questions college students frequently ask point up some of the common fears and misunderstandings: "My sister is mentally retarded. Does that mean that my children have a good chance of being retarded also?" "My aunt and my grandmother both died of breast cancer. The doctor says that I am much more likely to get breast cancer because it runs in my family. Is it true that I've got genes that will give me breast cancer?" "People have told me that diabetes and high blood pressure are genetic diseases, but other people say these diseases are caused by diet. Which is it?" "Everybody in my family is fat, so there must be something in my genes that makes me fat, too." After reading this chapter you should be able to determine which of these questions are answerable and the form of the answers for yourself.

To emphasize the difference between hereditary diseases or defects and other kinds of diseases, in this chapter we shall use the term *hereditary* instead of the term *genetic,* because genetic changes may also occur in somatic cells. There are two kinds of mutations: *germinal,* which are hereditary, and *somatic,* which are not. **Germinal mutations** occur in the chromosomes of the reproductive (sex) cells, while somatic mutations occur in all body cells except the sex cells.

For many human diseases and physical defects, it is often difficult to say whether or not the disease is caused by a person's genes. The assignment of hereditary diseases is complicated by the fact that not all are caused exclusively by mutant genes; many result from complex gene-environment interactions. However, all hereditary diseases do differ from environmentally caused diseases in that hereditary diseases cannot occur without some change in the genetic information.

About 5 percent of all newborn infants have some observable anatomical or physiological abnormality that can be established as a hereditary disease or defect. However, during pregnancy, exposure to infections, drugs, and other environmental agents may cause abnormal development of the embryo, resulting in a congenital defect—a physical abnormality that is detectable at birth. Congenital defects are not necessarily inherited, although genes may play a role. This chapter describes the basic rules that distinguish human hereditary diseases from the many other sorts of congenital defects.

Common Congenital Defects

The first opportunity to detect abnormalities in an individual is usually at his or her birth. Each newborn infant is examined and any physical abnormality is noted. Also, each newborn is tested for phenylketonuria (PKU) and for other hereditary diseases if there is some indication that the child is at risk for them. Some clearly recognized human hereditary diseases, their phenotypic consequences, and their modes of inheritance are listed in Table 13-1.

The most frequent chromosomal abnormality observed at birth is *Down's syndrome,* which affects 1 in every 10,000 to 15,000 children born in the United States (see Box 13-1). Down's syndrome usually results from an extra chromosome number 21, hence its other name, trisomy 21 syndrome. The extra chromosome 21 can arise by nondisjunction in either the mother's or father's sex cells at the time the parental gametes are formed, but the evidence indicates that most of the time the extra chromosome arises in the egg. Down's syndrome is not a typical hereditary disease because it is rarely passed on to progeny and does not occur generation after generation in the family lineage. Down's syndrome individuals, who typically are severely mentally retarded, usually do not mate. However, since the chromosomal abnormality is inherited from one parent or the other, Down's syndrome and other diseases caused by chromosomal abnormalities are classified as hereditary.

As we noted in Chapter 11, *phenylketonuria* is an inherited disease caused by a defect in the gene that codes for an enzyme that normally breaks down the amino acid phenylalanine. In PKU this gene's defect causes the enzyme to be wholly or partly inactive, so that the phenylalanine and its abnormal breakdown products build up in the developing brain, causing mental retardation during infancy.

PKU is a recessive trait because phenotypically normal pa-

Table 13-1. Some Well-Characterized Human Hereditary Diseases and Their Phenotypic Consequences

Disease	Major Consequences
Chromosome abnormalities	
Down's syndrome	Mental retardation
Klinefelter's syndrome	Sterility, occasional mental retardation
Turner's syndrome	Sterility, lack of sexual development
Autosomal dominant mutations	
Achondroplasia	Dwarfism
Retinoblastoma	Blindness
Porphyria	Abdominal pain, psychosis
Huntington's chorea	Nervous system degeneration
Neurofibromatosis	Growths in nervous system and skin
Polydactyly	Extra fingers and toes
Autosomal recessive mutations	
Cystic fibrosis	Respiratory disorders
Hurler's syndrome	Mental retardation, stunted growth
Xeroderma pigmentosum	Skin cancers
Albinism	Lack of skin pigment, vision difficulties
Phenylketonuria	Mental retardation
Progeria	Premature aging; early death
Maple sugar urine disease	Convulsions, mental retardation
Alkaptonuria	Arthritis
Galactosemia	Cataracts, mental retardation
Homocystinuria	Mental retardation
Tay-Sachs syndrome	Neurological deterioration
Sickle-cell disease	Anemia
X chromosome mutations	
Lesch-Nyhan's disease	Mental retardation, self-mutilation
Hemophilia	Failure of blood to clot, bleeding
Duchenne's muscular dystrophy	Progressive muscular weakness
Agammaglobulinemia	Defective immune system, infections
Testicular feminizing syndrome	Sterility, lack of male organs

rents can have a child who is afflicted with the disease. For PKU to manifest itself, both parents must carry the PKU (defective) gene on one of their chromosomes, and both genes must be passed on by chance when the sperm and egg unite to form the zygote. PKU can be detected during the first few weeks of life by means of a blood or urine test; such tests are now routine in many states. Once the disease has been diagnosed, it can be controlled by a low-phenylalanine diet.

Down's syndrome and PKU are classified as hereditary dis-

Mongolism:
How Prejudice Created a Description for a Hereditary Disease

BOX 13-1

In 1866 a London physician, J. Langdon H. Down, published a report entitled *Observations on an Ethnic Classification of Idiots*. Down had the opportunity to study a number of institutionalized retarded children who today would be recognized as being trisomic for chromosome 21. At that time, however, nothing was known about chromosomes or about the biochemistry of hereditary diseases, and Down ascribed the signs and symptoms to tuberculosis in the children's parents. Having observed the physical signs and mental defects of these children for a number of years, Down decided that about 10 percent of all children classified as congenital idiots appeared to be "typical Mongols." He described the characteristics of one such child in these terms:

The face is flat and broad and destitute of prominence. The cheeks are roundish, and extended laterally. The eyes are obliquely placed. . . . The lips are large and thick with transverse fissures. The tongue is long, thick, and is much roughened. The nose is small. The skin has a slight dirty, yellowish tinge and is deficient in elasticity. . . . There can be no doubt that these ethnic features are the result of deterioration.

Thus did the term *mongolism* come into general use to describe a particular kind of congenital idiocy in children. Moreover, the term *mongolism* implied that these individuals represented a degenerate ethnic group of human beings, specifically, Orientals.

Down was attempting to scientifically justify a common prejudice of the period: that Caucasians were evolutionarily the most advanced of the human races and that the English, in particular, were biologically far superior to all other people. Down also represented his ethnic classification as being in accord with Darwin's ideas of evolution. He reasoned that since English parents could give birth to children with mongoloid features, the mentally defective children must have represented throwbacks to some less evolved, "degenerate" form of the human species. Down closed his article with what might have been construed at the time as a liberal comment: ". . . the result of degeneracy among mankind, appears to me to furnish some arguments in favor of the unity of the human species."

Down may have been proposing biological unity among individuals or races, but it was obvious which ethnic group he believed to be most highly advanced and which groups he regarded as degenerate. This kind of *scientific racism* (the use of scientific facts or observations to support a prejudice) received widespread acceptance during the early 1900s. Such pseudoscientific nonsense about "inferior races" and a "master race" was ultimately used to justify genocide—the mass murder of religious and ethnic groups by the Nazis during World War II (see Chapter 19).

Some years ago a group of scientists recommended that the term *Down's syndrome* (or Down syndrome) replace the designation of mongolism in all scientific publications. Thus, the term *mongolism*, which carries with it the stigma of racial degeneracy, is no longer used by scientists and others who understand its historical significance.

Reading: Stephen Jay Gould, "Dr. Down's Syndrome" in *The Panda's Thumb*, New York: W. W. Norton, 1980.

eases because the genetic defects were present in the eggs or sperm of the parents before the child was conceived. By contrast, most congenital defects are due to a complex interaction of genetic information and environmental factors. Examples are *cleft lip* (harelip) and *spina bifida* (cleft spine), which are the results of developmental abnormalities in the formation of the oral cavity and the spine, respectively. Cleft lip has been known to

Figure 13-1 A victim of the teratogenic drug thalidomide, this armless boy has learned to use his toes and feet as substitutes for hands and fingers. (United Press International.)

occur in only one of a pair of identical twins, so factors other than genes must contribute to this abnormality.

Spina bifida (SB) afflicts 1 in every 1,000 newborns. It occurs when one or more spinal vertebrae fail to close and the spinal cord and associated nerves bulge through the cleft, forming an easily damaged fluid-filled sac. The protruding spinal nerves are vulnerable to paralysis-causing damage and life-threatening infection. Most SB babies also suffer from mental retardation caused by **hydrocephalus**, an abnormal accumulation of fluids in the brain. Today, spinal surgery can repair the exposed nerves and fluid-filled sac, and brain surgery can prevent hydrocephalus-caused mental retardation by the installation of tubes that continually drain the excess fluid. SB is more prevalent among persons of Northern European (white) ancestry than it is among Jews or blacks, suggesting that it has a hereditary as well as an environmental component.

Neither cleft lip nor spina bifida can be classified as hereditary diseases, although it is likely that the parents' genetic constitution contributes to the risk of progeny developing these defects. Because both cleft lip and spina bifida are detected at birth, they are classified as congenital defects.

While genes play a role in some congenital defects, many birth defects are caused solely by environmental agents called **teratogens** (from the Greek *teras*, "monster" and *gen*, "born"). Teratogenic agents that can adversely affect embryonic development include viruses, radiation, and many kinds of drugs. For example, if a woman is infected with German measles (rubella) while she is pregnant, the embryo may develop heart defects or suffer damage to its eyes or ears. Exposing the embryo's cells to radiation such as diagnostic x-rays can cause somatic mutations, which change normal development of tissues or organs. Finally, any of numerous drugs that a pregnant woman may take for one reason or another can affect embryonic development and cause congenital defects.

A particularly tragic example of a drug-induced congenital defect is that caused by **thalidomide**, a tranquilizer that was prescribed during the 1950s and 1960s to alleviate anxiety in expectant mothers. Thalidomide is a potent teratogen in humans, and thousands of women who took this drug early in pregnancy gave birth to children who had abnormal or missing arms and legs (Figure 13-1). These children are genetically normal, and they can have normal children when they reach reproductive age in the 1980s. Thanks to the alertness and efforts of Mrs. Frances O. Kelsey, a physician at the Food and Drug Administration in Washington, thalidomide was not prescribed in the United States; virtually all of the affected children were born in Europe, Asia, or in Canada.

Alcohol is another teratogen; it can cause serious physical defects, including some degree of mental retardation (Figure 13-2). Cortisone-like drugs are teratogenic in mice; they may or may not cause developmental defects in humans, but pregnant women are advised not to take them. Even as common a substance as the caffeine in coffee, tea, and many soft drinks is teratogenic in some animals, but there is no evidence yet that caffeine causes developmental defects in human embryos.

Because of the uncertainty about which substances are teratogenic, pregnant women are advised to forego taking drugs of any kind, including alcohol, coffee, aspirin, and tobacco. The early months are the most important for development of the embryo; by five to six months most tissues and organs, including the brain, have been formed, and from then until birth the fetus mainly increases in size. However, this does not mean that damage cannot occur from teratogenic substances during the last months of pregnancy.

What is important to remember is that congenital defects may or may not be due to genetic defects, and those developmental abnormalities that are not inherited will not be passed on to one's children. Developmental abnormalities may be caused by mutations in the chromosomes of the embryo's somatic cells. These kinds of abnormalities, though genetic, are not germinal mutations and cannot be passed on to offspring. Because developmental abnormalities and congenital defects can be caused by either hereditary or environmental factors or a combination of both, it is important to understand the criteria that establish whether a disease or defect is of hereditary origin.

When Is a Disease Hereditary?

Three well-established criteria are applied to determine whether a congenital defect or disease is due primarily to genetic factors

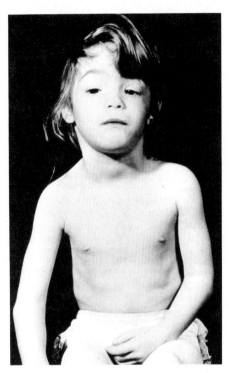

Figure 13-2 A child with fetal alcohol syndrome. Such children, born to women who consumed significant amounts of alcohol while pregnant, are undersize and have eyes and facial features that are abnormally proportioned or positioned. Many have heart defects, and most are mentally retarded to some degree. (Courtesy of Kenneth Lyons Jones, M.D.)

Figure 13-3 A family pedigree for color blindness. The Mendelian mode of transmission of a trait from generation to generation establishes that a particular defect or disease is inherited. Color blindness is a sex-linked hereditary defect that shows up almost exclusively in males. Can you figure out why the status of one of the females in the second generation is unknown?

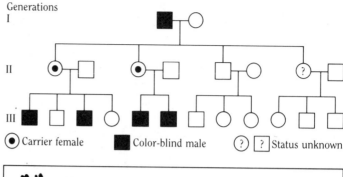

Generations

Figure 13-4 Karyotype for Down's syndrome. A karyotype analysis will show any gross chromosomal abnormalities in an individual's cells. Down's syndrome is characterized by an extra chromosome number 21 (trisomy 21). Chromosomal rearrangements and partial chromosomal abnormalities can also be detected by karyotype analysis.

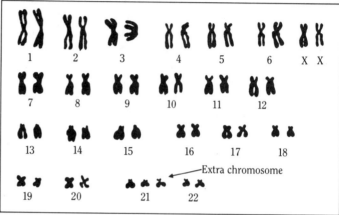

that have been passed on from parents—that is, whether the disease has been inherited. If any one of these criteria for inheritance can be demonstrated, we are justified in classifying the disease as hereditary.

The three criteria are:

1. A Mendelian pattern of inheritance; that is, a predictable pattern of inheritance over several generations that obeys Mendel's laws (Figure 13-3).

2. A chromosomal abnormality; that is, loss of a chromosome, presence of an extra chromosome, or rearrangements of chromosome segments from their normal locations, as shown by a cytological examination or karyotype analysis (Figure 13-4).

3. A biochemical defect that can be assigned to a particular gene, such as the mutant allele that produces abnormal hemoglobin molecules and causes sickle-cell anemia.

If a trait, defect, or disease can be characterized by one of these three criteria, its inheritance is established. Each of these criteria provides proof that a mutation of one sort or another was present in one or both parental gametes and was transmitted to the progeny. However, the fact that a trait does *not* meet these criteria does not in turn mean that the trait lacks a hereditary basis or genetic component; it simply means that the issue cannot be decided one way or another. As we shall see in the next chapter, there are important ethical, medical, and psychological

Figure 13-5 A carved stone showing the pedigree of horses raised by a breeder about 4,000 years ago. This stone, discovered in Asia, shows that breeders kept records of desirable traits in their animals. (Courtesy of Dorsey Stuart, University of Hawaii.)

reasons for requiring conclusive proof before a person is characterized as having a hereditary defect.

Pedigree Analysis

For centuries before the experiments of Gregor Mendel showed how traits are inherited, breeders of plants and animals kept some sort of record of desirable traits (Figure 13-5). Once Mendel had demonstrated that traits are determined by discrete genetic factors that are expressed in a dominant or recessive manner and that are transmitted in predictable ratios, it became possible to identify hereditary diseases by constructing a **family pedigree**—the pattern of a trait's occurrence in family members from generation to generation. To construct a pedigree, it is necessary to have medical histories of all the family members, or as many as possible, spanning several generations. Some hereditary defects may skip a generation completely because of their particular mode of inheritance, as in the case of male color blindness. Thus, the more extensive and detailed the pedigree, the easier it is to accurately determine the mode of inheritance.

In diagraming pedigrees, geneticists use a few universally recognized symbols (Figure 13-6). With the system used in the United States, females are always indicated by circles and males by squares. If an individual is affected (has the trait or the disease), the square or circle is specifically marked (in Figure 13-6, it is blackened). **Carriers** of the defect—that is, individuals who carry the mutant gene on one chromosome but are not themselves affected because they carry a normal allele on the other chromosome—are indicated by a different marking (in Figure 13-6, by partially blackened symbols). If a sufficiently detailed or complete pedigree can be constructed over several generations, a geneticist can determine not only whether the trait is inherited but also whether the gene for the trait is recessive, dominant, or sex linked.

Because the royal families of Europe kept detailed family

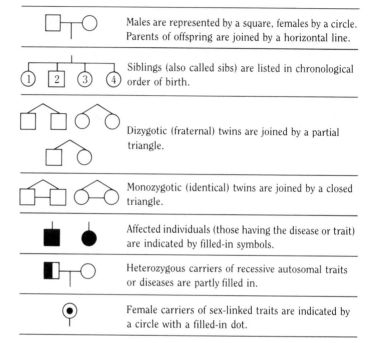

Males are represented by a square, females by a circle. Parents of offspring are joined by a horizontal line.

Siblings (also called sibs) are listed in chronological order of birth.

Dizygotic (fraternal) twins are joined by a partial triangle.

Monozygotic (identical) twins are joined by a closed triangle.

Affected individuals (those having the disease or trait) are indicated by filled-in symbols.

Heterozygous carriers of recessive autosomal traits or diseases are partly filled in.

Female carriers of sex-linked traits are indicated by a circle with a filled-in dot.

Figure 13-6 The meaning of the various symbols used in the United States in constructing pedigree diagrams.

records and because they tended to have large numbers of offspring, their pedigrees are particularly informative. Hemophilia—a sex-linked, recessive trait caused by a gene on the X chromosome—used to be called a royal disease because many males of the royal families of Europe and Russia suffered from this hereditary affliction. Hemophiliacs, or "bleeders," bleed profusely from the slightest scratch or injury because they lack one or more blood proteins called *clotting factors*. In the past, hemophilia was a life-threatening disease, and afflicted males had to be extremely careful not to be cut or bruised. Today, the disease, while still quite serious, can be controlled by transfusions and periodic injections of the missing clotting factor.

Queen Victoria of England (1819–1901) was a carrier of hemophilia, and it is possible that it arose in her family as a result of a mutation that she inherited (Figure 13-7). Be that as it may, among Victoria's nine children one son was a bleeder and two daughters, Alice of Hesse and Beatrice, were carriers. Alice, in turn, had seven children and passed the mutant gene on to at least two daughters, Irene and Alexandra (Figure 13-8). As a result of intermarriage with the German and Russian royal families, hemophilia was transmitted to other European royal families, and males in succeeding generations were affected. Because Edward VII was not affected, the disease could not be transmitted to future generations of the English royal family.

Investigation of royal pedigrees has uncovered another

Figure 13-7 Queen Victoria and her family. When her granddaughter Alexandra (above Victoria and to our left, with neck fur) married Nicholas II, Tsar of Russia, their son Alexis was a "bleeder." When granddaughter Irene (above Victoria and on our right, with neck fur) married Prince Henry of Prussia, the mutant gene was transmitted to the German royal family as well (see Figure 13-8). (Courtesy of Photography Collection, Humanities Research Center, University of Texas at Austin.)

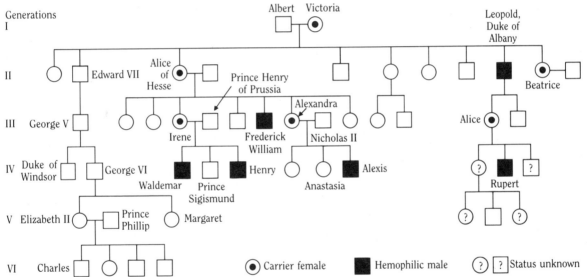

hereditary disease, **porphyria**, which is caused by a biochemical imbalance that sometimes produces insanity and a variety of other symptoms. Some historians think it likely that this hereditary disease, which can be traced through more than 200 years of British royalty (Figure 13-9), had an important impact on the War of Independence and, thereby, on American history.

King George III (1738–1820), Queen Victoria's grandfather, was England's monarch during a turbulent period of world history that included both the French Revolution and the American Revolution (1775–1783). During these years in which new governments and countries were created, George III was having serious health problems that affected his mental state and judg-

Figure 13-8 A pedigree of the European and Russian royal families showing the transmission of hemophilia. The disease can be traced to Victoria, who was a carrier of the X-linked mutant gene. Two of Victoria's daughters were also carriers.

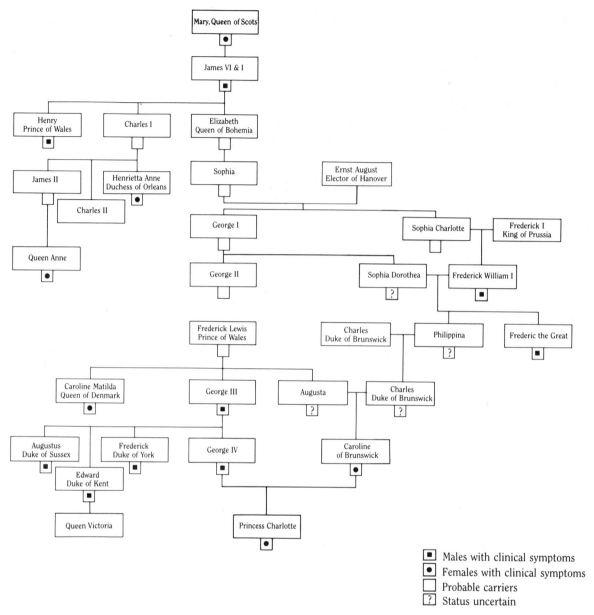

Mary, Queen of Scots

James VI & I

Henry
Prince of Wales

Charles I

Elizabeth
Queen of Bohemia

Sophia

Ernst August
Elector of Hanover

James II

Henrietta Anne
Duchess of Orleans

Charles II

George I

Sophia Charlotte

Frederick I
King of Prussia

Queen Anne

George II

Sophia Dorothea

Frederick William I

Frederick Lewis
Prince of Wales

Charles
Duke of Brunswick

Philippina

Frederic the Great

Caroline Matilda
Queen of Denmark

George III

Augusta

Charles
Duke of Brunswick

Augustus
Duke of Sussex

Frederick
Duke of York

George IV

Caroline
of Brunswick

Edward
Duke of Kent

Queen Victoria

Princess Charlotte

■ Males with clinical symptoms
● Females with clinical symptoms
□ Probable carriers
? Status uncertain

Figure 13-9 An abbreviated pedigree of the English royal families showing the transmission of porphyria. The disease can be traced all the way back to Mary, Queen of Scots. Because porphyria is caused by an autosomal dominant gene, both males and females are affected. However, because of extremely variable expression, not all individuals with the mutant gene showed clinical symptoms.

ment and certainly reduced his ability to manage colonial affairs. In 1765, at age twenty-six, he had his first documented bout of insanity; another more serious attack occurred in 1788, when he was fifty. Altogether his medical records show five periods of insanity that were described by his physicians as including severe physical symptoms as well as delirium and hallucinations. He was treated with accepted medical practices of the period: being shackled to a bed or chair while subjected to forced vomitings, cuppings, blisterings, and leechings. None of his doctors understood what was wrong, but they knew that insanity among members of the English royal family was quite common and that insanity had also affected the royal houses of Stuart, Hanover, and Prussia over the centuries.

It has now been established from medical records, recorded symptoms, and pedigree analysis that George III, as well as some of his ancestors and descendants, was afflicted with a hereditary disease known today as *acute intermittent porphyria*. The symptoms are acute pain in the abdomen and sometimes in the limbs and back, headaches, insomnia, nausea, hallucinations, depression, and delirium. All are caused by a dominant gene with variable expression, which simply means that the severity of the disease can range from mild symptoms that may go unnoticed to full-blown psychosis and even death due to respiratory paralysis.

Porphyrin molecules are present in most body cells and are essential for the functioning of many enzymes. They play a particularly important role in the transport of oxygen to body cells because they are part of the hemoglobin molecule, a component of all red blood cells. During the reign of George III, nothing was known about the genetics or biochemistry of hereditary diseases or about porphyrins, and so his insanity was presumed to have a psychological basis. Modern medical research has shown that all the symptoms of this disease, including insanity, are due to a toxic effect from the overproduction of porphyrins. The expression of the mutant gene is influenced by many factors, including diet, the use of alcohol and certain drugs, and the menstrual cycle—an example of the interplay often seen between gene expression and the external (and internal) environment.

American history might have been different if porphyria had not occurred in the British royal families during the two centuries from Mary Queen of Scots (1542–1587) through the lifetime of George III. It is fruitless to speculate how different our lives might be today if George III had not been afflicted with porphyria or if Alexis, the heir to the Russian throne, had not suffered from hemophilia. Aside from hereditary diseases that cause severe symptoms, each one of us has genes that, while producing no signs of disease, may yet affect our lives and livelihood (see Box 8-2).

X-Linked Inheritance

Genes that are located on the X chromosome are said to be *sex-linked* (X-linked) because generally only one sex (males) is affected by the mutant alleles. A female can also have X-linked diseases, but both her X chromosomes must carry the mutant alleles if she is to be affected (assuming that the genes are recessive, which they frequently are for serious hereditary diseases). Female hemophiliacs, for example, are rare, as their absence from royal pedigrees shows. Today, however, a few females who have hemophilia are known, and both their X chromosomes carry mutant genes.

Because chromosomes segregate at random into the eggs and sperm during meiosis, it is possible to calculate the probability that a daughter will be a carrier of an X-linked disease or that a son will be affected by an X-linked trait (Figure 13-10). On the average, if the mother is a heterozygous carrier, half of her daughters will be carriers and half of her sons will be hemophiliacs. Albert and Queen Victoria had nine children—five girls and four boys, very close to the expected 50–50 sex ratio. Of the five female offspring, two were definite carriers of the defective, hemophilia-causing allele, close to the expected frequency of one-half, although it is not certain that the others were not carriers. However, of the four male offspring, only one— Leopold, Duke of Albany—was affected, whereas two out of four would be expected to be affected. It is important to remember that probabilities become more accurate when the numbers are large; if the number of individuals (or events) is small, deviations from predicted numbers will often be observed.

In calculating the probability that a particular allele or group of alleles will be inherited, it may be helpful to think of the segregation of chromosomes as being analogous to the tossing of a coin. Just as a coin always comes up either heads or tails, eggs and sperm always receive either one or the other chromosome of each pair. On the average, tossing a coin a large number of times produces nearly equal numbers of heads and tails. However, it is possible to get runs of either heads or tails, and a run can badly skew the outcome of a limited number of tosses. So it is with inheritance: All the offspring may get the abnormal chromosome, all may get the normal one, or the frequency may approximate the expected 50–50 ratio.

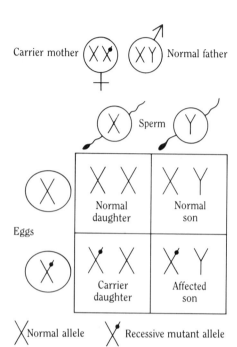

Figure 13-10 The pattern of inheritance of an X-linked trait. Both parents are phenotypically normal, but the mother is a carrier of a recessive gene on the X chromosome. All daughters of this couple will be phenotypically normal, but on the average half of them will be carriers. Half of the sons, on the average, will be affected, since there is one chance in two that they will receive the X chromosome carrying the mutant gene from their mother.

Autosomal Recessive Inheritance

A recessive gene is one whose abnormal function is of little or no consequence as long as cells contain a normal allele whose expression masks that of the recessive allele. This is why heterozygous carriers of harmful recessive genes are not themselves affected by the disease or defect; for offspring to be affected, both parents must be carriers of the recessive gene. Because of the random segregation of chromosomes during meiosis, on the average one-fourth of the offspring of two carriers will be affected and one-half will be carriers themselves (Figure 13-11). Some autosomal recessive diseases are listed in Table 13-1.

Autosomal recessive genes are usually recognized in human pedigrees by two distinguishing characteristics: (1) the trait is usually present in only one generation or it skips one or more generations, because as long as individuals marry outside of their immediate families it is unlikely that two carriers will mate; and (2) when the parents are heterozygous for the recessive gene, among a large number of children the ratio of nonaffected to affected individuals should be 3 to 1. Rarely do actual pedigrees conform to the predicted pattern, but it is usually possible to distinguish an autosomal recessive trait from an X-linked or autosomal dominant trait.

Autosomal Dominant Inheritance

To say that a gene is expressed in a dominant fashion means that the phenotype is determined by that gene regardless of what other allele is present. The number of recognized dominant genes that cause serious human hereditary defects is much smaller than the number of deleterious recessive genes because if the gene reduces a person's chances of survival, he or she is less likely to reproduce and pass the deleterious dominant genes on to the next generation. Some of the more common dominant genes that cause serious diseases have survived in the human **gene pool** (all of the alleles of all the individuals in a population) because they are not expressed until late in life, when the affected individuals have already had children.

Huntington's disease (also called *Huntington's chorea*) is one of the best-known examples of a dominant inherited disease. Woody Guthrie, the famous folk singer, died of this hereditary disease. At this writing, Arlo Guthrie, his son and a famous folk singer in his own right, does not yet know whether he carries the gene for Huntington's chorea, nor does he know whether he has passed the gene on to his own children. About half of the individuals with Huntington's disease do not show any signs of the disease until age forty, and obviously some individuals may die of other causes before the gene is ever expressed.

If only one parent has an autosomal dominant allele for a disease or physical defect, half of the couple's children will be affected if the gene is fully expressed (Figure 13-12). Porphyria is caused by an autosomal dominant allele, but its expression varies from person to person, and its presence does not always cause clinical symptoms. The symptoms of Huntington's disease vary markedly with age, and some persons with the mutant gene may even go undetected. However, since dominant genes are commonly expressed to some extent in individuals who possess them, there are no "carriers" of dominant traits as there are for recessive ones.

A rather common autosomal dominant gene that is fully expressed causes **achondroplasia**, or dwarfism characterized by a disproportionate shortening of arms and legs. This hereditary defect is due to abnormal growth of cartilage and has no other serious phenotypic effects. Persons with this autosomal dominant gene are fertile, which accounts in part for the many "little people" in our society with this inherited defect. Generally, for

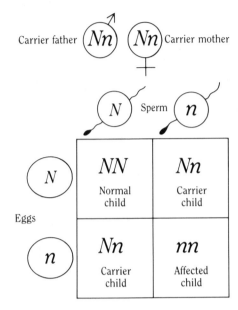

Figure 13-11 The pattern of inheritance of an autosomal recessive trait. Both parents are phenotypically normal and carriers of the same mutant gene. Half of the children will be heterozygous carriers of the trait like their parents and a quarter of the children will be affected.

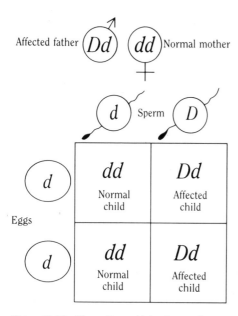

Affected father (*Dd*)   (*dd*) Normal mother

(*d*) Sperm (*D*)

Eggs

(*d*)

| *dd* | *Dd* |
| Normal child | Affected child |

(*d*)

| *dd* | *Dd* |
| Normal child | Affected child |

Figure 13-12 The pattern of inheritance of an autosomal dominant trait. One parent is affected and the other is phenotypically normal. Half of all children, on the average, will be affected and half will be phenotypically normal. This pattern assumes that the dominant trait is fully expressed in individuals carrying the mutant gene.

deleterious autosomal dominant genes to persist in a population, the expression of the mutant allele must occur later in life (as in the case of Huntington's disease) after individuals have had children, or it must not be serious enough to affect reproduction (as in the case of achondroplastic dwarfism).

Polygenic Inheritance

Although hundreds of human diseases and physical defects are caused by X-linked, autosomal recessive, or autosomal dominant mutant genes, many others have an ambiguous hereditary basis and do not obey any of the genetic rules we have discussed. Diseases such as hypertension (elevated blood pressure), diabetes, schizophrenia, and even allergies and cancers are often described as genetic diseases or as diseases that show familial patterns and thus by inference are assumed to have some genetic component. These particular human diseases, as well as such normal traits as stature, skin and hair color, intelligence, and muscularity, have a genetic basis known as **polygenic inheritance**. This means that these traits or diseases are caused by more than one gene, and the numbers, chromosomal locations, and degree of expression of these genes are unknown.

For many, if not all, polygenically inherited conditions, environmental factors are found to influence the expression of the genes. For example, height is influenced by nutrition and disease during the growing period of life, skin color is influenced by exposure to sunlight, and intelligence is influenced by the amount of stimulation, attention, and love received in the developing years. Even congenital polygenic disorders such as the more common forms of cleft lip and cleft palate are influenced by environmental factors. It has been shown that a woman's use of the tranquilizer Valium during pregnancy quadruples the chance that her baby will be born with a cleft lip or palate. Because of the observed influence of environment on the manifestation of polygenically inherited conditions, the expression *polygenic-multifactorial inheritance* is commonly used for these conditions. However, this technical phrase simply means that polygenic traits cannot be ascribed to any known genes and that it is difficult to say how much the traits are due to genetic or environmental factors.

One of the best-studied examples of polygenic inheritance is fingerprints. There are three basic types of fingerprints (Figure 13-13), and no two persons have the same set. Fingerprints are formed in the embryo by the twelfth week of pregnancy and remain unchanged throughout life. Their patterns are determined by an unknown number of genes interacting with the intrauterine environment. Strange as it may seem, even the fingerprints of identical twins, although fairly similar in their general pattern and number of ridges, are not identical, despite

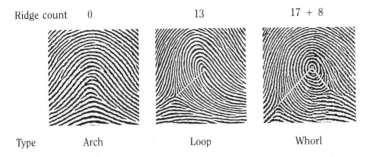

Ridge count	0	13	17 + 8
Type	Arch	Loop	Whorl

Figure 13-13 The three basic types of finger-prints: arch, loop, and whorl. The loop and whorl are further distinguished by the ridge count. The pattern of fingerprints is determined by polygenic inheritance. The environment is always important in polygenic inheritance—not even identical twins have identical fingerprints.

the fact that identical twins share identical genes and a common intrauterine environment. In fact, a finger-by-finger comparison of identical twins leads to the conclusion that during early development external factors play a major role in the formation of fingerprint patterns.

Another easily measured polygenic trait is stature, or height. Although researchers have identified numerous genes that can cause dwarfism in humans, they do not yet know how many or which genes are important in determining normal stature. As mentioned earlier, the environment plays an important role in determining stature. In one study of stature, for example, it was shown that Japanese men and women who migrated from Japan to Hawaii had children who were, on the average, 3 to 4 inches taller than they were (Table 13-2). It is clear that the genes in this population of Japanese had not changed in one generation; therefore, the increase in stature must have been due to environmental factors. It is interesting to note in Table 13-2 that the increase in stature was complete in males in one generation, whereas females continued to increase in height for at least two generations.

Presumably the factor that changed most significantly in these migrating families was their diet. Even in Japan the stature of persons born after World War II has increased as a result of increased milk in the diet. Other studies have shown that nutrition affects not only physical stature but also many other growth characteristics, including the growth of brain cells.

What can be said about the inheritance of such diseases as hypertension, diabetes, schizophrenia, and allergies? Can these diseases be regarded as hereditary in the same sense as those listed in Table 13-1? Clearly the answer is no. Most of them can be prevented, cured or controlled by appropriate changes in the affected person's environment. For example, blood pressure can often be brought under control by weight loss and diet; similarly, diabetes can be controlled in many instances by weight loss, reduced sugar consumption, and, if necessary, insulin injections.

Most people regard hereditary diseases as being incurable—a view that to a considerable degree is correct (see Box 13-2). That is why it is so important for people to understand the rules that

Table 13-2. The Effect of Environment on the Stature of Japanese-Americans in Hawaii

Generations	Stature (height in inches)	
	Women	Men
First	58.1	61.4
Second	59.8	65.6
Third	61.0	65.6

Data from: Froelich, J. W. "Migration and the Plasticity of Physique in the Japanese-Americans of Hawaii." *American Journal of Physical Anthropology*, 32 (1970) 429.

Gene Therapy:
The Limited Promise of Curing Genetic Diseases

BOX 13-2

Human hereditary diseases are not curable today. Because all genetic diseases are the result of abnormal genes that are already present in the egg or sperm at conception, there is no way to correct the defect that exists in every cell of affected individuals. All that medicine can offer at this time is treatment of the symptoms of those genetic diseases that can be managed by administering drugs, by continually supplying the missing gene product, or by continually removing the abnormal gene product. For example, PKU can be treated successfully by limiting the victim's intake of phenylalanine, through strict management of the individual's diet beginning immediately at birth. Similarly, some forms of hemophilia can be controlled by periodic injections of the clotting factor that is missing from the affected person's blood.

In order to actually cure a hereditary disease, the abnormal genes would have to be replaced by normal ones in all the cells of the affected person. This is impossible, but someday it might be possible to insert a sufficient number of normal genes (pieces of DNA) into the organs where those gene functions are normally expressed. It is quite unlikely that PKU or any of the other hereditary metabolic diseases could be helped by such *gene therapy*, since the functions of the abnormal genes in these disorders are expressed throughout the body. Moreover, in the case of many diseases the harmful effects of the abnormal genes have already occurred during development of the embryo. But for certain genetic diseases—especially blood diseases such as sickle-cell anemia and beta-thalassemia, in which one specific protein (hemoglobin in these two cases) is altered—it may be possible to reverse the defect by gene therapy.

In fact, in 1981 attempts were made to genetically change the bone marrow cells of two persons who were terminally ill with beta-thalassemia. The rationale for these experiments was that since the patients' bone marrow cells produced red blood cells with the abnormal hemoglobin molecules, inserting even a few normal hemoglobin-producing cells in the bone marrow might be sufficient to halt the effects of the disease. These first experiments in human gene therapy failed, and they were criticized by some scientists as having been premature. The critics argued that any technique of gene therapy should be shown to be effective in laboratory animals before it is tried on people—even terminally ill patients. The controversy over these experiments pointed up the moral and legal dilemmas that are inherent in any form of gene therapy attempted on people.

There is some evidence that gene modification in certain genetically caused blood diseases, and perhaps a few other genetic diseases, may eventually succeed. The DNA containing the gene for the synthesis of one portion of rabbit hemoglobin has been successfully inserted into mouse embryo cells. The embryos are implanted into female mice and give rise to baby mice that produce a small amount of rabbit hemoglobin in their red blood cells along with mouse hemoglobin. These preliminary experiments suggest that it may be possible to insert hemoglobin genes into animal cells and to have those cells become an integral and functional part of the animal. However, much remains to be learned before genes can be manipulated as readily in humans as they are in mice and rabbits.

Readings: Joan Arehart-Treichel, "Duchenne Muscular Dystrophy: A Cure in Sight?" *Science News*, January 15, 1983.
Julie Ann Miller, "Toward Gene Therapy: Lesch-Nyhan Syndrome." *Science News*, August 6, 1983.

establish whether or not a disease is hereditary. It is worth remembering that families share environments as well as genes; polygenic inheritance, in particular, involves diseases or traits in which the environment may well be the decisive factor. We cannot speak of the inheritance of polygenic diseases or traits with the same degree of assurance that we can speak of traits and diseases determined by single genes with a well-defined mode of inheritance. The significance of the distinction between heredity and environment for individuals and for society is discussed in depth in Chapter 19. Finally, it should be pointed out that modern medical techniques can detect an increasingly large number of hereditary diseases or defects in the developing embryo. The new techniques that permit diagnosis of prenatal defects and diseases create difficult decisions and ethical dilemmas for prospective parents—issues that are discussed in the next chapter.

Key Terms

hereditary (genetic) diseases   diseases resulting from abnormal chromosomes or mutant genes that are transmitted from one or both parents to offspring.

germinal mutation   a mutation that arises in the chromosomes of sex cells and that is inherited.

congenital defect   any physical abnormality that is detectable at birth.

hydrocephalus   an abnormal accumulation of fluids in the brain that causes mental retardation if it is not drained.

teratogen   an environmental agent that adversely affects the development of an embryo.

thalidomide   a teratogenic drug (formerly used as a tranquilizer) that interferes with limb development in human embryos.

family pedigree   the pattern of a trait's occurrence in family members from generation to generation.

carriers   individuals who carry a mutant gene on one chromosome but who are themselves unaffected because the allele on their other chromosome is normal.

hemophilia   a recessive sex-linked (X-linked) disease that causes afflicted persons to be "bleeders" because of the lack of a blood clotting factor.

porphyria   an autosomal dominant disease causing physical and mental symptoms and characterized by variable penetrance, which means that the mutant allele is not expressed in all individuals.

gene pool   all of the alleles in all the individuals in a population.

Huntington's disease (chorea)   an autosomal dominant disease causing neurological abnormalities; the gene is not expressed until late in life, after an individual may already have produced progeny.

achondroplasia   an autosomal dominant defect that affects cartilage growth, causing dwarfism.

polygenic inheritance   inheritance in which traits or diseases are caused by more than one gene and the number and chromosomal locations of the genes are undetermined.

Additional Reading

Holden, C. "Looking at Genes in the Workplace." *Science*, July 23, 1982.

Motulsky, A. G. "Impact of Genetic Manipulation on Society and Medicine." *Science*, January 14, 1983.

Notkins, A. L. "The Causes of Diabetes." *Scientific American*, November 1979.

Severo, R. "The Genetic Barrier: Job Benefit or Job Bias?" *The New York Times*, February 3–6, 1980.

Treichel, J. A. "Duchenne Muscular Dystrophy: A Cure in Sight." *Science News*, January 15, 1983.

Review Questions

1. Do genetic changes have anything to do with cancer?

2. What kind of mutation most likely causes PKU?

3. Why are teratogens dangerous?

4. Which of the following is a well-characterized hereditary disease: genital herpes, hemophilia, cancer, allergies, obesity?

5. Do more men or women suffer from sex-linked diseases?

6. What fraction of a couple's children will be affected if both parents are carriers of the same recessive autosomal mutation?

POLLINATION, THE ATTACHMENT OF POLLEN TO THE STIGMA IN A FLOWER. (215 ×)

14

GENETIC DISEASES

Screening and Counseling

*Scientists are real people, warts and all,
whose ambitions and morals, as well as
psychological and economic needs, span
the broad range encountered in
nearly all walks of life.*

GUNTHER S. STENT

Over a million persons are hospitalized each year in the United States for some hereditary or congenital disease. It is estimated that at least 10 of every 100 Americans suffer at some time in their lives from a serious disease that has a significant hereditary component. Among individuals with severe mental retardation—those who are incapable of functioning in society—many have an observable chromosomal abnormality, and most of the rest have either hereditary or developmental defects. Each year about 5,000 children are born with Down's syndrome, with an average life expectancy of about forty-five years. A majority of these people will sooner or later be relinquished by their families into foster homes or institutions.

Despite Nature's ability to eliminate genetically defective individuals before and at birth, about 5 percent of all live newborns have some inherited congenital defect. With recent advances in medical care, many thousands of infants now survive and grow up with handicaps that would have been fatal a few years ago.

This rather depressing presentation of the numbers, kinds, and consequences of congenital defects emphasizes the burden that these affected individuals place on the feelings of families and on the resources of society. Because hereditary diseases cannot be cured, remedial efforts must focus on their detection and prevention. Amniocentesis, a medical technique now widely used to detect and diagnose genetic or developmental defects early in pregnancy, is of particular benefit to women or couples who are at risk for bearing children with genetic defects. Genetic counseling both before and after amniocentesis can help high-risk prospective parents decide whether they want to abort the pregnancy, and it can help them to cope with their child if it is eventually born with a defect. Genetic counseling can also help them avoid having other children who may also be genetically handicapped. This chapter takes a look at what new medical techniques and genetic counseling programs can accomplish and discusses the ethical and moral problems associated with amniocentesis and abortion.

Determining Who Is at Risk

The first problem in trying to reduce the number of babies born with serious genetic diseases or defects is to determine which couples, and more particularly which pregnant women, are at a higher than average risk for giving birth to a genetically defective child. Certain factors are known to increase the risk sufficiently for some form of genetic counseling to be advisable:

1. The prospective mother is age thirty-five or over or the father is over age fifty. It is known that chromosomal abnormalities occur more frequently with increasing maternal age and to a lesser degree with increasing paternal age.

2. The couple have had a previous child with either a chromosomal abnormality or a neural tube defect or have had such an abnormality in a close relative. Such couples have an increased risk of producing another child with a defect.

3. The prospective mother is known to be or is suspected of being a carrier of a deleterious X-linked trait. It can be predicted that about half of the male children of carrier females will be affected.

4. Both prospective parents are known to be carriers of serious recessive autosomal genes, or one parent has a defective autosomal dominant gene. A high proportion of the children of such couples will be affected.

5. Polygenic-multifactorial traits are present in an earlier offspring or close relative. In such cases the risk of having an affected child is increased.

Because it is impossible to genetically screen or provide genetic counseling for everyone who decides to have children, these five criteria can help determine who is most likely to benefit from counseling and from the medical tests that are available.

THE AFP TEST FOR NEURAL TUBE DEFECTS

Among the most serious congenital defects—which may or may not involve genetic abnormalities—are those in which the neural tube (the brain and spinal cord) fails to develop properly in the embryo. Usually, if embryonic development of the nervous system is defective, other bodily and intellectual functions are also impaired. Approximately 1 in every 500 children born in the United States has a **neural tube defect**, of which there are three basic types: anencephaly, hydrocephalus, and spina bifida.

Anencephaly results from abnormal development of the brain; babies with this defect are usually stillborn or die soon after birth. As noted in Chapter 13, *hydrocephalus* is caused by the accumulation of excess cerebrospinal fluid in the brain. If the pressure exerted on the brain by this fluid is not relieved, brain damage and mental retardation result. If detected at birth or soon thereafter, this defect can usually be corrected by surgical implantation of tubes that drain the excess fluid. The most common form of neural tube defect is *spina bifida* (also discussed in Chapter 13), in which the base of the spine does not close completely during development. This congenital defect has a variety of consequences, including partial or complete paralysis from the waist down and, in severe cases, mental retardation.

Anencephaly and spina bifida can be detected in most cases by examining the pregnant woman's blood for the presence of a particular protein called **alpha-fetoprotein (AFP)**. During the development of the embryo, AFP is present in high levels in the uterus and enters the woman's bloodstream through the **placenta**, the organ that connects her blood supply with that of

The Question at the Heart of the Abortion Controversy: When Does Human Life Begin?

BOX 14-1

Killing another human being without just cause is a crime in all civilized societies. Once a live baby is born, everyone agrees that it is alive and that anything which is knowingly done to stop its life constitutes murder. But determining when (or if) an unborn fetus is alive is another issue entirely. Recent advances in medical care for premature babies, in amniocentesis, and in *in vitro* fertilization have created thorny medical, legal, and moral questions in this regard. The most vexing problem concerns abortion, which is inextricably linked to the question of when life begins.

Opinions differ enormously on this question. Some people argue that human eggs and sperm are alive and are already human because the fertilized egg will eventually develop into a human being. Other people insist that life begins at the moment of conception—when the sperm encounters the egg—since from that moment a human being begins to develop. Yet about half of all human conceptions abort spontaneously; are the pregnant women in which this happens guilty of involuntary manslaughter?

Some people argue that human life begins when the individual is born and able to survive on its own. Still others point out that even normal infants cannot survive without years of parental care. But infants born prematurely—after only six or seven months of development instead of the usual nine—can now be kept alive by heroic medical efforts, and most "preemies" survive into adulthood and live healthy, normal lives. Are "preemies" not to be considered living human beings?

There are even some people who are convinced that there is no such thing as a time when human life begins, because individual human life is part of a continuous process. They hold that the act of reproduction is merely a stage in an endless cycle of human lives. Some scientists point out, in what could be construed as support for this view, that for human life to "begin" at some moment it logically must have "stopped" at some other moment, so the whole issue of the beginning of human life is a red herring.

In yet another approach to

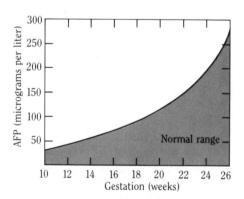

Figure 14-1 Normal alpha-fetoprotein (AFP) levels in pregnancy. Neural tube defects in the developing embryo are often signaled by abnormal levels of AFP circulating in the pregnant woman's blood. AFP levels that fall outside the normal range indicate that further diagnostic tests are needed.

the embryo. The level of AFP in the woman's blood increases during the first half of pregnancy but thereafter usually remains within certain normal limits (Figure 14-1). If the early pregnancy AFP levels are found to be abnormally high on repeated blood tests, the embryo may have some kind of neural tube defect. However, several other normal and abnormal factors can complicate the interpretation of high levels of AFP, including such a normal event as development of twins. Moreover, since the level of AFP is determined by the length of development of the embryo, the time of conception must be known with some degree of accuracy; otherwise, a correct interpretation of the AFP level may be impossible.

The use of the AFP blood test for detecting neural tube and other developmental defects is still controversial because the test's degree of accuracy leaves something to be desired. About 50 out of every 1,000 pregnant women test positively for high AFP levels. A second blood test eliminates about 30 of these positives. If the remaining 20 or so women agree to further tests, such as

the abortion problem, some people invoke the dictionary definition of the word *individual:* "an indivisible single entity." They argue that the embryo cannot be an individual because, beginning with the division of the zygote, the embryo's cells continue to divide during development. In this view, a person exists only after the umbilical cord is severed, and therefore the fetus cannot be "murdered" in a legal sense.

As the foregoing arguments indicate, the range of views about when life begins is a broad one indeed. The extremes of this range are perhaps best represented in the diametrically opposed positions presented by two scientists in testimony before a 1981 Congressional committee that was attempting to draft laws concerning abortion, amniocentesis, and genetic counseling. The French geneticist Jerome LeJeune held that abortion for any reason is immoral and illegal. He testified emphatically that "the human nature of the human being from conception to old age is not a metaphysical contention, it is

plain experimental evidence." In stark contrast was the testimony of the Yale University geneticist Leon Rosenberg: "I know of no scientific evidence which bears on the question of when 'actual human life' exists."

After all the arguments and facts have been heard, the legal issues of abortion remain unresolved. In the 1970s abortion laws in the United States were liberalized so that most women who desired to abort a fetus were able to do so legally and with medical assistance. In the 1980s strong pressure is being exerted by certain groups to revise the abortion laws so that all abortions will be illegal regardless of the circumstances, and it is this pressure that prompted the Congressional committee hearings. The legal issue of abortion has been well stated by Brian G. Zack of the Rutgers Medical School:

The issue is at what stage of development shall the entity destined to acquire the attributes of a human being be vested with

the rights and protections accorded that status. It is to the moral codes of the people that the law must turn for guidance in this matter, not to the arbitrary definitions of science. . . . Science may never make moral judgments; the law must."

As we saw in Chapter 1, even the distinction between living and nonliving material is a matter of definition, and science cannot provide the answer. Ultimately, any legal definition of when life begins must be provided through the political process, which is to say that (in a democratic society, at any rate) it must be a matter of social consensus. In our pluralistic society, a consensus on this issue is agonizingly difficult to reach, and it will never be unanimous.

Readings: Brian G. Zack, "Abortion and the Limitations of Science." Editorial in *Science,* July 17, 1981.
"Human Life" testimony, letters in *Science,* July 10, 1981.
John D. Biggers, "When Does Life Begin?" *The Sciences,* December 1981.

---

ultrasound scanning and amniocentesis, ultimately only 1 or 2 embryos out of the original 1,000 will actually be shown to have neural tube defects.

The AFP test is so simple that many pregnant women may elect to have it done. If the AFP levels are normal, they are reasonably reassured that their baby is developing normally. However, for those women who have abnormally high AFP levels, deciding whether to undergo further tests can be distressing. For those who do seek reassurance, further prenatal tests can determine whether the embryo is normal, usually in time to abort it if the parents wish to do so. The decision of whether or not to abort a pregnancy entails many moral questions and ethical judgments that are not easy for prospective parents to deal with (see Box 14-1).

ULTRASOUND SCANNING AND FETOSCOPY

Until recently, the only way to visualize an embryo in a womb was by radiological examination (x-rays). However, it is now generally

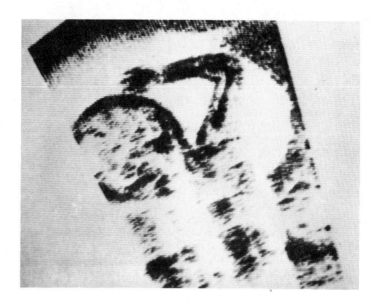

Figure 14-2 Image of a fetus obtained by ultrasound scanning. Such ultrasound scans reveal the position of the fetus and may also indicate certain physical abnormalities. (Courtesy of Willard Centerwall, University of California, Davis, Medical Center.)

recognized that x-rays are potentially harmful to the developing embryo, since such radiation may injure tissue and may induce mutations or other chromosomal abnormalities. Thus, pregnant women are now advised not to undergo abdominal x-ray examination unless the benefits to the fetus or woman clearly outweigh the risks. Fortunately, two other techniques are now available for visual examination of the fetus.

Within the last few years, **ultrasound scanning** has become an accepted and (presumably) safe technique for visualizing the fetus. Ultrasound scanning, or *sonography* as it is sometimes called, involves the use of high-frequency sound waves to outline the structural features of the embryo's body. The sound waves penetrate the womb and are reflected differently from embryonic tissues that differ in composition and density. The reflected sound waves are displayed on a screen, and the pattern of the picture is interpreted by the doctor.

Ultrasound scans can be used to detect multiple fetuses and to determine the orientation of the fetus in the womb. They can also ascertain the location of the placenta, which is particularly important if amniocentesis is to be performed. And they can gauge the embryo's head size, thereby providing an independent means of determining the age of the embryo and of ascertaining normal or abnormal brain development (Figure 14-2).

Fetoscopy is another diagnostic technique used to visually examine the developing embryo, but it is more hazardous to the mother and fetus because it requires an instrument (the fetoscope) to be inserted through a small surgical opening in the abdomen and into the amniotic sac. However, developmental defects that would go unnoticed by ultrasound examinations,

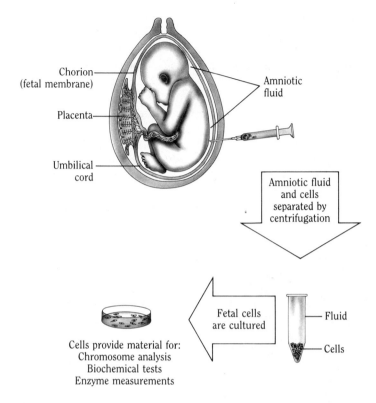

Figure 14-3 Amniocentesis. This procedure is used to test for hereditary or developmental defects in the fetus. Around the fifteenth week of pregnancy, a sample of amniotic fluid containing fetal cells is removed. The cells are cultured *in vitro* to increase their quantity. The cultured cells are then used for a karyotype (chromosome) analysis and for any biochemical or enzyme tests that are deemed necessary.

such as deformed limbs or extra toes or fingers, can be seen with the fetoscope. If necessary, this instrument can also be used to remove a sample of fetal blood that can then be analyzed for blood diseases or abnormal cellular enzyme levels. Because of the danger of harming the fetus and inducing a miscarriage, fetoscopy is used only in the case of serious conditions that cannot be diagnosed by simpler techniques.

AMNIOCENTESIS

The medical technique that has most revolutionized prenatal genetic counseling in recent years is **amniocentesis**. This procedure, which is now widely used in the analysis of hereditary diseases and developmental defects, involves removing a small quantity of fluid from the amniotic sac by means of a long needle and syringe (Figure 14-3), thereby making the amniotic fluid available for genetic and biochemical tests. The extracted amniotic fluid and cells are analyzed for abnormal enzyme levels. Even more significantly, fetal cells in the fluid can be cultured *in vitro,* and the chromosome constitution can be analyzed to determine whether any chromosomes are abnormal. Amniocentesis has been developed to the point where it is quite safe for both the pregnant woman and the embryo. More than 100 hereditary diseases can be detected by amniocentesis, and more are being added yearly.

Amniocentesis is usually performed around the fifteenth week of pregnancy, about the earliest time to safely and consistently obtain a sufficient amount of fluid and cells. At this stage of fetal development there is still enough time to culture the cells and analyze the results (three to four weeks) and to safely abort the pregnancy, should that option be elected by the prospective parents if the fetus is abnormal. Even earlier analysis would be better, however, and researchers are striving to improve the technique of amniocentesis so that the results can be obtained more quickly.

Amniocentesis is recommended only if there is reason to suspect that the embryo is at risk for a particular hereditary disease or developmental defect, although in some countries it is available to anyone who requests it. Before amniocentesis or any of the other prenatal tests are performed, the prospective parents must be told about the risks of the procedures and about the various options. The reasons for performing amniocentesis, the slight risks involved, and stipulations to the effect that the success of the cell and chromosome analyses cannot be guaranteed are spelled out in the consent form that must be signed by the patient (Figure 14-4). If the analyses reveal a genetic defect, the prospective parents receive further genetic counseling so that they will have the best possible informational basis on which to decide whether to terminate or continue the pregnancy.

Genetic Counseling

Just what is **genetic counseling**? The American Society of Human Genetics has formulated this definition:

> *Genetic counseling is a communication process concerning the risks of occurrence of a genetic disorder in a family. It involves an attempt to help the person or family comprehend the medical facts, appreciate the hereditary nature and recurrence risks in specific relatives, understand the options dealing with the risk, choose the most appropriate course of action, and make the best possible adjustment.*

Because genetic counseling may lead to the abortion of a fetus, counselors must undergo considerable training in order to be effective. They must also be sensitive and tactful in conveying information to their clients and apprising them of the various options.

Giving advice or making recommendations that affect the life of another person inevitably involves moral decisions—decisions that are influenced by the adviser's own views, beliefs, prejudices,

---

INFORMED CONSENT FOR AMNIOCENTESIS

I have been informed that my physician has recommended that a chromosome and/or biochemical test be performed on me to provide further information about my pregnancy. I understand that the cells of the fetus required as a basis for such an analysis are obtained in a procedure called transabdominal amniocentesis which involves penetration of my abdominal and uterine walls by a needle and the withdrawal of fluid from the sac in which my unborn child is contained. This fluid is called amniotic fluid.

The reasons for this procedure as well as the limitations and complications, which are listed below, have been explained to me:

1. That although transabdominal amniocentesis is a proven technique which has been used extensively and hazard to me or the fetus is considered to be small, it cannot be guaranteed that the procedure will not cause damage to me or the fetus including infection, bleeding or initiate premature labor possibly resulting in spontaneous abortion.

2. That any particular attempt to obtain amniotic fluid by transabdominal amniocentesis may be unsuccessful and repeat amniocentesis may be required.

3. That any attempt to obtain a viable tissue culture from the cells or any particular sample of amniotic fluid may be unsuccessful or the chromosome preparations and/or biochemical analyses may be of poor quality and unusable.

4. That although the likelihood of a misinterpretation of the chromosome analysis and/or biochemical analyses in this case is considered to be extremely small, a complete and correct diagnosis of the condition of the fetus based on the karyotypes or biochemical analyses obtained, cannot be guaranteed.

5. The results provided of normal chromosomes or normal biochemical status of the fetus does not eliminate the possibility that the child may have birth defects and/or mental retardation because of other disorders.

6. In the case of presently undiagnosed twins, the results provided pertain only to one of the twins.

I acknowledge that I have had an opportunity to ask questions and become fully informed about this procedure.

I request that _____ M.D., and/or his associates, assistants of his choice and personnel assigned by the hospital or medical group attempt to perform a chromosome and/or biochemical test (transabdominal amniocentesis) on me.

Date: _____ Patient: _____

Date: _____ Witness: _____

Figure 14-4 A typical consent form for amniocentesis. By signing the form, the patient attests that she has had counseling and understands the risks and limitations of the procedure.

and convictions. For example, an advisor's religious beliefs will influence the advice given, as will his or her sensitivity to the fears, social and cultural attitudes, and religious beliefs of the prospective parents.

One of the ethical dilemmas that arises in genetic counseling is that of deciding whether a genetically defective fetus should be aborted. The advice and recommendations that are conveyed to the prospective parents vary with the genetic counselor's own views on the question. Ideally, in the better counseling situations, the personal views of the counselor should not influence the individuals' or the families' decisions. Clients should feel free to arrive at their own decision after careful consideration of all medical and legal options that have been explained to them.

Even among educators who have carefully considered the issues, views differ. Joseph Fletcher, a theologian and philosopher, has stated his convictions on the matter:

I would say that it is always unjust and therefore unethical or if you like, immoral, to knowingly and deliberately victimize innocent third parties or the innocent others. And I think that to deliberately and knowingly bring a diseased defective child to the world injures society, very probably injures the family, and certainly injures the individual who is born in that condition.

This forthright statement expresses the viewpoint of many people but is disputed by others. First, it does not take into account the various degrees of disability that the child might have. Some defects are much more debilitating and handicapping than others. For example, children with hemophilia now usually survive to adulthood with medical assistance, and some lead relatively normal lives. Hereditary diseases that cause severe and irreversible mental retardation, on the other hand, are usually accompanied by serious emotional problems. They also create serious burdens, including financial ones, for the rest of the family—burdens that can cause emotional problems for other family members and can damage family life in other ways. Still, even in the case of Down's syndrome, some parents elect to have the genetically abnormal child even though they know from the results of amniocentesis that the child will be mentally retarded and suffer from the other disabilities of the syndrome. For some persons, the anguish caused by abortion would be greater than the anguish of caring for a defective child.

Over the years I have discussed with students the ethical issues involved in the birth of genetically handicapped children. Neither they nor I have ever resolved the issues, but discussing them has sharpened our understanding. Because many young people will eventually face these issues in their own lives, it is important that they come to grips with their thoughts and feelings about them. Here are one student's views:

The entire process of conception and childbirth is a very fulfilling private experience for a couple that has planned and desired a child. Modern medical technology can provide the information parents need to make intelligent decisions about the present health and future well-being of their unborn child. I do not believe anyone has the right to tell a couple whether or not they can, or even whether they should, have a child. Perhaps prospective parents should be informed about the possible risks of having a child, but once that information has been conveyed to them, it should be entirely up to them whether they will have the child or not. I do not believe that researchers, scientists, or doctors should decide what is right for that child or his parents.

Although genetic counselors strive to be objective, the counseling process is often, subtly and unintentionally, colored by the counselors' own views and prejudices. For example, prospective parents who carry recessive genes can be told that any child that they bear will have one chance in four of being abnormal, or they can be told that the odds are three to one that the child will be normal. Both statements express the same truth about the probabilities, but the prospective parents might well interpret the two statements differently. A counselor with strong antiabortion beliefs may counsel a couple in such a way that an abortion will not seem advisable, whereas a counselor with different attitudes and beliefs may advise clients in such a way that abortion seems to be the only sensible way to cope with the abnormal embryo. So, although genetic counseling begins with objective calculations of degree of risk (which may approach certainty) that an abnormal fetus is present, from that point on subjective elements—value judgments—inevitably come into play and inevitably influence what the counselor tells the clients.

Genetic Screening

Genetic counseling is essentially a communications process that informs prospective parents about the nature of genetic disorders, about their risk of having a genetically defective child, and about the options available to them in dealing with that risk. **Genetic screening**, by contrast, is a routine diagnostic procedure devised to detect those few individuals who are carriers of, or who are themselves affected by, a hereditary disease. Genetic screening applies to populations rather than to individuals.

The most widespread application of genetic screening in the United States is for phenylketonuria (PKU). The vast majority of newborn babies in the United States and in some other countries are screened for PKU by a blood test. A drop of the infant's blood is checked for the presence of excess phenylalanine, one of the twenty amino acids. If a PKU infant is detected shortly after birth, the serious effects of the hereditary disease can be prevented by providing the infant with a special diet very low in phenylalanine. About 1 in 20,000 newborns is found to have PKU. Because the PKU test is simple, reliable, and inexpensive, it is practicable to screen an entire population—millions of newborns—for this hereditary disease.

Although numerous other inherited diseases involve the metabolism of amino acids, sugars, or nucleic acids and thus could be detected by genetic screening programs, this is not done to any great degree. Some metabolic disorders, although they produce serious diseases, are so rare that it is simply too costly to screen all newborns for those very few individuals who may have inherited the diseases. For other detectable metabolic diseases, there is no known means of preventing their harmful effects, so

Table 14-1. The Probabilities of Hereditary Diseases in Particular Ethnic Groups or Nationalities

Hereditary disorder	Ethnic group with highest risk	Probability that individual is a carrier	Probability that individual's child will inherit the disease
Sickle-cell anemia	Black-Americans	1 in 10	About 1 in 400
Beta-Thalassemia	Italian-Americans	1 in 10	About 1 in 400
	Greek-Americans		
Tay-Sachs disease	Jews (Ashkenazic)	1 in 30	About 1 in 4,000
Adult lactose intolerance	Orientals	Almost all	Nearly 100%
	Blacks	Most	About 7 in 10
Phenylketonuria	No ethnic differential	1 in 80	About 1 in 20,000
Cystic fibrosis	No ethnic differential	1 in 25	About 1 in 2,500
Mediterranean fever	Armenians	1 in 45	About 1 in 8,000

there is not the same value or incentive for early detection. Before instituting any large-scale genetic screening programs, authorities must evaluate the costs of the programs, the benefits to society, and the benefits for the affected individuals.

ETHICAL ISSUES IN GENETIC SCREENING

Aside from the logistical and economic problems involved in widespread screening for hereditary diseases, other questions have been raised concerning the right of any government or society to impose genetic screening for a particular trait. Are personal freedoms and individual rights violated by mandatory screening programs? Advocates of such programs have sometimes been accused of racism or even worse, because certain hereditary diseases tend to occur more frequently in particular ethnic groups (see Table 14-1). For example, about three-fourths of all cases of Tay-Sachs disease occur among Ashkenazic Jews, beta-thalassemia in the United States is primarily a disease of Italian-Americans and Greek-Americans, and sickle-cell anemia occurs almost exclusively among black Americans. Singling out a particular hereditary disease or a particular ethnic group for mandatory genetic screening is often interpreted as governmental interference with the right of people to have children with whatever partner they choose.

It should be pointed out that although screening tests are available for the three hereditary diseases mentioned above, such tests have never been mandatory. And even for the disorders for which screening tests are now mandatory, no restrictions have been placed on the breeding practices of the affected individuals or carriers who are detected. However, the concern of some people is that someday such infringements on personal liberties will come about. The concern of others is that no adequate checks are available to curb the production of predictably defective and untreatable children, to the disadvantage both of the innocent affected ones and of society. Such are the ethical and legal dilemmas that can arise from advances in modern medical science.

Accusations of prejudice in genetic screening programs and of interference with people's right to bear children are often dealt with by comparing these programs to the way society deals with communicable diseases (ones transmitted from person to person). Few people question the right of county, state, or federal governments to prevent epidemics of serious infectious diseases such as plague, cholera, typhoid fever, and smallpox by mandating sanitation measures, pasteurization of milk products, and foreign travel immunizations. Most people would agree that the spread of sexually transmitted diseases is also a matter of public concern and that it would be desirable to reduce the incidence of syphilis, gonorrhea, and genital herpes infections. Throughout history, societies have used quarantine and other strict measures to control the spread of infectious diseases in populations.

In certain respects serious hereditary diseases can be compared to communicable infectious diseases. Genetic diseases are passed from one person to another, as are communicable diseases. Both kinds of disease are subject to environmental influences. Both differ in frequencies in different populations. Both also cause death and disability. And some of both types of disease can be successfully treated or prevented.

Why do some people accept and support efforts to eradicate infectious diseases yet angrily oppose proposals to employ similar measures to reduce the incidence of hereditary diseases? One reason is that despite feeling that laws mandating genetic screening for serious hereditary diseases are useful and beneficial, they worry that such laws may eventually lead to screening for less serious traits and even to the acceptance of the ideas of eugenics, which advocates the breeding of people with desirable traits. Eugenics is a notion that has fostered an ugly history of racism and persecution and that, in its worst manifestations, has spawned the horrors of genocide (see Chapter 19).

GENETIC SCREENING AND DOWN'S SYNDROME: AN EXAMPLE

A better understanding of the controversies surrounding genetic counseling and genetic screening can be gained by examining one example: Down's syndrome. This syndrome causes a number of physical defects and abnormalities, the most serious of which is mental retardation. All Down's individuals require special care; none will ever function normally in society. Even many couples who initially want to care for their Down's child usually cannot do so beyond the early years; they find the financial and/or emotional costs too high and end up placing the child in professional care.

About 5,000 Down's syndrome babies are born each year in

Table 14-2. Yearly Incidence and Approximate Prevalence of Some Hereditary Diseases in the U.S.

Disease	Approximate annual incidence	Approximate prevalence	Genetic cause	Detected by:
Down's syndrome	5,100	44,000	Chromosomal abnormality	Amniocentesis: chromosome analysis
Muscular dystrophy	Unknown	200,000	Hereditary: often recessive inheritance	Apparent at onset
Spina bifida and/or hydrocephalus	6,200	53,000	Hereditary and environmental	Amniocentesis; prenatal x-ray; ultrasound; maternal blood test; examination at birth
Cleft lip and/or cleft palate	4,300	71,000	Hereditary and/or environmental	Visual inspection at birth
Cystic fibrosis	2,000	10,000	Hereditary: recessive inheritance	Sweat and blood tests
Sickle-cell anemia	1,200	16,000	Hereditary: incomplete recessive—most frequent among blacks	Blood test
Hemophilia	1,200	12,400	Hereditary: sex-linked recessive inheritance	Blood test
Phenylketonuria (PKU)	310	3,100	Hereditary: recessive inheritance	Blood test at birth
Tay-Sachs disease	30	100	Hereditary: recessive inheritance—most frequent among Ashkenazic Jews	Blood and tear tests; amniocentesis
Thalassemia	70	1,000	Hereditary: incomplete recessive inheritance	Blood test
Galactosemia	70	500	Hereditary: recessive inheritance	Blood and urine tests: amniocentesis

Source: The National Foundation for Birth Defects.

the United States (Table 14-2). Down's syndrome occurs in all populations, but women over thirty-five years of age are particularly at risk for bearing a Down's child (Figure 14-5). From age thirty-five and up, a woman's risk of having a child with Down's syndrome increases markedly. A quarter of all the Down's babies born each year could be detected if all pregnant women over age thirty-five were examined by amniocentesis. In fact, it is now virtually obligatory for physicians to inform any pregnant woman over age thirty-five of her increased risk of bearing a child with Down's syndrome and of the availability of prenatal detection by amniocentesis. If a physician prefers not to have anything to do with prenatal testing that might lead to abortion, he or she must refer the at-risk woman to other persons who can provide the information. Failing to do so opens up the possibility of a financially catastrophic lawsuit if the uninformed woman subsequently has a child with Down's syndrome.

No law compels a pregnant woman over thirty-five to undergo amniocentesis, however, and many couples simply prefer to take their chances. Then, too, many people feel that what they do with their bodies and their childbearing is their business, not the state's, and so it is unlikely that laws will be enacted that will make genetic screening or amniocentesis obligatory. Thousands of Down's babies will continue to be born each year that could be prevented by genetic counseling, amniocentesis, and abortion. Perhaps as more people become aware of the serious consequences of hereditary diseases and learn that defects can be detected early enough to take remedial actions, their attitudes regarding genetic screening programs and genetic counseling will change. Such preventive measures are the only real hope in the case of many genetic diseases; it appears that only a few of these diseases may ever be curable (see Box 13-2).

Genetic Load

Most of us are born healthy and do not have to worry about hereditary diseases in ourselves or in our children. However, even the healthiest people carry three to five lethal alleles in their chromosomes and many more alleles that are deleterious but sublethal. In a population of individuals, the decrease in overall reproductive capacity that results from the presence of deleterious genes, whether expressed or not, is called the **genetic load** of the population.

A **lethal gene** is one that causes the death of the fetus before or after birth if it is expressed; a *sublethal gene* permits the individual to survive but produces some form of serious disease or disability if expressed. The reason that people survive with lethal genes and can pass them on to progeny is that the expression of recessive lethal and sublethal alleles is masked by the presence of dominant normal alleles. This occurs because, as we saw earlier in the book, people carry two copies of every gene on the pairs of homologous chromosomes (except for the X and Y chromosomes in males). For most traits, expression of a dominant normal allele is sufficient to maintain normal function and phenotype in the individual.

The number of people who are carriers for a particular deleterious mutant gene can be calculated by knowing the number of persons that are actually affected (the method of the calculation is discussed in Chapter 17). Even for hereditary diseases in which affected persons are extremely rare (1 per 1 million), the number of people who carry the recessive allele for that disease is quite frequent—about 1 per 500. For hereditary diseases that occur more frequently in the population, the number of carriers of the mutant alleles becomes quite large (Table 14-3). Altogether these deleterious genes constitute the genetic load of the human population. Obviously, as the genetic load

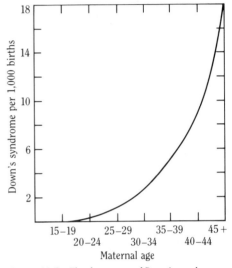

Figure 14-5 The frequency of Down's syndrome births in relation to the age of the mother. Women over age thirty-five are advised to undergo amniocentesis to be certain they are not carrying a Down's child.

Table 14-3. Frequency of Disease-Causing Recessive Alleles in a Human Population

If the number of persons affected is	The number of carriers is
1 in 10	1 in 2.3
1 in 100	1 in 5.6
1 in 1,100	1 in 16
1 in 10,000	1 in 51
1 in 100,000	1 in 159
1 in 1,000,000	1 in 501

increases, fewer individuals survive, and if it becomes sufficiently great, the survival of the population as a whole may be threatened.

Proposals for reducing the genetic load of human populations have generated heated controversy and much resistance. Although a strong case can be made for attempting to reduce the number of babies born with Down's syndrome or PKU, the arguments become less convincing when relatively minor defects such as cleft palate and club foot are included in estimates of the genetic load. And when it is suggested that in order to reduce the genetic load all individuals classified as mentally retarded not be permitted to reproduce, most sensible people become alarmed about just how "mental retardation" is to be assigned.

It is virtually impossible to draw the line between severe human hereditary defects that almost everyone agrees should be eliminated and those defects that are less severe. Deciding where to draw the line would require conflicting value judgments to be reconciled: what some people would consider an unacceptable defect others would consider quite tolerable. Moreover, trying to change the frequency of genes in the population of a modern society by selective breeding would be impracticable. As shown in Table 14-3, the number of people who carry recessive deleterious genes is enormous, even for traits that are quite rare. No counseling or breeding program could change these numbers appreciably. Furthermore, mutant genes are continually being introduced into human chromosomes by the occurrence of mutations that cannot be prevented.

The Future of Genetic Screening

Except in the case of identical twins, every person carries a unique set of genes, making him or her genetically and biochemically distinct from every other individual. The proteins produced in the cells of each organ in the body are unique to that particular organ, and many proteins vary from person to person. Because proteins are the products of genes, analyzing the kinds of proteins found in particular cells enables researchers to determine the specific alleles that different people carry in their chromosomes.

Biochemical techniques have been developed that allow the separation of more than 2,000 different human proteins from a sample containing just a few hundred cells. The separation of all of these proteins by techniques of electrophoresis is so accurate and reproducible that the position and quantity of each protein can be recorded in a computer by automated protein scanning devices (Figure 14-6). Using this modern electrophoretic technique for the biochemical analysis of proteins (and indirectly of genes), researchers can compare any protein of an individual to the normal average value of that protein in a healthy human population. If the value should be much too low or much too

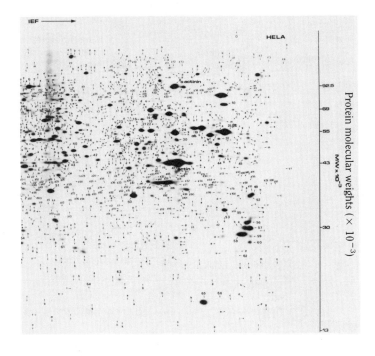

Figure 14-6 Biochemical analysis of human proteins. Two-dimensional electrophoresis can separate thousands of different proteins extracted from a few hundred human cells. Each protein is assigned a number, and its position and quantity is entered into a computer. This newly developed technique allows protein comparisons to be made between cells, individuals, or species. In this example, human cancer cells from a line called HeLa were grown *in vitro* and the proteins were analyzed. (Courtesy of J. E. Celis, Aarhus University, Aarhus, Denmark.)

high, abnormal gene expression and abnormal cellular functions can be detected.

These advances may eventually make it possible to establish a computerized file of each person's particular proteins. If this file were recorded at birth, a comparison of each newborn infant's protein profile with some standard profile might provide an early indication of abnormalities that would not otherwise be detectable. Recording a baby's protein profile may eventually become as routine as recording its sex and weight at birth. So far, the cellular functions of only a few of the protein spots shown in Figure 14-6 have been identified; most spots represent proteins (and genes) whose cellular functions are unknown. Applications of this new technique of protein identification are still several years in the future; nevertheless, the use of protein profiles for genetic screening and for overall human health assessment appears promising.

Another recent technical development permits the automated separation of individual human chromosomes, which can facilitate the mapping of genes and may also help in detecting chromosomal defects in fetal cells obtained by amniocentesis. At present, fetal cells must be cultured *in vitro* to obtain enough material to examine the chromosomes for defects. Microscopic examination can uncover only gross chromosomal abnormalities, and smaller gene defects go undetected. Having a large quantity of any one of the forty-six human chromosomes provides material for a more detailed chromosome analysis.

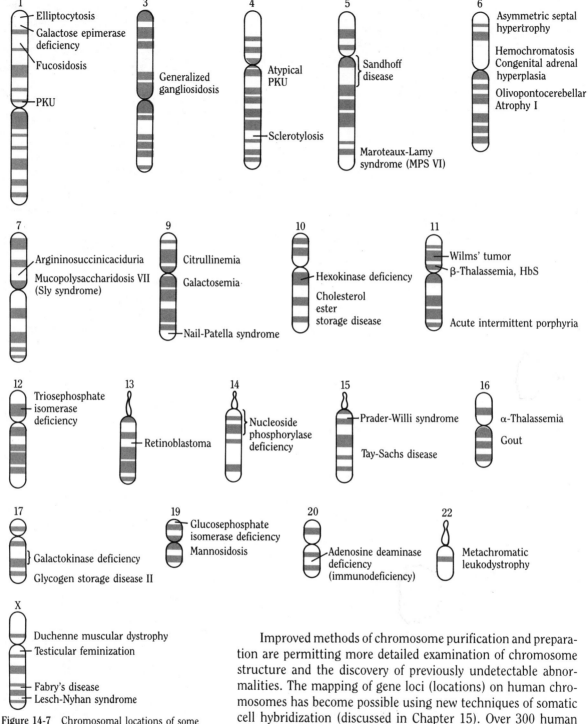

Figure 14-7 Chromosomal locations of some hereditary human diseases. Knowing the positions of the gene loci on the chromosomes facilitates genetic analysis of hereditary diseases and may lead to prevention or prenatal detection of serious diseases.

Improved methods of chromosome purification and preparation are permitting more detailed examination of chromosome structure and the discovery of previously undetectable abnormalities. The mapping of gene loci (locations) on human chromosomes has become possible using new techniques of somatic cell hybridization (discussed in Chapter 15). Over 300 human genes and approximately 50 hereditary diseases have already been assigned to specific chromosomes, and the number of gene assignments to human chromosomes is expected to increase rapidly (Figure 14-7).

Frozen Embryo Implantation In Animals

BOX 14-2

Experiments on reproduction that cannot ethically be performed with humans can nevertheless be carried out successfully using other animals, especially domestic cattle. For example, many kinds of valuable animals have been produced by the process of *frozen embryo transfer*. First, hormones that cause *superovulation* (the release of many eggs) are administered to cows. Next, these eggs are fertilized by artificial insemination. About a week after fertilization, the embryos are removed from the cows. The normal embryos are identified and frozen, in which state they may be stored almost indefinitely. Finally, they are implanted into other female cows.

Frozen embryo technology allows rare and valuable cows and bulls to be produced and raised in surrogate mother cows of much less value. Instead of the prize-winning cow producing only one or two calves each year, the number of valuable calves—that is, calves with her genotype and therefore many of her prized traits—she can produce is limited only by the number of eggs she can generate by superovulation. Moreover, it may soon be possible to determine the sex of frozen embryos so that either male or female animals can be produced, depending on the use for which the animal will be raised.

Another fascinating application of the reproductive techniques developed in cattle is propagation of rare and nearly extinct species of animals. Recently, a Holstein cow was implanted with the *in vitro* fertilized egg of a Gaur, a rare wild ox native to India that is threatened with extinction. Because of the biological closeness of Gaur and cow, the surrogate Holstein mother was able to give birth to a baby Gaur.

It may soon be possible to propagate other threatened animal species by *in vitro* fertilization using eggs and sperm from zoo animals and establishing a pregnancy in a biologically related surrogate female.

Animal husbandry will, in the future, be dramatically altered by advances in reproductive technology and frozen embryo transplantation. Rare breeds of cattle, sheep, horses, and other animals will likely become increasingly common. The sex and other characteristics of animals will be determined by appropriate choices of egg and sperm. Herd-threatening disease outbreaks will be reduced, because frozen embryos can be shipped around the world without danger of transmitting diseases—often a serious problem when live animals are shipped from one place to another. Even the construction of exotic new breeds of animals by *in vitro* fertilization and embryo transfer may soon become possible.

Reading: George E. Seidel, "Superovulation and Embryo Transfer in Cattle." *Science*, January 21, 1981.

In Vitro Fertilization and Surrogate Motherhood

On July 25, 1978 a healthy baby girl, Louise Joy Brown, was born to an elated couple in Bristol, England. What was unique about the birth was that it was the first reported instance in which a human egg had been successfully fertilized by a human sperm *in vitro* (in a test tube) and successfully reimplanted in the female uterus. The fertilized egg was inserted into the uterus of Mrs. Brown, where it attached and developed normally until Louise was born nine months later. Since that widely publicized birth, several dozen others have been reported that resulted from such **in vitro fertilization**.

This technique is used when reproductive problems in women who want to bear a child prevent their doing so unaided. The children resulting from *in vitro* fertilizations are popularly, but incorrectly, referred to as "test tube babies." The only procedure that takes place in a "test tube" is the initial fertilization of the egg; the subsequent development of the embryo can proceed only in the uterus. With techniques currently available, human embryonic development cannot continue for more than a few days (several cell divisions) outside the human body. However, in other species of animals some remarkable manipulations of embryos have been accomplished (see Box 14-2).

The first successful *in vitro* fertilization of a human egg was achieved in 1969 by three English scientists. Nine years later, two of them, R. G. Edwards and P. C. Steptoe, implanted the egg fertilized *in vitro* that became Louise Joy Brown. Human *in vitro* fertilization has been attempted many times since then, but many more of the attempts have failed than have succeeded, because the procedure is difficult. First, several eggs must be surgically removed from the woman's ovaries without damaging them. Next, the eggs must be fertilized *in vitro* and cell division must be initiated. Then a single fertilized egg must be selected and implanted into the woman's uterus and a pregnancy established. The implantation is the most difficult procedure to control because the uterus must be prepared for pregnancy by hormones that are given the woman prior to implantation. In natural pregnancies, the uterine changes are initiated by the release of the egg from the ovary and by the fertilization that usually takes place in the Fallopian tube.

Another option for infertile couples is to contract for a surrogate mother to bear them a baby. Advertisements such as the one below now appear quite frequently:

> *WANTED: surrogate mother. Infertile married couple willing to pay fee to white single woman to bear child for them. Conception to be achieved through artificial insemination supervised by medical doctors. Child to be given to married couple through adoption court. All expenses paid plus fee. All replies confidential.*

Such advertisements point up the willingness of many infertile couples to have another woman bear their child. The husband's sperm is used to artificially inseminate a fertile woman who agrees to deliver a child to the couple, generally for a fee. Although it has not yet been reported, it is technically possible for the *in vitro* fertilized egg derived from one couple to be implanted in another woman, who could go through pregnancy and give birth to the couple's child. In the future, single women might make a successful career of bearing other couples' children for a fee. Some people are morally outraged by the prospect of made-to-order babies and surrogate motherhood; others view such prospects as beneficial and logical outgrowths of the recent improvements in reproductive technology.

The ability to manipulate eggs, sperm, and genetic material

in both animals and humans has advanced tremendously in the past decade. Two things can safely be predicted for the future: (1) These techniques will continue to advance and to be used in animals and people, and (2) the ethical, legal, and social problems that accompany these advanced reproductive technologies will also increase.

_____ Key Terms _____

neural tube defects   defects in the formation of the brain or spinal cord in developing embryos.

alpha-fetoprotein (AFP)   a protein produced during embryonic development, the level of which in the blood can be used to detect neural tube defects.

placenta   the organ that connects a pregnant woman's blood supply with that of the embryo.

ultrasound scanning   production of a visible image of a fetus in the uterus by using sound waves.

fetoscopy   visual observation of a fetus by inserting an optical instrument into the uterus.

amniocentesis   a procedure in which a sample of amniotic fluid and fetal cells is removed from the uterus and examined for genetic or biochemical defects.

genetic counseling   a process of communicating the risks of having a defective child to prospective parents.

genetic screening   a diagnostic procedure that detects individuals in a population who carry defective genes.

genetic load   the decrease in fitness in a population due to the number of carriers of deleterious genes.

lethal gene   a gene that causes the death of the individual (before or after birth) if it is expressed.

in vitro **fertilization**   fertilization of one or several eggs by sperm outside of the body, such as in a test tube.

_____ Additional Reading _____

Biggers, J. D. "When Does Life Begin?" *The Sciences,* December 1981.

DeYoung, H. G. "Stalking the Silent Killers." *High Technology,* March 1983.

Epstein, C. J. and M. S. Golbus. "Prenatal Diagnosis of Genetic Diseases." *American Scientist,* November/December 1977.

Fletcher, J. *The Ethics of Genetic Control.* New York: Anchor Press, 1974.

Fuchs, F. "Genetic Amniocentesis." *Scientific American,* June 1980.

Hubbard, R. "Human Embryo and Gene Manipulation." *Science For The People,* May/June 1983.

Mitchell, M. L. and H. L. Levy, "The Current Status of Newborn Screening." *Hospital Practice,* July 1982.

Motulsky, A. G. "Brave New World: Current Approaches to Prevention, Treatment, and Research of Genetic Diseases Raise Ethical Issues." *Science,* August 23, 1974.

Schimke, R. N. "The Physician's Experience." *Hospital Practice,* March 1983.

Walters, W. A. W. and P. Singer (eds.). *Test Tube Babies.* New York: Oxford University Press, 1982.

_____ Review Questions _____

1. About what percentage of babies born in the U.S. have a hereditary defect?

2. Which congenital defects are detected by measuring the level of alpha-fetoprotein?

3. About how many hereditary diseases can be detected by amniocentesis?

4. What protein is affected in people afflicted by β-thalassemia?

5. What hereditary disease is screened for in all newborns in the U.S.?

6. Are chromosomal defects likely to be corrected by gene therapy in the future?

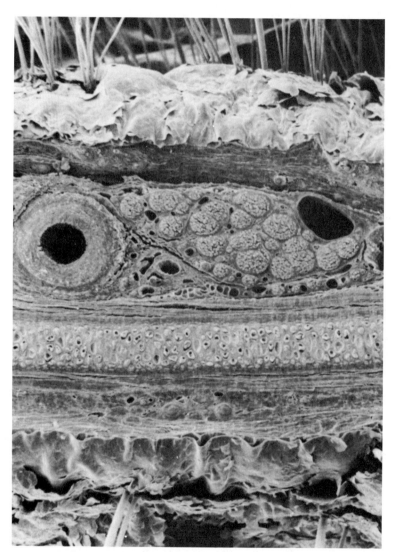

A CROSS SECTION OF HUMAN SKIN. (150 ×)

15

SOMATIC CELL GENETICS

Manipulating Genes in the Laboratory

Every true scientist should undoubtedly muster sufficient courage and integrity to resist the temptation and the habit of conformity.

ANDREI SAKHAROV

The development of any animal involves the progressive specialization of cells, particularly of the different somatic cells that form the tissues and organs of the developing embryo. As we saw in Chapter 9, each somatic cell of an adult animal contains all of the genetic information that was originally present in the fertilized egg that gave rise to that animal. Differences in the regulation of gene expression in the various cells account for the diverse specialized functions of the body's tissues and organs. However, the precise mechanisms that regulate this gene expression, particularly during development of the embryo, are still unknown.

Scientists would like to understand how somatic cells differentiate and perform special functions—how, for example, genes direct the synthesis of a human brain consisting of hundreds of billions of nerve cells that are interconnected and that function together to produce human thoughts and abilities. How are particular somatic cells of the developing embryo instructed to become bones, lungs, blood, muscles, skin, and brain? These questions cannot be answered by the study of intact animals or even of embryos.

Recognizing the impracticality of studying animal development and cell differentiation in whole animals, scientists have turned to somatic cells, which can be grown in large quantities *in vitro*. If cells from different tissues and organs can be grown and analyzed genetically, perhaps the molecular mechanisms that govern gene expression in the different kinds of somatic cells can be understood. But growing somatic cells from animals, especially humans, is not as easy as growing bacteria or yeast cells, which are much less complex and which do not differentiate into tissues with specialized functions. The nutritional requirements of growing animal cells are much more complex than those of bacteria, and animal cells tend to die when grown for more than a few generations *in vitro*. Bacteria and yeast, on the other hand, keep growing until nutrients are used up, and they resume growing when fresh nutrients are added.

This chapter discusses how somatic cells are grown *in vitro* and how they can be fused together to form hybrid cells that provide information about the location of genes on chromosomes and about the mechanisms that regulate gene expression and cellular differentiation.

Growing Somatic Cells *in vitro*

The cells in an animal's body are able to grow because they receive nutrients supplied by the blood. The blood vessels supply each living cell with the essential substances needed for growth and reproduction, and at the same time the bloodstream removes cellular waste products. Unlike bacteria, the cells of hu-

Table 15-1. A Nutrient Medium for Growing Animal Somatic Cells *in vitro*

Amino acids	Minerals	Vitamins	Other components
Alanine	$CaCl_2$	Biotin	Glucose
Arginine	$Fe(NO_3)_3$	Pantothenic acid	Sodium pyruvate (for energy) Succinic acid
Asparagine	KCl	Choline	Phenol red (to monitor pH)
Aspartic acid	KH_2PO_4	Folic acid	Streptomycin } (to prevent Penicillin } bacteria contamination)
Cystine	$MgCl_2$	Inositol	
Glutamic acid	$MgSo_4$	Nicotinamide	Blood serum (additional growth factors)
Glutamine	NaCl	Pyridoxine	
Glycine	$NaHCO_3$	Riboflavin	
Histidine	$NaHPO_4$	Thiamine	
Isoleucine			
Leucine			
Lysine			
Methionine			
Phenylalanine			
Proline			
Serine			
Threonine			
Tryptophan			
Tyrosine			
Valine			

mans and other animals cannot synthesize all the essential chemical substances—amino acids, vitamins, nucleotides—they need; many nutrients must be obtained from the food that the animal eats, digests, and absorbs into the blood. The addition of these nutrients (Table 15-1) to somatic cells growing *in vitro* will not, by itself, enable most animal cells to grow; they also require the addition of blood serum. Serum, the clear fluid obtained from blood after all of the red blood cells have been removed by filtration or centrifugation, is a complex mixture of molecules consisting of hundreds of different proteins and other growth factors.

The more highly differentiated the somatic cell—that is, the more specialized its functions—the more difficult it is to grow *in vitro*. Embryonic cells generally grow more readily than cells taken from the organs of an adult animal because they are less specialized. The inference drawn from this fact is that in fully differentiated cells some essential growth genes may be

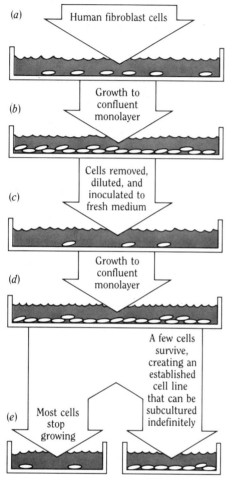

Figure 15-1 Growth of somatic cells *in vitro*. (a) A few thousand human fibroblast cells are added to a culture medium in a plastic dish. (b) The cells multiply until the surface is covered with a monolayer of cells. At this stage the cells stop dividing. (c) If a few thousand cells are removed, a fresh dish can be inoculated, and (d) again the cells will grow into a confluent monolayer. (e) After several passages of this sort, most of the cells stop growing and die. However, sometimes a few cells will survive. These cells, which are genetically changed from the original ones, can be subcultured indefinitely and are called an established cell line.

permanently switched off. However, with varying degrees of difficulty, cells from skin, muscle, heart, lung, and other tissues can be grown *in vitro*. By combining the improved techniques for growing somatic cells *in vitro* with the ability to transfer nuclei from one cell to another, some fascinating animal experiments are now possible and practical (see Box 15-1).

Growing somatic cells *in vitro* is still more an art than a science; even after all of the necessary growth substances have been determined, the cells may not grow, or they may grow quite poorly. Most normal somatic cells will not begin to grow and divide until they have attached to the surface of a glass or plastic dish. If several thousand somatic cells from human connective tissue (fibroblast cells) are placed in a Petri dish and covered with an appropriate nutrient medium, some of them will attach to the surface and begin to grow and subsequently divide by mitosis (Figure 15-1). They will continue to grow and reproduce until a more or less continuous layer only one cell thick (a confluent monolayer) covers the bottom of the dish. At this stage the cells stop growing and further cell division is inhibited by a process known as **density-dependent growth**; that is, when somatic cells growing *in vitro* physically touch other cells, the cell–cell contact prevents further cell division.

The mechanisms of density-dependent growth inhibition are unknown but are thought to reflect an important biological function of cells. Imagine what would happen in a living animal if tissues failed to stop growing when a certain organ size was reached. Mechanisms that must be dictated by the genetic information of the cells regulate the eventual size, shape, and functions of all the organs in the body. Density-dependent growth has another essential function: it regulates wound healing. When a tissue such as the skin is damaged by injury or disease, cells begin to grow and multiply rapidly in the injured area. When the healing process is complete, cell proliferation is once again switched off.

The density-dependent growth observed *in vitro* is thought to reflect the mechanisms that operate *in vivo* to regulate cell proliferation. Growth-inhibited cells can be made to resume growth by removing them from the culture dish and reinoculating them into fresh nutrients. To accomplish this, the depleted nutrient medium is poured off, along with unattached cells, and the growth-inhibited cells are gently scraped from the surface of the dish. The cells are resuspended in a solution and are again separated into single cells, either by gentle stirring or by treatment with enzymes. A few thousand cells are again placed in a Petri dish, where they attach and begin to grow if supplied with fresh nutrients. This process can be repeated several times, but eventually most somatic cells become too old (senescent) to resume growth and reproduction. Most *in vitro* cells age like *in vivo* cells and eventually die.

Bring Back the Woolly Mammoths!

BOX 15-1

With the dramatic technical advances achieved in growing somatic cells *in vitro* and in manipulating eggs that contain transplanted nuclei, a variety of science-fiction-like experiments have become possible—at least in genetics texts. Here is one of my favorite fantasy experiments that is actually a quite reasonable extension of experiments that have already been performed.

Woolly mammoths were elephantlike creatures that became extinct many thousands of years ago. Intact woolly mammoth specimens have been found in the frozen wastes of Siberia, where their bodies and cells have been preserved by the extreme cold that has prevailed there since the last Ice Age. It is impossible to revive the frozen creatures; they undoubtedly were dead for some time before they froze. However, it *is* possible that some of the somatic cells have been so well preserved that they could be made to grow *in vitro*.

Here is how a woolly mammoth might be resurrected. First, nuclei would be extracted from well-preserved mammoth somatic cells. These cells would presumably contain all the genetic information for directing the development of the entire animal. At the same time, superovulation (Box 14-3) could be induced in a female elephant and several eggs removed from her ovary. These elephant eggs would then be enucleated, and the somatic cell nuclei that had been obtained from the mammoth cells would be injected into the elephant eggs.

Next would come the delicate step (but remember that it has been successfully accomplished

with cattle, mice, and humans): An elephant egg containing a mammoth nucleus would be allowed to develop *in vitro* for a few cell divisions and would then be implanted into the uterus of a female elephant that had been prepared for pregnancy by hormone treatment. With luck, the egg would attach to the lining of the elephant's uterus, and development of a mammoth would begin according to the genetic information in the nucleus. Because the elephant and mammoth are biologically related, it is quite conceivable that development would proceed normally and that a baby mammoth would be born some months later.

Admittedly, each of the steps in this experiment would have only a slight chance of success. But a few years ago no one would have predicted that the cloning of mice would be accomplished so soon. The woolly mammoth experiment is only one of many that might be used to re-create extinct creatures or to construct unusual allophenic animals.

Reading: Roger Rapaport, "The Case of the Woolly Mammoth." *Science 80*, January/February 1980.

With care and persistence, however, some somatic cells can usually be made to grow slowly even in an aging culture, and over a long period of time—often many months—the rate of growth of these few cells will increase until they become what is called an *established cell line*. Such cells can be subcultured *in vitro* indefinitely; they do not age or die. Established cell lines are almost always genetically different from the cells that were initially used to start them. They are said to be **transformed cells**; during their months of growth *in vitro*, genetic changes have made them better adapted to growth in the synthetic medium. The karyotype of cells that have been established *in vitro* usually shows marked changes, such as loss of chromosomes, extra chromosomes, or translocation of DNA segments from one chromosome to another.

Established cell lines are valuable to researchers for a number of reasons. For example, somatic cells that have been grown *in vitro* can be frozen and stored for many years without loss of viability; the cells can be thawed and will resume growth after inoculation into a nutrient medium. In addition, samples of the frozen cells can be shipped from laboratory to laboratory so that the experimental results of different investigators can be compared, since they are using cells that, for the most part, are genetically identical. Many transformed cell lines can produce tumors when the *in vitro* grown cells are injected into animals. Studying the properties of transformed cells has thus provided insights into the process of cancer formation in people and other animals.

Fusion of Somatic Cells

Although somatic animal cells reproduce *in vitro* until the density or contact of cells prevents their further growth, but they almost never fuse with one another. Each somatic cell remains isolated from the others by its membrane. About twenty years ago, scientists began to look for ways to fuse somatic cells with one another while they were growing *in vitro*. They realized that if this could be accomplished—especially with cells that were derived from different tissues or from different species of animals—it would become possible to map genes on particular chromosomes and to learn how genes are regulated and expressed in different types of somatic cells. Performing *in vitro* genetic experiments with most species of animals is out of the question because of the long generation times involved—about twenty years in humans, for example. By contrast, genetic experiments with somatic cells *in vitro* provide an enormous advantage because somatic cells have a generation time of one or two days when grown this way.

In 1960 scientists in France discovered that an influenza-like virus, the Sendai virus, would facilitate fusion of two different lines of mouse cells. When two cells fuse, first their cytoplasms are united within a common membrane and eventually the nuclei of the cells fuse to form a single nucleus initially containing all of the chromosomes of both cells. This is not a stable condition for most fused cells, and some chromosomes are lost. As we shall see later, because of this chromosome loss, it is possible to map genes on particular chromosomes.

The first step in cell fusion is to inactivate the DNA of the Sendai viruses so that they cannot destroy the cells they infect. The inactivated viruses are still able to attach to the cell mem-

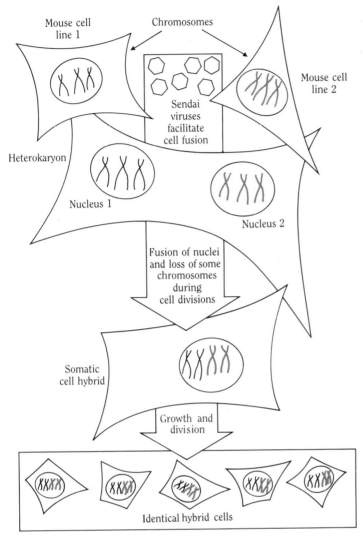

Figure 15-2 Fusion of mouse cells *in vitro* from different established lines. When two cells fuse, they first form a heterokaryon—a single cell with two nuclei containing part or all of the genetic information of the original cells. Further growth allows the nuclei to fuse, producing a somatic cell hybrid. During the fusion of the nuclei, some chromosomes are lost from one or both cell lines. If the somatic cell hybrid is genetically stable, millions of genetically identical cells can be derived from the original hybrid.

branes and to injure them sufficiently so that, in some instances at least, the membranes of two cells fuse together and the two cells become one. At this stage the fused cell is called a **heterokaryon** because it contains two distinct nuclei (visible under a microscope) and two different sets of genetic information derived from the different cells (Figure 15-2).

As heterokaryons grow, the nuclei fuse together in some of them, forming a true **somatic cell hybrid**—a single cell with a single nucleus that contains all, or at least some, of the chromosomes derived from each of the two original cells. Before the hybrid cell is actually isolated, it undergoes several mitotic and cellular divisions. During these divisions, some chromosomes fail to segregate into the daughter cells, and these chromosomes are missing from all subsequent generations of cells.

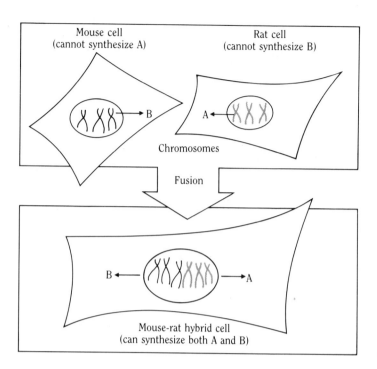

Figure 15-3 Genetic complementation. Fusion of cells from different species facilitates selection of somatic hybrid cells. In this example, the line of mouse cells chosen cannot synthesize compound A, which is essential for cell growth *in vitro*, but it does have a gene that synthesizes compound B. The line of rat cells chosen cannot synthesize compound B but is able to synthesize compound A. If the mouse and rat cells are fused and grown in a medium that lacks compounds A and B, they are able to grow and multiply, because of genetic complementation. The somatic cell hybrids synthesize both compounds A and B because they contain chromosomes from both cell lines.

Following the original discovery that mouse cells could fuse to form somatic cell hybrids, other researchers developed techniques for fusing cells and for selecting the hybrid cells from the majority of unfused cells. By 1967 scientists in the United States were successful in fusing cells from established mouse lines with those from human cell lines. Once this was accomplished, the mapping of a human gene on a particular human chromosome became a fairly routine matter. Before we get to exactly how human genes are assigned to specific human chromosomes by this method, we must understand how somatic cell hybrids are selected from a mixture of fused and unfused cells.

Genetic Complementation

One way to select the relatively infrequent hybrid cells from the rest of the unfused cells in an *in vitro* culture is by the technique of **genetic complementation**. This procedure generally begins with two cell lines, each of which lacks the capacity to synthesize some product that is essential for growing the cells *in vitro*. The genetic deficiency may occur naturally because the gene coding for the product is missing from that species, or mutant cells may be selected that have lost a particular function. In either case, the cells will grow *in vitro* only if the medium is supplemented with the necessary gene products.

Consider, for example, a line of mouse cells that cannot synthesize compound "A" (an amino acid, vitamin, or base) and a line of rat cells that cannot synthesize compound "B" (a different amino acid, vitamin, or base). The mouse cells can grow *in vitro* if compound A is added to the medium, while the rat cells can grow if supplied with compound B. If a mixture of mouse and rat cells is inoculated into a medium that is deficient in both compounds, neither kind of cell will grow or divide. However, if the mouse and rat cells are fused prior to inoculation in the medium, any somatic mouse-rat hybrid cells that contain the genes for synthesizing both compounds A and B will be able to grow (Figure 15-3). The hybrid cells grow because the missing genes and their products are complemented by functional genes present on the chromosomes. Thus, genetic complementation in somatic cells refers to provision of essential functions by genes on chromosomes from different cell lines. In a more general sense, complementation is a test applied by geneticists to determine whether mutations occur in different functional genes or affect the same functional gene on two different chromosomes.

To prove that the hybrid cell actually contains genetic information derived from both rat and mouse cells, it is necessary to examine the chromosomes present in the nucleus of the hybrid (Figure 15-4). Many hybrid somatic cells are chromosomally unstable, and chromosomes are sometimes lost as the hybrid cells undergo mitosis and cell division. However, by genetic complementation and other techniques that select for the presence of specific functions in hybrid cells, certain essential chromosomes can be maintained in the subsequent cell line.

Gene Expression in Hybrid Cells

Although somatic cells from different animal species synthesize many proteins that are structurally and functionally quite similar, these proteins have slightly different amino acid compositions, which in turn cause them to have different net electrical charges. These charge differences enable scientists to separate the various proteins from one another by electrophoresis (described in Chapter 6).

Protein extracts from cells of interest are spotted onto a gelatinous material (called a *gel*) that is electrically charged. The resulting interaction of the electrical charges on the protein's amino acids with the electrical charge on the gel causes the proteins to migrate through the gel. Their rate of movement depends on how highly charged they are: the greater the charge, the faster they move. After the proteins have been separated by electrophoresis, a chemical or dye is applied to the gel that stains the particular proteins, making it possible to identify their positions in the gel.

The proteins coded for by different alleles carried on different chromosomes will usually have different net electrical charges as a result of having different amino acid compositions.

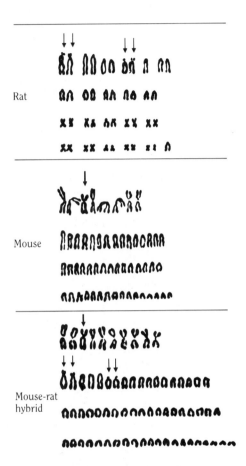

Figure 15-4   Karyotypes of chromosomes from mouse, rat, and mouse-rat hybrid cells. Marker chromosomes that have distinctive structures (indicated by arrows) are used to prove that somatic cell hybrids contain chromosomes derived from both mouse and rat cells.

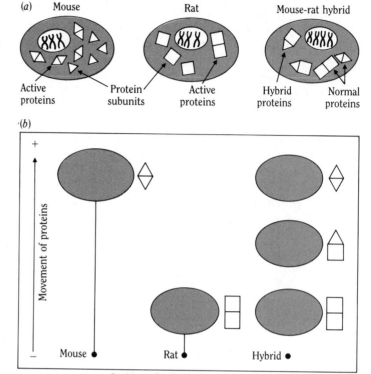

Figure 15-5 Use of electrophoresis to separate functionally similar proteins that have different structures and different net electrical charges. (a) The functionally active protein consists of two subunits that are different in mouse and rat cells. A mouse-rat hybrid cell synthesizes three kinds of proteins because of the random assembly of mouse and rat subunits. (b) The mouse protein moves faster in the electrically charged gel than the rat protein, so the two proteins can be distinguished from each other. The mouse-rat hybrid cell produces three distinguishable protein spots because of random assembly of the two types of subunits, shown alongside each protein spot.

(Even one amino acid difference may change the net electrical charge.) In the case of proteins synthesized in somatic cell hybrids, the presence of certain chromosomes can be shown by the presence of certain gene products—namely, the proteins that are characteristic of the individual chromosomes in the two cell lines.

Many functional proteins are composed of two or more subunits that join together to produce the active protein (Figure 15-5a). In addition, a protein that is synthesized by a gene on a mouse chromosome may differ in structure and net electrical charge from the analogous protein synthesized by the corresponding gene on a rat chromosome. Because of the differences in electrical charge, the mouse and rat proteins migrate at different rates in the gel and can be separated from one another by electrophoresis (Figure 15-5b). In the mouse-rat hybrid cell, not only are the characteristic mouse and rat proteins visible but a new hybrid protein is detected that is due to the random joining of mouse and rat subunit proteins. Analysis of proteins by electrophoresis provides an indirect means of demonstrating the existence of particular chromosomes in the somatic hybrid that have been derived from either the mouse or rat cells.

Other kinds of somatic hybrid cells can be used to study mechanisms of gene regulation by measuring the protein activities of various genes. For example, the nuclei in red blood cells of chickens contain all of the chicken chromosomes—that is, the complete chicken genome—yet none of the genes on the chromosomes are active because no messenger RNA is synthesized in these nuclei. Chicken red blood cells are so functionally specialized that expression of the entire genome has been shut off. These blood cells function with the proteins that were synthesized before the genes were switched off. (In human red blood cells, the nuclei actually disintegrate by the time the red blood cells are released into the bloodstream. Thus, human red blood cells do not contain any chromosomes and therefore do not grow or reproduce; new red blood cells are continuously synthesized in bone marrow and released into the blood.)

Some of the inactive genes in chicken red blood cells can be reactivated by somatic cell hybridization. If chicken red blood cells are fused with other kinds of chicken cells or even with mouse or human cells, mRNA synthesis resumes from genes on the previously inactive chromosomes. How the chicken cell genes are switched off in red blood cells and switched on in hybrid cells is not yet understood, but cell hybrids provide a means for studying the regulation of gene expression in somatic cells, since the gene switching on must be due to chemicals that are present in the hybrid cells but absent in the red blood cells.

Mapping Human Genes

When researchers constructed hybrid somatic cells by fusing mouse and human cells, they found that the mouse-human hybrids that grew best were those that contained mostly mouse chromosomes. This property of mouse-human hybrids has led to the mapping of hundreds of human genes on specific human chromosomes. Overall, about 3,000 human genes have been identified by various genetic techniques, and about 300 of these have been mapped on the twenty-three different chromosomes. Among these 300 are many of the 50 or so genes that, when present in a mutant form, cause hereditary diseases.

Several characteristics of human chromosomes aid greatly in the mapping of genes by somatic cell hybridization. One is their relatively large size. Mouse and rat chromosomes are quite small and difficult to distinguish from one another, whereas human chromosomes are generally larger and can be distinguished in cells on the basis of their size and particular structures. The three basic chromosome structures are: *telocentric,* in which the **centromere** is located at the extreme end of the chromosome; *acrocentric,* in which the centromere is located so that one pair of arms is much longer than the other; and *metacentric,* in which the centromere is located approximately in the middle of the chromosome (Figure 15-6). The size and structures of the human chromosomes allow them to be distin-

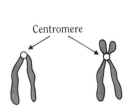

Centromere

Telocentric   Acrocentric   Metacentric

Figure 15-6 Three basic structures of chromosomes. When the centromere is located at the extreme end, the structure is telocentric. When the centromere is further in, the structure is acrocentric. When the centromere is approximately in the middle, the structure is metacentric.

The Fragile X Chromosome

BOX 15-2

One of the goals of genetics is to be able to correlate observable chromosomal abnormalities with heritable human diseases; the classic example is the correlation between an extra chromosome 21 and Down's syndrome. A *cytogeneticist* studies chromosomes by examining chromosome structure at different stages in a cell's life cycle. Metaphase is the most useful stage in which to examine chromosomes because this is when they have condensed in the nucleus and are most visible.

Some chromosomal aberrations—for example, presence of an extra chromosome, loss of an entire chromosome, or translocation of large chromosomal segments between nonhomologous chromosomes—are easily detected. But detection of chromosomal abnormalities becomes increasingly difficult as the genetic defects become physically smaller. If a few bases are added to or deleted from DNA, cytogenetic techniques cannot detect the changes. However, modifying the conditions in which the cells are grown *in vitro* may assist in the detection of small chromosome aberrations that otherwise go unnoticed.

By varying cell culture conditions, cytogeneticists in several laboratories have been able to

identify a human chromosomal abnormality known as the *fragile X chromosome,* which is found in the cells of certain mentally retarded males. In the photomicrograph shown here, the single abnormal X chromosome is located in the upper righthand corner and is indicated by an arrow. Observe that this X chromosome has two faintly stained knoblike structures that are situated at the

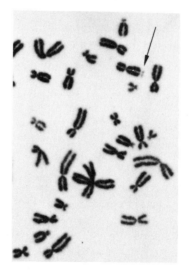

end of the long arm of the X chromosome but that are only faintly connected to it—hence the name *fragile X*. The significance of these tiny structures at the ends of the X chromosome is not known, but they are not present in normal X chromosomes. Because some (but not all) mentally retarded males have this abnormal X chromosome, it is assumed that they inherited it from their mothers. But the mothers are not mentally retarded, indicating that the other X chromosome in these women is normal and, in some way, masks the effects of the abnormal X chromosome.

Special cell growth conditions are necessary to expose the fragile X chromosome in a karyotype analysis. As better cytological techniques are developed, it may become possible to detect this fragile X chromosome in women who are carriers. Genetic counseling and amniocentesis could then be offered to those women who are at high risk for having mentally retarded male offspring. It should be emphasized that, as yet, there is no proof that having a fragile X chromosome is the cause of mental retardation in some males, since many environmental factors can also cause or contribute to mental retardation.

(Courtesy of Patricia Jacobs, University of Hawaii.)

guished from the resident chromosomes in mouse, rat, and other kinds of cells.

Recent advances in the visual examination of chromosomes have uncovered more subtle differences in human chromosome structures (see Box 15-2). Perhaps the most distinguishing feature of human chromosomes that allows them to be individually identified are the **chromosome banding** patterns. When human chromosomes are stained with certain dyes, characteristic bands

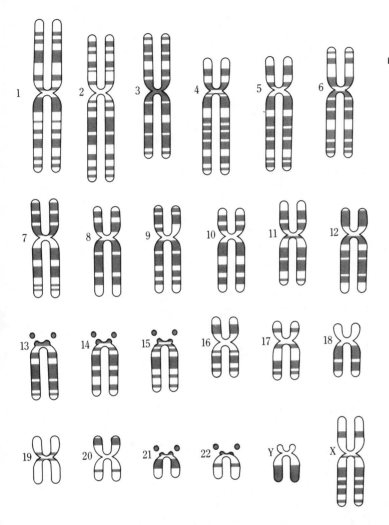

Figure 15-7 The characteristic banding patterns of individual human chromosomes that have been stained with giemsa and quinacrine dyes. The banding patterns permit identification of individual chromosomes.

appear that permit unambiguous identification of each chromosome (Figure 15-7). Two different dyes are used that produce different banding patterns: giemsa dye gives what are referred to as *G bands;* quinacrine dye produces *Q bands.* It is not known what the bands on the chromosomes represent, but they do serve to identify particular human chromosomes and regions on chromosomes.

Various techniques that employ mouse-human hybrid somatic cells have been used to map human genes. For genes that produce enzymes whose activities can be measured, the use of genetic complementation is quite effective. Suppose that the growth of the hybrid cell depends on its ability to synthesize an essential amino acid, such as proline. A mouse somatic cell is selected that is defective in proline synthesis; thus, mouse-human hybrid cells will grow only if the human chromosome that contains the genes responsible for proline synthesis is maintained in the hybrids. After fusing proline-deficient mouse cells with human cells, functional mouse-human hybrid cells are isolated that contain only one of the twenty-three human chromosomes (the remainder of the hybrid's chromosomes are derived from the mouse cell). Because these mouse-human hybrid cells are able to grow without proline added to the medium, the human chromosome they contain must carry the genes for proline synthesis. This chromosome can then be stained and identified.

In some cases the loss of a particular protein function can be correlated with the simultaneous loss of a single human chromosome in the hybrid. Consider the case of an isolated hamster-human hybrid cell that still contains eight human chromosomes in addition to the hamster chromosomes. As the hamster-human hybrid grows and divides over several cell generations, the number of human chromosomes is gradually reduced. Initially, this hamster-human hybrid also synthesizes a human protein that is easily detected by biochemical tests. Whenever a human chromosome is lost from the hybrid cells, all the cells in the culture are immediately tested to determine whether this particular protein is still present or whether it has disappeared along with the human chromosome (Figure 15-8). If a particular protein and a human chromosome disappear simultaneously from the hybrid cells, the gene that synthesizes that protein is assigned to the lost chromosome.

Normal human cells can also be fused with cancer cells to probe the genetic changes that convert a normal cell into a cancer cell (discussed in Chapter 8). Sometimes these somatic hybrids grow normally *in vitro* and are unable to initiate the formation of tumors when they are injected into an experimental animal. These normal cell–cancer cell hybrids show that whatever genes are responsible for causing a tumor are no longer active in the hybrid cell. In a way that is not yet understood, some genetic activity of the normal cell's chromosomes is able to turn off the cancer-causing genes on the cancer cell's chromosomes.

It should be evident from these brief examples that hybrid somatic cells constructed by fusing cells from different species or by fusing cells with different properties can provide insights into the organization and expression of genes—insights that cannot be obtained by other methods. The value of using somatic hybrids for human genetic studies depends on three properties of the somatic hybrid cells: (1) human chromosomes tend to segregate (be lost) from somatic hybrids when cells divide; (2) human chromosomes are readily identifiable by their size, structure, and particular banding patterns; and (3) human proteins can be separated and distinguished from similar proteins produced by other species.

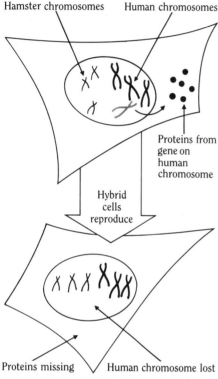

Figure 15-8 A technique for mapping human genes. In a hamster-human hybrid line of cells, loss of a human chromosome is correlated with loss of a protein activity. The gene for this protein must therefore be on the chromosome that was lost.

A Mouse With Four Parents?

Somatic cells removed from genetically different mouse embryos can be combined (not fused) *in vitro* to construct allophenic mice. This genetically useful technique involves the joining together of embryonic somatic cells derived from two genetically different mouse embryos and the reimplantation of the composite embryo into a female mouse. For example, if true-breeding mice with completely black or white fur are used in this experiment, allophenic mice can be constructed that have alternating black and white stripes in their fur coats. These stripes are indicative of their mixed parentage (Figure 15-9). Because each of the embryonic cell types is produced by two parents, allophenic mice have four biological parents instead of two.

To construct allophenic mice, a researcher begins by mating true-breeding strains of mice, in this case a male and a female having all-black coats and a male and a female having all-white coats. The embryos are removed from the two pregnant females' oviducts shortly after fertilization, generally around the eight-cell stage (Figure 15-10). The embryonic cells are then separated by means of enzymes, and the cells from the two embryos are mixed together. New embryos will form that are mixtures of the embryonic somatic cells from the black and the white embryos. Several of these embryos are then implanted into the uterus of a female mouse that has been hormonally prepared for pregnancy by having been mated with a sterile male. The implanted embryos develop normally and are born allophenic.

Why construct allophenic animals? Because they provide valuable insights into how genetic instructions govern the development and differentiation of tissues in animals. For example, by studying the patterns of the black and white fur stripes that appear in the progeny mice, researchers can determine how skin tissue is organized during embryonic development. If allophenic mice were born with half their body black and the other half white, or if they were covered all over with black and white spots, such a pattern would suggest a different developmental organization than the one that produces black and white stripes.

Other genetically instructive experiments that can be performed with allophenic mice involve the acceptance or rejection of skin grafts. Understanding how tissues are rejected has important application in medical problems that occur with organ transplants (discussed in Chapter 16). Normally, an animal will reject any foreign tissue that is grafted onto it (Figure 15-11a). However, allophenic mice will accept skin grafts from any one of their four parents (whereas the parents themselves will reject skin grafts from one another). Patches of white fur can be

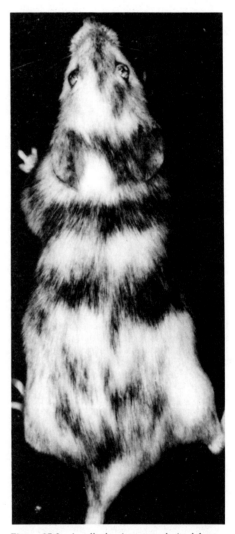

Figure 15-9 An allophenic mouse derived from four parents. (Courtesy Beatrice Mintz, Fox Chase Cancer Center, Philadelphia.)

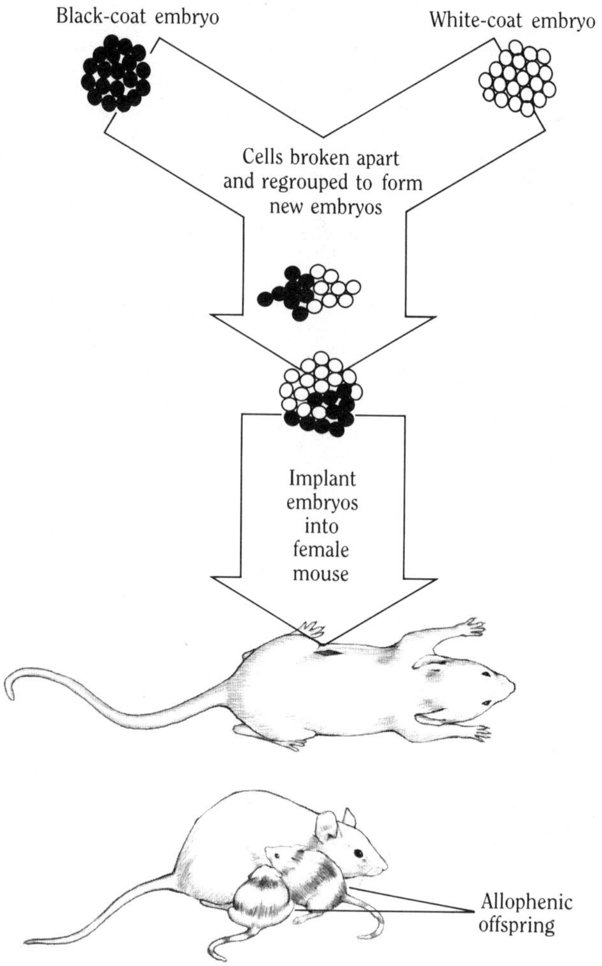

Black-coat embryo   White-coat embryo

Cells broken apart
and regrouped to form
new embryos

Implant
embryos
into
female
mouse

Allophenic
offspring

Figure 15-10 Procedure for constructing allophenic mice. Somatic cells of black-coat embryos are mixed with somatic cells of white-coat embryos. The mixed-cell embryos are then implanted into the uterus of a female mouse, which produces allophenic offspring.

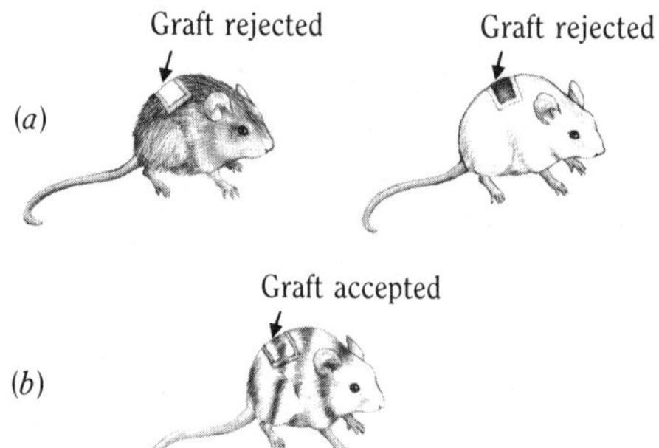

Graft rejected   Graft rejected

(a)

Graft accepted

(b)

Figure 15-11 Skin grafts in allophenic mice. (a). True-breeding black-coat mice and true-breeding white-coat mice will not accept skin grafts from one another. (b). Their allophenic offspring will accept a black graft onto a white-coat area or a white graft onto a black-coat area.

grafted onto the black patches in allophenic mice, even though the black and white cells are genetically different and the graft would normally be rejected (Figure 15-11b). This shows that immunological tolerance is a property that is acquired by cells during development and is not just due to the genetic information carried in the cells.

Mice have now been cloned by taking nuclei from somatic cells and injecting them into enucleated eggs—techniques similar to those used in the cloning of toads (see Chapter 9). So far, successful nuclei transplants have been from somatic cells of embryos, not of adult animals. From several hundred such eggs that have been implanted into female mice, a few baby mice have been born that are clones—that is, that are genetically identical. The ability to manipulate genes, cells, and embryos *in vitro* provides powerful tools for studying the mechanisms of gene expression and cellular differentiation during animal development. Future success in this area may be regarded as exciting or frightening, depending on one's point of view.

Key Terms

density-dependent growth   growth of somatic cells *in vitro* that stops when a certain cell density is reached or when neighboring cells come into contact with each other.

transformed cells   a line of established cells that can be cultured indefinitely *in vitro* and whose properties are quite different from those of the normal cells from which the established line was derived.

heterokaryon   a single cell with two separate nuclei, formed by the fusion of two cells *in vitro*.

somatic cell hybrid   a somatic cell with a single nucleus containing chromosomes from two genetically different cells that fused to form a single one.

genetic complementation   in somatic cell hybrids, the provision of essential functions by genes on different chromosomes from different cell lines. More generally, a test for whether mutations occur in different functional genes or on two different chromosomes.

centromere   a region on chromosomes that partially determines their structure and that assists in segregating them during mitosis and meiosis.

chromosome banding   the patterns of bands that become visible when chromosomes are stained with dyes.

allophenic mice   mice that are formed by mixing embryonic somatic cells from two genetically different embryos and implanting the composite embryo into the uterus of a female mouse.

Additional Reading

Burke, D. C. "The Status of Interferon." *Scientific American,* April 1977.

Ephrussi, B. and Weiss, M. C. "Hybrid Somatic Cells." *Scientific American,* April 1969.

Gold, M. "The Cells that Would Not Die." *Science 81,* April 1981.

McKusick, V. A. "The Anatomy of the Human Chromosome." *Hospital Practice,* April 1981.

Ruddle, F. H. and R. S. Kucherlapati. "Hybrid Cells and Human Genes." *Scientific American,* July 1974.

Shepard, J. F. "The Regeneration of Potato Plants from Leaf-Cell Protoplasts." *Scientific American,* May 1982.

Wolkomir, R. "Aristocratic Mice Are a Keystone of Genetic Study." *Smithsonian,* May 1983.

Review Questions

1. Which chromosomes are usually retained in mouse-human hybrid cells?

2. Do all human cells contain chromosomes?

3. How are specific human chromosomes identified?

4. How many parents contribute genetic information to allophenic mice?

5. In principle, is it possible to re-create extinct organisms?

6. Is it possible for a single cell to contain proteins derived from more than one species?

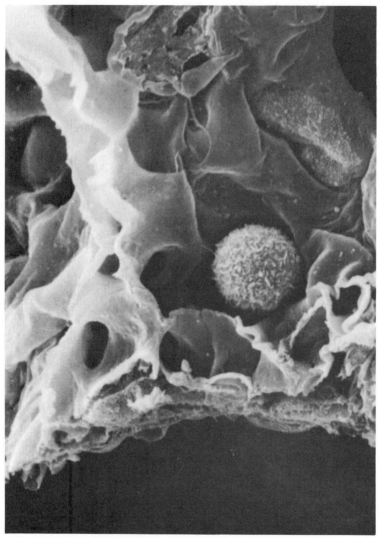

A PHAGOCYTE CELL IN THE LUNG DEVOURS HARMFUL PARTICLES THAT ARE INHALED. (1,630 ×)

16

IMMUNOGENETICS AND ANTIBODIES

Protection Against Disease

*A scientist should never attempt to judge
the value of his own achievements, whether
significant or not, but especially when not.*
ANDRE LWOFF

If geneticists were asked to name the most important and interesting organ system in the human body, many would choose the immune system. Other scientists might select the nervous system, which includes the brain, as being more interesting. But from a genetic, evolutionary, and medical point of view, the immune system has provided the most insights into the organization, expression, and evolution of hereditary information in humans and other vertebrate animals. The human immune system is capable of distinguishing the body's own cells from those that are foreign to it either because the foreign cells have invaded the body from outside or because some of the body's own cells have changed, as when cancer cells arise. Our health—even survival—depends on our immune system's ability to recognize and destroy foreign microorganisms and substances before they cause disease or otherwise interfere with the body's normal functions. When functioning properly, then, the immune system accurately discriminates between "self" and "nonself." But it is not infallible.

Like any complex biological system, the immune system may go awry. When it mistakenly attacks "self"—normal body cells—autoimmune diseases, such as certain kinds of rheumatoid arthritis or lupus erythematosus, may result. In *arthritis,* the immune system mistakenly attacks cells in the body's joints; in *lupus erythematosus,* the immune system may attack cells almost anywhere in the body. We do not yet know why these disease-causing changes occur in the functioning of the immune system of certain individuals (see Box 16-1).

Allergies are undesirable reactions of the immune system that are poorly understood. Allergic reactions of the nose, skin, lungs, eyes, and other parts of the body are caused by inappropriate responses of certain cells of the immune system to harmless foreign substances such as dust, pollen, or food. Allergies result when environmental substances stimulate particular cells of the immune system, which, in turn, produce chemical substances that cause the discomforting rashes, wheezes, and sneezes characteristic of allergic reactions.

The most remarkable aspect of the immune system from a geneticist's point of view is that the millions of specialized cells of this system are genetically programmed during embryonic development to recognize any of the innumerable kinds of viruses, bacteria, and other harmful substances that might be encountered by people during their lifetimes. Because the entire genome of a person probably does not contain more than a million genes, it is a great mystery how the enormous amount of genetic information seemingly needed by a person's immune system can be encoded in the forty-six human chromosomes. In fact, in the past few years it has become apparent that only about 300 different genes provide all the genetic information necessary

Differences Between Male and Female Immune Systems

BOX 16-1

Agammaglobulinemia, an extremely rare hereditary disease in males, was first described in 1951. Victims of this disease have no detectable plasma cells in their blood and produce almost no antibodies. They are extremely susceptible to infections and must remain in completely sterile environments lest they succumb to an infection and die. All infants are protected for some months after birth by the antibodies that they have received from their mothers, but once these antibodies disappear, the babies must synthesize their own. Because male infants affected with agammaglobulinemia cannot do so, they need the protection of a sterile chamber.

Five such hereditary immune system diseases have been described that occur almost exclusively in males. This suggests that most antibody-producing genes are carried on the X chromosome. It has been noticed that women have a lower incidence of viral and bacterial infections, have higher serum levels of antibodies,

and develop certain forms of cancer much less frequently than males. Another fact also points to female immunological superiority: Although about 6 percent more males are born than females, the male survival rate at all ages after birth is lower. By reproductive age the numbers of men and women are equal, but eventually women outlive males by almost eight years in the United States. It seems that the immune systems of women provide a better defense against diseases than those of men.

It has been suggested that the enhanced immunocompetence of females is necessary to compensate for the unavoidable immunological interactions that occur between the mother and fetus during pregnancy. Immunologically speaking, fetal tissue is foreign to the mother, yet it is not rejected by her immune system. Some sci-

entists speculate that the extra X chromosome in women may have arisen because of the need for a higher degree of immunocompetence; if important antibody-producing genes governing female-fetus immunological interactions were located on the X chromosome, they might have contributed to the natural selection of females with two X chromosomes.

The enhanced functioning of the female immune system also carries with it some disadvantages. Autoimmune diseases, in which antibodies begin to attack the person's own cells, occur much more frequently among women than among men. The incidence of lupus erythematosus, for example, is almost ten times higher in women. Thus, although women appear to derive certain advantages from their extra X chromosome and extra antibody production, they also pay a price in terms of autoimmune diseases.

Reading: W. J. Cromie. "Genetic Tradeoffs: The Hidden Costs of Survival." *Sciquest,* January 1981.

for synthesizing as many as 18 *billion* different antibodies, the proteins synthesized by certain cells of the immune system that recognize foreign substances.

How the antibody-determining genes of the immune system are organized, rearranged, and expressed during embryonic development is now partially understood and is the subject of this chapter. Also discussed here are some of the many research and medical applications that are beginning to emerge from our increased understanding of immunogenetics.

What Is the Immune System?

In vertebrates, the **immune system** consists of numerous mechanisms that recognize foreign substances and alien cells in

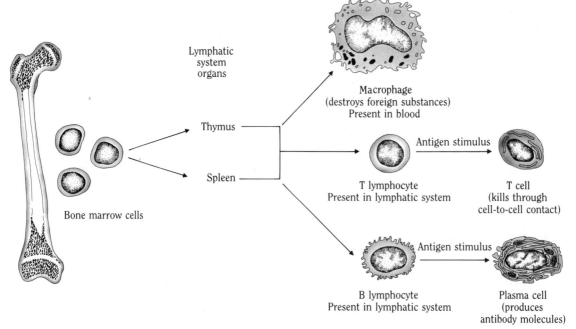

Lymphatic system organs

Thymus

Spleen

Bone marrow cells

Macrophage
(destroys foreign substances)
Present in blood

T lymphocyte
Present in lymphatic system

Antigen stimulus

T cell
(kills through
cell-to-cell contact)

B lymphocyte
Present in lymphatic system

Antigen stimulus

Plasma cell
(produces
antibody molecules)

Figure 16-1 Cells of the immune system. Bone marrow cells are processed by organs of the lymphatic system and become specialized in their functions. The macrophages, T lymphocytes, B lymphocytes, and plasma cells all play specific roles in the immune system.

the body. The diverse biological functions of the human immune system are carried out by a variety of white blood cells, or **leukocytes,** that circulate in the blood as well as in the fluid of the **lymphatic system.** This system consists of several organs, including the lymph nodes, tonsils, spleen, and thymus. The lymphatic system of humans and other vertebrate animals performs many functions, one of which is to develop immunities to potentially harmful substances and microorganisms that invade the body.

The term *white blood cell* is used to distinguish leukocytes from the **red blood cells,** which owe their color to the hemoglobin molecules that transport oxygen from the lungs to all body cells. Both red and white blood cells are manufactured in bone marrow and are secreted into the circulatory system. Specific kinds of white blood cells are secreted into the fluid of the lymphatic system as well and for this reason are referred to as **lymphocytes.** White blood cells that are synthesized in bone marrow are processed by various organs of the lymphatic system, such as the spleen, tonsils, and thymus, in such a way that they become more specialized in their functions (Figure 16-1). The tonsils are now thought to play a role in the immune system, which is why tonsils are no longer removed from children unless the tonsils are so badly infected that they are causing a serious disease.

Three main kinds of immune system cells circulate

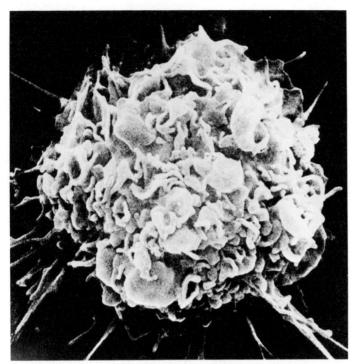

Figure 16-2 Electron micrograph of a macrophage. These large cells engulf and destroy foreign substances. Macrophage cells in the stomach, such as this one from a mouse stomach lining, protect animals from foreign particles contained in food. (Courtesy of L. Lin, Cetus Corp.)

throughout the body and protect it in various ways. The **macrophages** are large phagocytic cells that engulf and destroy viruses, bacteria, and other foreign particles (Figure 16-2). **T lymphocytes**, which are processed in the thymus, are special white blood cells that regulate the responses of other cells of the immune system in ways that are not fully understood. They also assist other cells in the recognition of foreign substances. The functions of the **B lymphocytes** are rather well defined; these white blood cells carry the genetic information for directing the synthesis of particular antibodies. When B lymphocytes detect the presence of a foreign substance or alien microorganism in the body—referred to as an **antigen**—a specific B lymphocyte cell that is genetically preprogrammed to respond to that particular antigen becomes activated. The antigen-activated B lymphocyte cell replicates and ultimately produces many **plasma cells**, which in turn synthesize the millions of antibody molecules that eventually inactivate the antigens. Because each B lymphocyte can direct the synthesis of many plasma cells, which together produce one specific kind of antibody, there must be as many different types of B lymphocytes in the body as there are different harmful antigens in nature. If a person were not capable of destroying all potentially harmful microorganisms or substances, he or she would probably not survive long (Figure 16-3).

Before we address the problem of how the body can gener-

Figure 16-3 A photo of David on his eleventh birthday, September 21, 1982. As a result of a sex-linked immune deficiency disease, David was born with no thymus gland, lymph nodes, or tonsils, and he lacks the ability to manufacture antibodies. David was born in a germ-free environment and has lived his entire life in sterile rooms either at the Baylor College of Medicine or, more recently, at home. When he goes out, he wears a "spacesuit" similar to those worn by astronauts to protect him from the environment and harmful organisms. (Courtesy of Baylor College of Medicine.)

ate every conceivable antibody using only a small portion of its genetic information, we need to examine the structure of antibody molecules—a structure that allows them to recognize and inactivate specific antigens.

Antibodies

Antibodies are a large and enormously varied class of proteins found in the blood. Each antibody is a large protein molecule consisting of four polypeptide chains that are present in identical pairs, referred to as the heavy and light chains. Various antibody proteins have specific shapes that permit them to recognize and inactivate pairs of antigens whose shapes they recognize (Figure 16-4).

The four polypeptide chains of an antibody molecule are held together by disulfide bridges. These bridges are formed by chemical bonds between sulfur atoms in certain of the amino acids located in the four polypeptide chains (Figure 16-5). The two heavy (H) chains are so named because of their larger size; the light (L) chains consist of smaller polypeptides. The two heavy chains of an antibody molecule are identical in amino acid sequence, as are the two light chains. Each light and heavy chain has a *variable* amino acid sequence at one end and a *constant* sequence in the rest of the chain. The constant sequence of amino acids usually does not change within the same class of antibodies that recognize different antigens. Antigen recognition is principally determined by the *hypervariable regions* located within the variable regions at the ends of both the light and heavy chains of each antibody. These hypervariable regions consist of amino acid sequences that are different in each different kind of antibody.

This antibody structure was elucidated by analyzing the most common class of antibodies found in the blood, but it probably applies to the other classes of antibodies as well. Immunoglobulins, the scientific term for antibody proteins, are divided into five major classes (Table 16-1). These immunoglobulin classes—IgG, IgA, IgM, IgD, and IgE—are distinguishable by their different kinds of heavy chains as well as by their specific immunological functions and their characteristic concentration levels in blood serum. Most immunoglobulin molecules found in blood serum are of the IgG type; they recognize a wide range of antigens because of their unique structures. The functions of the other immunoglobulins are less well understood; however, it has been shown that synthesis of small amounts of IgE can cause allergic reactions. Immunoglobulins are found in saliva, tears, stomach secretions, and most other body fluids as well as in the blood.

If each different heavy and light polypeptide chain in the different IgG immunoglobulins were coded for by a different

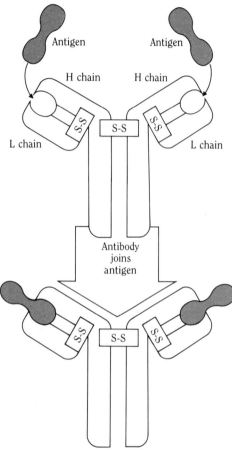

Figure 16-4 Basic antibody structure and function. There are two identical large polypeptides called heavy (H) chains and two identical small ones called light (L) chains. Each antibody recognizes a specific antigen and can inactivate a pair of them. The sulfur-sulfur (S-S) bridges hold the four chains together.

Table 16-1. The Five Major Classes of Immunoglobulin (Ig) Molecules

Immunoglobulin class	Relative amount in blood serum	Functions
IgG	1.0	Inactivates viruses, bacteria, and toxins
IgA	0.25	Inactivates toxins
IgM	0.1	Inactivates viruses, bacteria, and toxins
IgD	0.03	Normal function is unknown
IgE	0.0001	Activates allergic responses; normal function is unknown

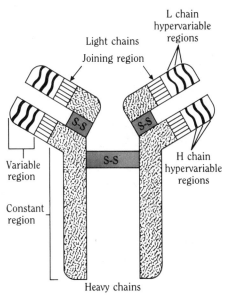

Figure 16-5 An antibody molecule. The constant, joining, and variable regions of each molecule provide its specificity. The constant region consists of a similar amino acid sequence found in different kinds of antibodies. The variable region contains amino acid sequences that are different in each different kind of antibody; within this region are segments that are hypervariable. The joining region is the location where different combinations of polypeptides are linked together.

gene, millions of genes would be needed. But, as we noted earlier, it has been determined that only about 300 human genes are devoted to the synthesis of antibodies. Thus, the puzzle that confronted immunogeneticists for many years was: How can a virtually unlimited number of different antibodies be synthesized from a small number of genes?

ANTIBODY DIVERSITY

It was by determining the amino acid sequences of many different IgG antibodies that researchers discovered the variable and constant regions of amino acids in the heavy and light polypeptide chains. Researchers also noted certain recurring patterns in the amino acid sequences of both the heavy and light chains from one antibody to the next. Although antibodies often had the same sequences of amino acids in the constant regions, different antibodies always had different sequences of amino acids in the variable regions. Eventually it became apparent that different antibodies could be constructed from a relatively few kinds of heavy and light chains whose amino acid compositions were identical in one part (the constant region) but quite different in another part (the variable region).

Using recombinant DNA techniques, researchers isolated the genes that code for light and heavy chains and determined the sequence of bases in the heavy- and light-chain genes. The amino acid sequences in the chains could then be compared with the base sequences. Each light chain, we now know, is constructed of short polypeptides synthesized from three families of genes: genes that code for the constant region, genes that code for the variable region, and genes that direct the joining of any constant region with any variable region. Because several different copies of these genes are present in the genome, when they are joined in different combinations an enormous number of different light chains can be generated.

The heavy chains are constructed in a more complex manner. The particular class of the antibody molecule and its antigen specificity are determined by four families of genes that direct the synthesis of heavy chains: genes that code for the constant region, genes that code for the variable region, genes that direct the joining of the constant and variable regions, and a fourth group of genes that enhances antibody diversity. Since genes from the four families may be joined in any combination, an enormous number of different heavy chains can be synthesized. Each antibody molecule is assembled from two identical heavy chains and two identical light chains that are constructed from polypeptides synthesized from several different genes.

The puzzle as to how the enormous diversity of antibodies arises can now be explained. Individual genes from each of the seven families of genes are brought together by recombination during development of the embryo and wind up in the chromosomes of millions of genetically different B lymphocytes. If any gene from one family can be joined to any gene from one of the other families, billions of antibody combinations are possible. Calculations show that even a few hundred different genes can be rearranged by somatic recombination in B lymphocyte cells to generate all conceivable antibodies. For example, by recombining only 100 different genes from the seven gene families that have already been identified, more than 10 billion gene combinations (and presumably the same number of different antibodies) are possible. Even limiting the mechanisms by which the different genes can be recombined would still produce millions of different antibodies. One important prediction from such a scheme is that each B lymphocyte or plasma cell derived from it should synthesize one, and only one, kind of antibody.

CLONAL SELECTION EXPLAINS DIVERSITY

In the early 1950s, Macfarlane Burnet, an Australian immunologist, proposed the **clonal selection model**, which explains how cells of the immune system develop and how the genetic information is rearranged in these cells so that they can recognize any conceivable antigen that might be encountered. Burnet's clonal selection model also accounts for the fact (although it does not account for the mechanism) that the immune system destroys foreign tissues and cells that enter the body yet does not harm the body's own cells or tissues. In other words, the clonal selection model proposes how the immune system distinguishes "self" from "nonself."

According to this model, very early in development embryonic cells that are destined to become part of the immune system undergo extensive recombination (or mutation) among the genes that will eventually be used to synthesize antibodies. From these embryonic cells of the immune system, a vast number of genetically different bone-marrow cells are generated that are capable of synthesizing all possible antibodies (Figure 16-6). During development of the embryo, any immune system cell that recognizes a normal antigen of the body is inactivated or destroyed. This ultimately leaves a population of immune system cells capable of recognizing foreign antigens only.

The clonal selection model was viewed with scepticism at the time it was proposed. In his original hypothesis Burnet suggested that the genetically different immune-system cells were the result of **hypermutability**—an extremely high rate of mutation—of these cells. That certain cells could mutate at extremely high rates seemed unreasonable to many scientists. However, Burnet's idea was close to the truth. We now know that antibody diversity results primarily from an unusually high rate of recombination among genes in immune system cells during development, but mutations also contribute to the diversity of immune cells.

If the clonal selection model is correct, a single immunocompetent cell should produce one and only one kind of antibody. This important prediction was tested in an experiment using two immunologically different strains of bacteria, strain A and strain B. A mixture of the two different strains was injected into a rat (Figure 16-7). After a week, during which the specific immune cells that reacted with the bacterial antigens proliferated, the rat's spleen was removed and homogenized into individual cells. Some of these cells were the plasma cells that synthesized antibodies directed against the injected bacterial antigens. From the homogenized mixture, individual spleen plasma cells were placed into tiny dishes, and a mixture of both strains of bacteria was added to each dish. If the spleen cell produced no antibody, both strains of bacteria were unaffected and continued to swim. If anti-A antibody was produced by the spleen cell, bacteria of type A were inactivated and stopped swimming. If anti-B antibody was produced, bacteria of type B stopped swimming. No dish was observed in which *both* strains of bacteria stopped swimming.

Because no spleen cell was able to stop the movement of both types of bacteria, this experiment showed that both types of antibodies are never produced by a single spleen cell. It is now firmly established that each plasma cell synthesizes only one kind of antibody and that the diversity of antibodies derives from the genetic diversity of immune system cells. Antibody diversity is generated by recombination and by mutations during development of the immune system. Thus, the clonal selection model proposed by Burnet about thirty years ago has been shown to be essentially correct.

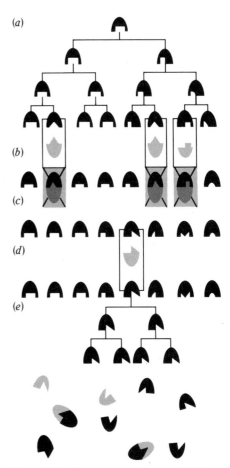

Figure 16-6 The clonal selection model. (*a*) During embryonic development, genes in immune system cells are rearranged to give every possible combination of light and heavy chains as well as constant and variable regions of each chain. (*b*) Antigens on normal body cells are recognized by cells of the immune system as the embryo develops and those cells are destroyed. (*c*) What remains are millions of immune system cells that can recognize any foreign substance. (*d*) If a foreign antigen is detected by cells of the immune system, one specific class of immune cell begins to proliferate. (*e*) These clonally derived cells synthesize antibodies that inactivate the antigen.

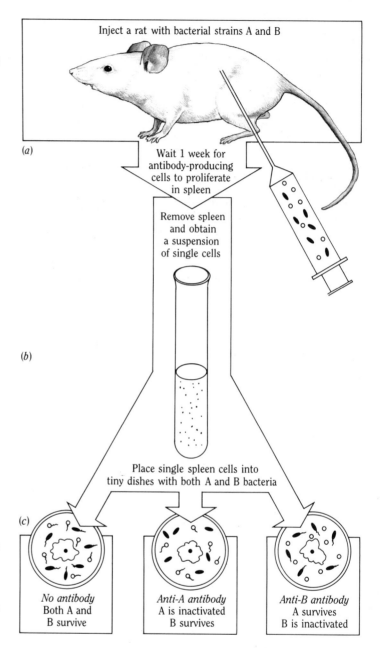

Figure 16-7 Test of the clonal selection model. The model predicts that each immune system cell can produce only one kind of antibody for one kind of antigen. (*a*) A rat is injected with two different strains of bacteria. (*b*) After one week the spleen, which contains the B lymphocytes, is removed and homogenized into single cells. (*c*) Individual cells are incubated with a mixture of both bacterial strains. Only one of the two strains is inactivated in any dish—never both. Individual spleen cells produce only one kind of antibody.

What prevents antibodies from attacking normal cells in the same way that they recognize and attack foreign cells? According to the clonal selection theory, as normal body cells and their antigens are formed, any immune system cells that might synthesize antibodies against these cells are themselves inactivated or destroyed. The mechanism that selects against these unwanted and potentially destructive immune system cells is still unknown, but whatever the mechanism, it is clearly essential to an organism's survival. The immune system of every person is able to distinguish its own tissues from those of any other individual; if cells from one person are injected or implanted into another, a strong immunological response is generated and the foreign cells are destroyed.

Why Are Tissue Transplants Rejected?

Skin transplantation experiments with inbred strains of mice led to the concept of histocompatibility; that is, the condition that determines whether a tissue (histo means "tissue") transplanted from one animal to another will be accepted or rejected by the recipient animal. Strains of mice in which all individuals are genetically identical can be produced by continuous inbreeding, such as by mating brothers and sisters generation after generation. After many generations these inbred strains of mice will accept skin grafts from one another but will reject grafts from mice of other inbred strains.

Genetic crosses can be made between male and female mice of two separate inbred strains to produce F_1 hybrid mice (essentially the same experiment that Mendel performed with his inbred strains of peas). Skin grafts between an F_1 hybrid mouse and its parent mice produce a characteristic pattern of graft acceptance and rejection (Figure 16-8). The F_1 progeny mouse will accept skin grafts from either parent, but neither the black nor white parent will accept a skin graft from its progeny. This pattern of skin graft acceptance or rejection in mice can now be explained in genetic terms.

GENETIC BASIS OF HISTOCOMPATIBILITY

Particular genes synthesize the proteins that appear on the surface of each cell and that can act as antigens. In the inbred strains of black and white mice, these genes are different and they produce different cell surface antigens. The F_1 progeny mice synthesize cell surface antigens corresponding to both the black and white parental mice since they have inherited chromosomes and genes from both parents. Thus, the cell surface antigens characteristic of both the black and white mice are a normal part of the F_1 hybrid mice; these mice are thus able to accept skin from either parent because their own antigens are compatible with both parents. However, skin from the F_1 mouse has cell surface antigens that are recognized as foreign by the immune systems of both parents; thus, each parent rejects skin grafts from the F_1 progeny. These experiments show that histocompatibility is hereditary and that cell surface antigens are determined by genes.

The genetic rules governing histocompatibility in humans are quite similar to those discovered in mice. Histocompatibility plays a critical role in the transplantation of organs from one

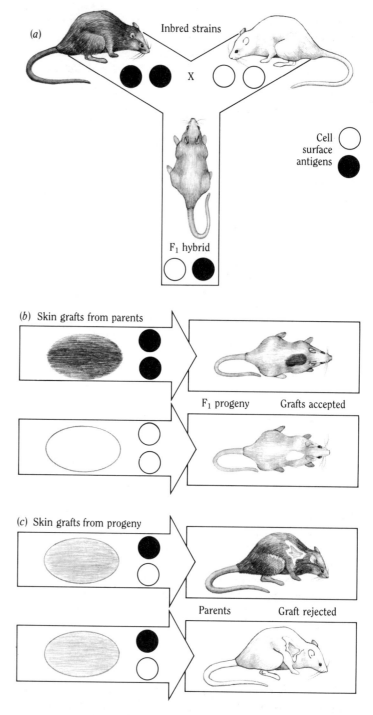

Figure 16-8 Pattern of skin graft acceptance or rejection in inbred strains of mice. (*a*) A male and female mouse of different inbred strains are mated and produce an F_1 hybrid offspring. (*b*) When skin from parental mice is grafted onto the F_1 progeny, the offspring accept the skin grafts. However, (*c*) the parents reject skin grafts from their offspring.

person to another. The more genetically related two individuals are, the more successful tissue transplants from one to the other will be. Identical twins can accept organ transplants from each other because they are genetically identical. Siblings other than identical twins share 50 percent of their genes and may be more similar in their histocompatibility antigens than unrelated persons.

When *tissues* are transplanted, the donor is usually unrelated to the recipient. In these cases, histocompatibility antigens of the recipient and of potential donors must be determined in order to obtain as close a donor-recipient match as possible. Inevitably, however, some antigens will differ between the donor and the recipient (unless they are identical twins). The recipient must therefore be given drugs that suppress the responsiveness of his or her immune system; otherwise the tissue is likely to be rejected.

HLA GENES

Human histocompatibility is determined by four families of genes that are located close to one another on human chromosome 6. These four groups of genes are called *HLA* (*h*uman *l*eukocyte *a*ntigen) genes because the antigens were originally identified on the surfaces of leukocytes. However, it is now known that the antigens coded for by three of the HLA genes (A, B, and C) are present on the surfaces of all body cells and that the D antigen appears only on certain cell types.

There is an enormous diversity of HLA genes and of HLA antigens in human populations. In fact, except in the case of identical twins, the likelihood that any two persons will have the same set of HLA genes or histocompatibility antigens is vanishingly small. Many alleles for each gene in the HLA complex exist in the human population. At least eight, and in some cases as many as forty, different alleles for each of the four HLA genes have been identified, and the number of HLA antigen combinations that can be constructed from these various alleles is in the billions.

Some years ago, it was noticed that particular HLA alleles are more commonly associated with certain human diseases than would be expected by chance (Table 16-2). These observations raise the possibility that persons at risk for particular diseases might be identified before onset of the disease by determination of their HLA type in infancy or childhood. However, except in the cases of sufferers of ankylosing spondylitis and Reiter's disease, both rare conditions, the correlation between a particular HLA allele and a particular human disease is not significant enough to warrant genetic screening. It is important to keep in mind that carrying a particular HLA allele does not *cause* a disease, although it may increase a person's susceptibility to that disease. For example, most people with the HLA B27 allele do not get ankylosing spondylitis even though they are at higher risk than the general population. Environmental factors initiate the disease process in ankylosing spondylitis as well as in many other diseases.

Any tissue that carries histocompatibility antigens that are

Table 16-2. Frequency of HLA Alleles Associated with Particular Human Diseases

Disease	HLA allele	Frequency among patients	Frequency in control population (%)	Increased risk factor
Ankylosing spondylitis	B27	71–100%	3–12%	90.1
Reiter's disease	B27	65–100	4–14	36.0
Myasthenia gravis	B8	38–65	18–31	4.4
	Dw3	23–36	14–19	2.3
Juvenile diabetes mellitus	B8	19–55	2–29	2.4
	Dw3	50	21	3.8
	Dw4	42	19	3.5
Multiple sclerosis	B7	12–46	14–30	1.7
	Dw2	47–70	15–31	4.3
Adult rheumatoid arthritis	Dw4	38–65	18–31	4.4

different from those in the animal receiving the tissue will be recognized by the recipient's immune system as foreign and will stimulate an immune response. Yet a common situation exists in which foreign tissue is not rejected: pregnancy. The embryo contains HLA genes contributed by both the mother and father. From what we now understand about the immune system, paternally derived antigens on the surfaces of the embryo's cells should be recognized as foreign by the female's immune system. But somehow the embryo is protected from being destroyed by antibodies synthesized by the mother-to-be. Despite the progress that has been made in understanding how the immune system works, this and many other aspects of the system are still not understood.

Blood Types

In the early part of this century, a blood transfusion was often fatal to the recipient because the transfused blood would clot. We now know that such clotting occurs because red blood cells have several different antigens on their surfaces that evoke strong immunological reactions in a recipient. In the early 1900s, Karl Landsteiner established the rules that allow blood transfusions to be made safely between individuals.

Reactions between transfused blood and the recipient's immune system are a special example of tissue rejection. A pair of genes consisting of pairs of three different alleles determines the red blood cell (RBC) surface antigens. The kinds of antigens in turn determine whether the transfused blood will be accepted or clotted (Table 16-3). These three alleles are referred to as the **ABO blood group**, and each person carries two of the three possible alleles: A, B, and O.

Persons with type O blood carry identical alleles that pro-

Table 16-3. How Blood Transfusions Are Determined According to ABO Blood Groups

Blood group	Genotype	Antigens on red blood cells	Transfusions cannot be accepted from	Transfusions are accepted from
O (universal donor)	OO	none	A, B, AB	O
A	AA, AO	A	B, AB	A, O
B	BB, BO	B	A, AB	B, O
AB (universal recipient)	AB	A, B	none	A, B, AB, O

duce neither antigen A nor antigen B on their RBCs; for this reason type O people are *universal blood donors*. Because neither antigen A nor antigen B is present on the RBCs, they do not provoke an immune response when transfused into persons carrying any of the four possible ABO blood groups. Persons who are type AB carry both the A and B alleles and have both antigens on their RBCs. Since these antigens are normal components of their tissues, their immune system recognizes them as "self" and does not react to them. For this reason type AB persons are *universal recipients:* they can accept blood transfusions from persons carrying any of the four ABO blood groups.

There are many other kinds of proteins on the surfaces of RBCs that can act as antigens and that differ from one person to the next. Generally, however, these other RBC antigens do not stimulate an immune response strong enough to cause the RBCs to clump together in the recipient's blood when the donor and recipient are genetically mismatched for them. The genetic basis for the ABO blood types, as well as for other less immunologically reactive blood groups, is now well understood, but prejudice and ignorance concerning human blood have had serious social consequences over the years (see Box 16-2).

Another RBC antigen, known as the **Rh factor,** is important, particularly during pregnancy. About 15 percent of all women are Rh negative—that is, their RBCs do not have the Rh antigen. If the woman's mate is Rh positive and she becomes pregnant, the child will inherit an Rh positive gene from the father and its RBCs will have the Rh antigen on their cell surfaces. The Rh positive antigens cause no problems in the first pregnancy. But during that pregnancy some of the embryo's blood cells may enter the woman's bloodstream and cause her immune system to produce antibodies directed against the Rh antigen. If the woman becomes pregnant a second time and if the embryo is again Rh positive, antibodies in her blood can enter the embryo's blood system and destroy its red blood cells. This reaction can produce severe, even fatal, anemia in the fetus.

By determining the Rh blood type of both the mother and the father, physicians can control the potential problems of Rh

Hereditary Traits Are Not Passed On In Blood

BOX 16-2

The idea that personal traits and even social status are determined by the kind of blood one has is well established in many societies. Reference is often made to the "royal blood" of kings and queens, to "pure blood lines" of nationalities and races, and to "bad blood" carried by criminals or poor people. The fact is that human red blood cells circulating in the bloodstream carry no hereditary information. Red blood cells are unique in that they are the only cells in the body that have no nuclei and therefore no chromosomes. When human red blood cells are synthesized in bone marrow, they do contain a complete human genome like all other cells, but by the time they enter the circulatory system the nuclei have disintegrated.

Many people have suffered because they have been presumed to carry "inferior" blood. In the early part of this century, some people attributed America's social and labor troubles to "undesirable blood" in workers who struggled for better working conditions. In 1924, Albert Wiggam, a popularizer of eugenic ideas (discussed in Chapter 19), wrote:

Heredity has cost America a large share of its labor troubles, its political chaos, many of its frightful riots and bombings, the doings and undoings of its undesirable citizens. Investigation proves that an enormous proportion of its undesirable citizens are descended from undesirable blood overseas. America's immigration problem is mainly a problem of blood.

Prejudices over blood were aggravated during World War II, when blood transfusions came into wide use and the American Red Cross established blood banks. During that war and for a period afterward, some American hospitals, particularly in the South, segregated blood donated by blacks from blood donated by whites. We now know that what is important in blood transfusions is matching the antigenic determinants on the cells; successful transfusions are not influenced by a person's race or ancestry. In some states blood donors are still asked to give information on their race or ethnic origin. However, this request does not arise from prejudice but rather from the fact that certain rare blood phenotypes occur only among Afro-Americans, Amerindians, Mexican Americans, and Orientals. Locating the appropriate blood for certain individuals who need transfusions can be a costly and time-consuming process if such information is not known about donors.

Reading: W. J. Miller, "Blood Groups: Why Do They Exist?" *BioScience,* September 1976.

---

incompatibility by immunological treatment. Nowadays, any woman who is Rh negative and who is carrying an Rh positive child is treated with antibodies immediately after delivery in order to destroy any Rh antigens in her bloodstream. In this way her immune system is prevented from responding to the embryo's Rh antigens. Because her blood does not subsequently produce high levels of Rh antibodies, it is quite safe for her to have additional pregnancies.

Monoclonal Antibodies

When a person is infected by viruses or bacteria, many different antibodies are produced by his or her immune system. Even simple single-celled organisms such as bacteria have dozens of antigens on their cell surfaces that can stimulate the synthesis of many different antibodies. Human blood serum contains thousands of different antibodies that are present in different concentrations and that continually change depending on a person's exposure to particular antigens. This antibody heterogeneity makes it extremely difficult to isolate and purify specific antibodies and, until only a few years ago, it was impossible to isolate and grow cells *in vitro* that produce only one particular type of antibody molecule.

In 1975 Georges Köhler and Cesar Milstein developed a way to select a particular kind of somatic hybrid cell called a **hybridoma** that produces only one kind of antibody that is specific for one kind of antigen. Such specially selected somatic hybrid cells produce **monoclonal antibodies,** or antibodies of one kind synthesized by genetically identical cells. The techniques of monoclonal antibody production have revolutionized immunological research and have already spawned a new industry that manufactures specific antibodies that can be used in diagnosing and treating diseases as well as in many other applications (see Box 16-3).

The basic procedure for producing monoclonal antibodies begins with the injection of an animal, usually a mouse, with either a specific antigen or a mixture of the antigens of interest (Figure 16-9). After several days the animal's spleen, which contains particular B lymphocytes and plasma cells that have been synthesized in response to the injected antigens, is removed and homogenized into single cells. Then the plasma cells are mixed with special cancer cells, called myelomas, that can be cultured *in vitro* indefinitely. The spleen cells are fused with the myeloma cells in a special nutrient medium that permits growth of the somatic hybrid cells but that does not permit growth of either the spleen cells or the myeloma cells by themselves (discussed in Chapter 15).

After the antibody-producing somatic hybrid cells have been grown *in vitro* for several generations, individual hybridomas producing the desired antibody are isolated from all the other kinds of antibody-producing hybridomas. The desired hybridomas are then cloned, and each clone is tested for the presence of the particular antibody that reacts only with the original antigen (or with one of the antigens if a mixture of antigens was used). If a hybridoma clone is found that produces the particular antibody, it can be grown *in vitro,* where it will continue to synthesize significant amounts of the antibody. Even larger amounts of the monoclonal antibody can be harvested from fluid in tumors that are induced in mice injected with the hybridomas. Finally, the hybridoma cells can be frozen and then revived whenever more of the particular antibody is needed.

Monoclonal antibodies are currently being used to determine people's blood types and to diagnose serious bacterial and

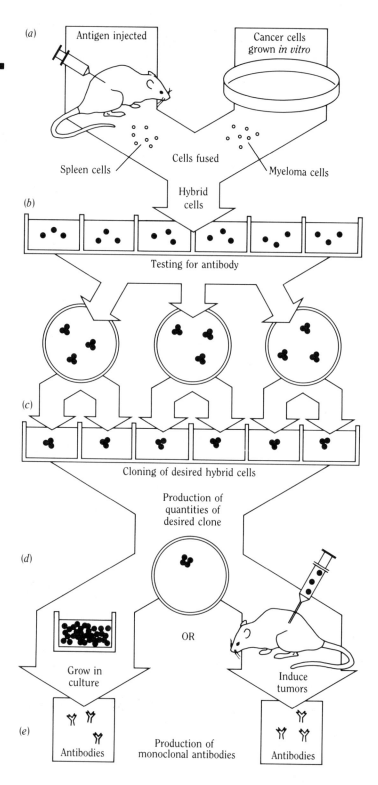

(a) Antigen injected

Cancer cells grown *in vitro*

Spleen cells

Cells fused

Myeloma cells

Hybrid cells

(b) Testing for antibody

(c) Cloning of desired hybrid cells

Production of quantities of desired clone

(d)

Grow in culture

OR

Induce tumors

(e) Antibodies

Production of monoclonal antibodies

Antibodies

Figure 16-9 The basic technique for producing monoclonal antibodies. (*a*) A mouse is injected with a specific antigen or antigens. After a week the animal's spleen cells are fused with an established line of mouse cancer cells called myeloma cells. (*b*) Somatic hybrid cells are selected and tested for production of the desired antibody. (*c*) Cells producing the antibody are cloned and finally the best clone is selected. (*d*) Quantities of these cells are grown *in vitro* or in an animal and (*e*) the desired antibodies are isolated.

Synthetic Antigens: Vaccines of the Future?

BOX 16-3

In the nearly 200 years since Edward Jenner first injected a child with cowpox viruses and showed that the injection immunized him against smallpox, vaccination has produced enormous improvements in human health. According to statistics kept by the World Health Organization, smallpox, a viral disease that formerly killed tens of millions of people, has been completely eradicated world-wide by vaccination. Polio has been reduced to a handful of cases in the United States by vaccination with either the Salk or Sabin polio vaccines. Other effective vaccines protect us from measles, cholera, typhoid fever, and other viral and bacterial diseases.

Vaccines consist of inactivated microorganisms that, while unable to cause disease, still evoke a characteristic immune response when injected into the body. However, some serious viral diseases still cannot be prevented by vaccination because the immune system does not respond to the viral vaccine preparations. Hepatitis B and rabies in humans, leukemia in cats, and hoof-and-mouth disease in cattle are among those serious diseases that cannot be prevented by vaccination. However, in the future it may be possible to use synthetic antigens instead of inactivated virus particles as vaccines for these diseases. Synthetically produced antigens may prove to be uniquely safe, effective, and inexpensive vaccines that will protect people and animals from viral and other kinds of infections.

The chemical construction of a synthetic antigen that can be used to develop a vaccine begins with identifying particular proteins on the surface of viruses that act as antigens. It appears that even a small fragment of a viral protein is sufficient to stimulate production of antibodies that can recognize and inactivate intact viruses. Knowing the amino acid sequence of a protein antigen and having some knowledge of its structure thus allow scientists to construct small peptides that mimic the antigenic properties of the complete protein.

Testing a large number of potentially useful synthetic peptide viral antigens is easily done with laboratory animals such as rabbits or mice. If a synthetic antigen is discovered that proves effective in protecting the mouse or rabbit, it may subsequently be possible to use it to develop a safe and effective human vaccine. Because mice and rabbits do not get liver disease when they are infected by the hepatitis B virus, vaccines for this viral disease must first be tested in chimpanzees—an expensive and time-consuming process. Nevertheless, based on promising results with synthetic animal vaccines, it appears likely that within a few years useful vaccines for human viral infections will be produced that consist of small, chemically synthesized peptide antigens.

Reading: R. A. Lerner et al., "The Development of Synthetic Vaccines." *Hospital Practice,* December 1981.

viral infections much more rapidly than can be done with other techniques. For example, identification of a pathogenic (disease-causing) bacterium can be accomplished in a few hours using monoclonal antibodies specific for certain pathogens, as compared to the several days necessary using other techniques. Monoclonal antibodies are also being used to identify particular hormone, pain, and other receptor sites on cells. Once the receptor sites on cells have been identified and chemically characterized, it may be possible to design more effective drugs—drugs that can interact with specific cellular receptor sites and produce the desired pharmacological effect with a minimum of undesirable side effects. For example, it may be possible to construct effective pain-relief drugs that are nonaddicting.

Some medical scientists believe that monoclonal antibodies may prove useful for carrying drugs to cancers in hard-to-treat areas of the body. Because most cancer cells have cell surface antigens that can be distinguished from those on normal cells, it is hoped that monoclonal antibodies can eventually be generated that will recognize specific antigens on cancer cells. If powerful drugs were chemically attached to monoclonal antibodies, perhaps they could be carried directly to the cancer cells and could destroy them without causing harm to other body cells. However, much still remains to be learned about the isolation and properties of monoclonal antibodies before they can be used effectively and safely to treat human diseases.

Because most of the monoclonal antibodies to date have been isolated from mouse cells, it is likely that these mouse-derived proteins would be recognized as foreign antigens if they were injected into people. The problem now is how to develop human monoclonal antibodies by using procedures that have proved successful in mice.

Key Terms

autoimmune disease   a disease such as some forms of rheumatoid arthritis or lupus erythematosus, in which the immune system reacts to normal body cells and damages or destroys normal tissues.

immune system   the numerous mechanisms in vertebrates that recognize and eliminate foreign substances and alien cells.

leukocyte   any kind of white blood cell

lymphatic system   a complex organ system consisting of the lymph nodes, tonsils, spleen, and thymus; it performs many functions, one of which is immunity.

red blood cell   a cell in the blood that contains hemoglobin and that transports oxygen.

lymphocyte   a particular type of white blood cell that is associated with the immune system.

macrophages   large phagocytic white blood cells that engulf and destroy foreign particles.

T lymphocyte   a white blood cell that regulates responses by other cells of the immune system.

B lymphocyte   a white blood cell that carries the genetic information for synthesis of particular kinds of antibodies.

antigen   any foreign substance that stimulates an antibody response in an animal.

plasma cell   a specialized cell derived from B lymphocytes that synthesizes antibodies.

antibody   an immunoglobulin molecule capable of combining with specific antigens and thereby inactivating them.

immunoglobulin   the scientific term for antibody.

clonal selection model   a theory explaining how antibody diversity is generated during embryonic development and how the immune system distinguishes "self" from "nonself."

hypermutability   a rate of mutation much higher than is normally observed.

histocompatibility   the condition that determines whether tissue transplanted from one animal to another will be accepted or rejected.

ABO blood group   three alleles that direct the synthesis of red blood cell antigens, which determine whether transfused blood will be accepted or rejected by a recipient.

Rh factor   a red blood cell antigen that can cause anemia in a fetus.

hybridoma   a somatic cell hybrid between a B lymphocyte and a cancer cell; it produces monoclonal antibodies.

monoclonal antibody   an antibody of a single type that is produced by genetically identical somatic hybrid cells (clones).

352

Additional Reading

Alper, J. "Vaccine Research Gets New Boost: Safer, Wider-Ranging Vaccines May Soon Eliminate Many of Today's Leading Infectious Killers." *High Technology,* April 1983.

Buisseret, P. D. "Allergy." *Scientific American,* August 1982.

Chisholm, R. "On the Trail of the Magic Bullet: Monoclonal Antibodies Promise Perfectly Targeted Chemicals." *High Technology,* January 1983.

Cromie, W. J. "Genetic Tradeoffs: The Hidden Costs of Survival." *Sciquest,* January 1981.

Edelman, G. M. "The Structure and Function of Antibodies." *Scientific American,* August 1970.

Hapson, J. L. "Battle at the Isle of Self." *Science 80,* March/April 1980.

Leder, P. "Genetic Control of Immunoglobulin Production." *Hospital Practice,* February 1983.

Miller, W. J. "Blood Groups: Why Do They Exist?" *BioScience,* September 1976.

Rose, N. R. "Autoimmune Diseases." *Scientific American,* February 1981.

Yelton, D. E. and M. D. Scharff. "Monoclonal Antibodies." *American Scientist,* September/October 1980.

Review Questions

1. Is it possible for a person to survive without functional antibodies?

2. What class of antibodies produces allergic reactions?

3. How many different antigens are recognized by an antibody molecule?

4. What function of DNA is primarily responsible for antibody diversity?

5. What kinds of cells produce antibodies?

6. How many genes determine histocompatibility in humans? How many genes determine ABO blood type?

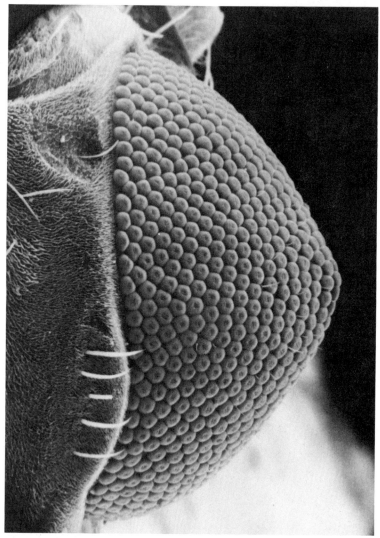

THE EYE OF A MOTH IS CONSTRUCTED OF HEXAGONAL-SHAPED FACETS. (280 ×)

17

EVOLUTION

Natural Selection and Populations

Evolution is a mixture of chance and necessity—chance at the level of variation, necessity in the working of selection.

STEPHEN JAY GOULD

Figure 17-1   Charles Darwin. Darwin's works established the scientific basis of evolution. (Courtesy of National Library of Medicine.)

Probably no area of biology arouses more interest or generates more controversy than evolution, which deals with the origin of living organisms, the genetic diversity of individuals and species, and the biological history of our planet. Evolution is concerned both with existing organisms and with those that have become extinct but whose earlier existence can be documented from fossil records. Evolution is the branch of biology that attempts to explain how the tens or possibly hundreds of millions of different species of plants, animals, and microorganisms arose during the several billion years of earth's history. And it is the various forces of evolution that cause species to adapt, to evolve (meaning to change), and to eventually become extinct. Evolution and genetics are inextricably interconnected because the biological changes that occur in organisms over time are due to changes in their hereditary information—changes in their genes.

Although various evolutionary ideas have been proposed over the centuries, the first generally accepted scientific explanation for the origin, adaptation, diversification, selection, and extinction of species was provided by the detailed observations and brilliant insights of the English naturalist Charles Darwin (Figure 17-1). In his famous book *On the Origin of Species by Means of Natural Selection,* published in 1859, Darwin provided a wealth of documentation for evolutionary change by the process he called **natural selection**. Darwin argued persuasively that individuals who are best suited genetically to survive and reproduce in particular environments are the ones that pass on their genetic information to succeeding generations (an idea often referred to as "survival of the fittest"). He believed that natural selection is the primary force of evolution—that it is responsible for the origin of new species, for their development over long periods of time, and for their eventual extinction.

Another of Darwin's important insights was that in each generation a certain amount of heritable variability is introduced into individual organisms and that natural selection operates on this genetic variability to continuously select new types. Darwin was unaware of Mendel's experiments, so he had no way of incorporating Mendel's discoveries on the random segregation of chromosomes into his evolutionary theories. Nor could Darwin (or anyone else at that time) know that ultimately all genetic diversity can be traced to mutations that arise spontaneously and randomly in DNA. But Darwin did not jump to his controversial conclusions about evolution; quite the contrary. Because he realized that his ideas were novel and controversial, he spent twenty-eight years mulling over his data, from the time he sailed on the *H.M.S. Beagle* in 1831 until he finally published his pioneering work on evolution in 1859.

The unification of Darwin's ideas with those of Mendel and the geneticists who followed eventually resulted in what is now called the **modern synthesis** of evolutionary theory (also referred to as *neo-Darwinism*). This modern evolutionary synthesis unites the ideas of natural and sexual selection with those that elucidate the basic hereditary properties of genes and chromosomes. And—quite remarkably—it accounts for both the continuity of genetic information generation after generation and the genetic diversity of individuals and species.

This modern evolutionary synthesis, which began in the 1930s and has continued to the present day, is one of the truly great accomplishments of the biological sciences. Indeed, it is the major unifying principle of all biology. On the one hand, molecular genetics explains the overall continuity of the hereditary information by showing how each DNA molecule can be an exact copy of the one from which it was replicated. On the other hand, evolutionary and population genetics explain how ever-so-slight genetic changes (mutations) can provide the variability that is essential for the ongoing natural selection of new types of individuals, new species, and new populations of organisms.

Evolutionists attempt to reconstruct biological history, and it is in this area that much of the controversy arises, especially in the recurrent confrontations between creationists and evolutionary scientists about what the schools should teach regarding biological history and human origins (see Box 17-1). Because it is an indisputable fact that humans along with an as yet undetermined number of millions of other species presently inhabit the earth, it is important to understand how and when they arose. Estimates from fossil records suggest that a hundred times the present number of species lived in the past and became extinct.

To appreciate the facts, theories, speculations, and controversies of evolution, it is necessary to first understand how evolutionary data are obtained and analyzed and how evolutionary hypotheses are tested. Only then can one decide which information and which ideas make sense and which do not. The discussion in this and the following chapter deals with the basic scientific discoveries of evolutionary biology and with some of the speculations that the scientific facts engender.

What Is Evolution?

The word *evolution* simply means an unfolding, a process of development and change. Evolution applies to the formation and development of the entire physical universe—atoms, molecules, mountains, planets, stars, galaxies—as well as to living organisms. Geneticists, however, limit their concerns to biological evolution (generally termed evolution for short) and generally define it in a stricter sense.

One of the ideas introduced by the so-called scientific creationists—those who believe in neither the facts nor the theory of evolution—is that of the "Master Designer." This change in term from God to Master Designer is supposed to make the creationists' views more scientific and acceptable to the general public. However, an all-powerful, all-knowing Master Designer still sounds like God to most people. In fact, the argument for the existence of God based on the perfection of His designs was advanced by a theologian, William Paley, in a book published several years before Darwin's *Origin of Species*.

Gary Parker is a scientific creationist and one of the authors of the biology books published by the Creation Research Institute, located in San Diego, California, of which he is a member. In his lectures Parker attempts to debunk evolution by using a can of Dr. Pepper as an example. In his view, "All the time, all the chance in the world, all the natural reactions of aluminum with other kinds of elements are never going to result in a little blue can with 'Dr. Pepper' written on it." But cans are not cells, and this fanciful criticism of the evolutionists' view is no more scientific than the explanations given in the *Berry's World* comic strip reprinted here.

Who Created Dr. Pepper?

BOX 17-1

All members of the Creation Research Society subscribe to a Statement of Belief whose first three paragraphs read as follows:

1. The Bible is the written Word of God, and because we believe it to be inspired throughout, all of its assertions are historically and scientifically true.

2. All basic types of living things, including man, were made by direct creative acts of God during Creation Week as described in Genesis. Whatever biological changes have occurred since Creation have accomplished only changes within the original created kinds.

3. The great Flood described in Genesis, commonly referred to as the Noachian Deluge, was an historical event, worldwide in its extent and effect.

The right of any individual to *hold* such beliefs is not at issue. Rather, the issue in many parts of the United States is whether these beliefs should be *taught in the classroom*. It should be pointed out that these creationist beliefs are not embraced by most other equally devout Christians, who do not believe these ideas should be taught in public schools. Nor do the majority of Catholics, Presbyterians, Methodists, Lutherans, and members of other Christian denominations find any contradiction between their religious beliefs and the theory of evolution.

Science constructs its theories—its views of how nature works—from observations and experiments that can be substantiated and repeated by other scientists. The hypotheses and theories of science are continually being modified, verified, or falsified by new information. Beliefs that can neither be tested nor falsified are not scientific. Belief in God (or in the idea of a Master Designer) can neither be proved nor disproved; therefore "scientific creationism" is not a science, it is a religion.

Readings: Joel Gurin, "The Creationist Revival." *The Sciences*, April 1981.
"The Genesis of Equal Time." *Science 81*, December 1981.
James Gorman, "Creationists vs. Evolution." *Discover*, May 1981.
Michael Ruse, "A Philosopher at the Monkey Trial." *New Scientist*, February 4, 1982.

As with all scientific theories, biologists devise experiments that will test the theory of evolution. To a geneticist, **evolution** is the changes in gene frequencies that arise and accumulate over time in populations of organisms. (It is populations, not individuals, that evolve.) It is these cumulative changes in gene frequencies that are subject to natural selection, as that process was originally proposed by Charles Darwin. According to Darwin, it is natural selection that determines the reproductive success of one individual as compared to another and thereby determines which individual's genes will be represented more frequently in subsequent generations. In Darwin's own words:

As many more individuals of each species are born than can possibly survive; and as, consequently, there is a frequently recurring struggle for existence, it follows that any being, if it vary however slightly in any manner profitable to itself, under the complex and sometimes varying conditions of life, will have a better chance of surviving, and thus, be naturally selected.

Darwin reached his conclusions by observing the kinds of slight phenotypic variations that occur among individuals of closely related species. The most famous of Darwin's many examples was the fourteen species of finches that he studied while visiting the Galapagos Islands (Figure 17-2). He concluded that the slight differences he observed in the physical appearances of the finches on the different islands were the consequences of adaptations that had accumulated over long periods of time in response to the different environments on the islands. The adaptations, Darwin concluded, were the result of heritable changes that ultimately resulted in the development of the different species of finches that were physically similar but that were reproductively isolated from one another by the water that limited each reproductive population to a particular island.

The term **species** refers to reproductive groups that interbreed among themselves but that do not mate or exchange genes with other reproductive groups in nature, even when different species share a common environment. Chickens, geese, and ducks may all share the same barnyard, but they mate only with other birds of their own species. **Race** is a rather arbitrary subclassification of a species based on physical or genetic differences. However, individuals of different races can and do interbreed, since they all belong to the same species. In some species, such as dogs, the term *breed* is used instead of *race* to denote a subspecies. All breeds of dog belong to the same species, and, in principle, any dog can mate successfully with any other dog. However, because of the artificial selection of specific characteristics by dog breeders over the centuries, physical and

Figure 17-2 The 14 species of finches on the Galapagos and Cocos islands. These finches are closely related yet have evolved into separate species. Some of the species of finches live in trees and feed on insects; others live on the ground and feed on seeds. Darwin's observations on the finches were important in the development of his evolutionary ideas. (From "Darwin's Finches" by David Lack. Copyright © 1953 by Scientific American, Inc. All rights reserved.)

reproductive problems can arise in matings between certain breeds of dogs; obviously, the problems of breeding a male Great Dane with a female dachshund, for example, are formidable.

The assignment of organisms to species and races is a method of classification that helps scientists organize the otherwise bewildering diversity of organisms into groups with some shared characteristics. Races of plants or breeds of animals can be organized by size, color, hair or skin texture, presence or absence of certain genes, or any other physical or genetic characteristics that can be measured. Assignment of persons to human races has often carried with it implications of social or intellectual inferiority or superiority. Historically, persons of different human races have been prevented from interbreeding, but the obstacle has been social and cultural, not biological. In nature, biological mechanisms ensure that any member of a species is fertile with any other member of the same species, regardless of how physically similar or dissimilar the individuals may appear. By the same token, other biological mechanisms ensure that matings between individuals of different species are infertile.

As the dog example shows, dramatic differences in appearance do not necessarily mean that individuals belong to different species. It also happens that individuals may appear similar or even identical to the nonexpert yet belong to different species. In the Hawaiian Islands several hundred species of the fruit fly *Drosophila* have been identified, each adapted to a particular island environment. Flies of one *Drosophila* species will mate only with flies of the same species, although to most people every fruit fly looks pretty much like any other.

Reproductive Isolation Mechanisms

Nature has numerous ways, often called **reproductive isolation mechanisms**, that prevent matings or reproduction between individuals of different species. Formation of species is important for evolution, since it provides a means of stabilizing genetic information in a way that optimizes the chances for survival and reproduction of organisms in particular environments. Advantageous genetic information is maintained by matings between individuals of the same species.

Reproduction between animals of different species is usually prevented by the absence of matings. Most animals exhibit sexual behaviors that influence male-female attraction and mating. Many species of birds, for example, display elaborate mating behaviors. Male birds generally have a more striking physical appearance: they tend to be larger and more extravagantly colored than females. *Sexual selection* obviously influences which genes are passed on to subsequent generations; hence, sexual selection is also an evolutionary force. Bigger, more colorful

Figure 17-3 A peacock displays his tail feathers to attract females. (Courtesy of Janis Hansen, Kaaawa, Hawaii.)

male birds are more successful in attracting females than their smaller, drabber male competitors. Thus, when peacocks fan their gorgeous tail feathers, they are announcing their sexual intentions to females (Figure 17-3).

Darwin regarded sexual selection as only slightly less important than natural selection as a force in evolution. In *The Descent of Man and Selection in Relation to Sex*, which deals specifically with the significance of sexual selection in humans and other animals, he wrote:

> *It cannot be supposed that male Birds of Paradise or Peacocks, for instance, should take so much pains in erecting, spreading, and vibrating their beautiful plumes before the females for no purpose.*

However, just because male animals are usually more colorful and aggressive in mating rituals does not mean that male behaviors necessarily determine which genes are passed on to progeny. For many species of birds and other animals, it is the female that chooses the male—decides which one is to be her mate—and thereby determines which male genes will be passed on to her progeny.

Reproductive isolation can result from mating behaviors; it can also result from geographical separation. If species of animals, plants, or microorganisms inhabit territories that are physically separated from one another, individuals are forced to mate within their particular territory. As mentioned earlier, animals isolated on islands that are widely separated from one another are forced to mate with other animals of that species on the same island. As Darwin observed with the finches, geographical isolation can eventually give rise to new species that are adapted to the particular isolated environments.

Unusual Birth in a Georgia Zoo

A few years ago, quite by accident, a mating took place between two apes of different species in the Grant Park Zoo in Atlanta, Georgia: a female classified by the zoo as a siamang was impregnated by a male classified as a gibbon. To everyone's surprise, a healthy hybrid ape was born some months later. This unique primate was named a *siabon,* since it was a cross between a *sia*mang and a gib*bon.*

Even though siamangs and gibbons occupy overlapping territories in the jungles of Southeast Asia, the two species are physically very different and presumably never mate in nature. Not only do the two species have different numbers of chromosomes— siamangs have fifty, whereas gibbons have forty-four—the organization of their genes on the chromosomes would be expected to be quite different. And, in fact, staining the chromosomes in cells from a siamang and a gibbon showed that the chromosome banding patterns are very different, an indication that the genes are indeed arranged differently on the chromosomes and, therefore, that the two ape species are only distantly related.

Despite the physical and genetic differences between these apes, the birth of a viable, normal infant hybrid ape as a result of their mating means that all the genetic information required for

BOX 17-2

normal development and differentiation was present in the hybrid cells of the embryonic siabon. This is all the more surprising because the cells of the siabon have an odd chromosome number (2N = 47), twenty-two chromosomes from the gibbon father and twenty-five from the siamang mother. Because of the odd number of chromosomes and the impossibility for all of them to

(Photo by Sister Moore, Atlanta, Georgia, courtesy of Richard H. Myers, Boston University Medical Center.)

pair with homologous partners during meiosis, it is likely that the siabon will be sterile (at this writing it is not yet mature enough to produce gametes).

The fact that such an unusual hybrid ape is viable and is growing normally raises the possibility that unusual matings might have played a role in primate evolution in the past. In 1982, a detailed comparison of the bands in orangutan, gorilla, chimpanzee, and human chromosomes showed that eighteen of the twenty-three human chromosomes are virtually identical to ones found in all three ape species. Human and chimpanzee cells differ in chromosome number by only one, and the banding patterns of both species' chromosomes are very similar.

More and more evidence suggests that slight genetic rearrangements or subtle changes in genetic regulation may have caused the evolution of the different species of primates. These genetic studies certainly raise the question—disturbing to some people—as to whether humans are indeed the culmination of primate evolution. Is it possible that other primates may yet evolve and replace the present human species, even as we replaced other, now-extinct hominids?

Reading: Richard H. Myers, "The Chromosome Connection." *The Sciences*, May/June 1980.

Biological mechanisms that affect fertilization and reproduction also contribute to the reproductive isolation of different species. Infertility between individuals of different species is often due to the wide variations in chromosome numbers of species. If individuals of different species are able to mate, the union of dissimilar gametes generally produces abnormal numbers of chromosomes or abnormal arrangements of genes on chromosomes in the hybrid. Occasionally, viable animal hybrids are formed between different species—the mule, for example, is produced by the mating of a horse and a donkey, but it is always sterile (see Box 17-2).

Many species of plants also have biological mechanisms to ensure that eggs will be fertilized only by pollen from the same species. However, plant breeders can often circumvent normal plant reproductive barriers, and they have been able to construct many kinds of hybrid plants that do not occur in nature. An example of a useful plant hybrid is *Triticale,* a new type of grain that was constructed by crossing wheat (Triticum) with rye (Secale). Wheat has a haploid chromosome number of forty-two and rye has a haploid number of fourteen. The reason that hybrid plants can be propagated is that some hybrid plants are fertile, whereas hybrid animals such as mules are usually infertile.

Population Genetics

Because the underlying basis for evolution is genetic diversity, one of the main ways of studying evolution is to measure the frequency and diversity of genes in populations and to determine how the frequencies of different genes change from generation to generation. This is the subject of **population genetics.**

As noted earlier, geneticists define evolution as the cumulative changes in gene (allele) frequencies in populations over time. In order to calculate how a particular pair of alleles changes in frequency from one generation to the next, we can make some simplifying assumptions. We will assume that initially our two alleles, *A* and *a,* are distributed equally among the members of the population ($\frac{1}{2}A$, $\frac{1}{2}a$). We will assume also that any individual can mate with any other individual in the population, a process called **random mating.** With these two simplifying assumptions, we can easily calculate the frequencies of the alleles in succeeding generations. (As pointed out earlier, sexual selection means that real matings are not random, so we are discussing an idealized population here.)

Because eggs and sperm combine at random in this idealized population, individuals of three distinct genotypes will be created in the next generation in predictable ratios (Figure 17-4a). One-fourth of the individuals will be homozygous for the dominant alleles (*AA*), one-half will be heterozygous (*Aa*), and one-fourth will be homozygous for the recessive alleles (*aa*). Similarly, if the genotypic frequencies of individuals in a population are known, the allelic frequencies can be calculated by counting alleles in the gametes (Figure 17-4b). It is apparent that in this simple example the frequency of alleles (*A* and *a*) in

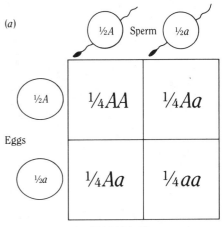

(b) Next generation ¼AA ½Aa ¼aa

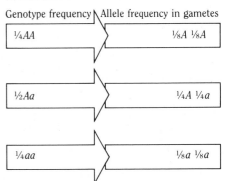

Total allele frequency in gametes:
$\frac{1}{8}A + \frac{1}{8}A + \frac{1}{4}A = \frac{1}{2}A$
$\frac{1}{8}a + \frac{1}{8}a + \frac{1}{4}a = \frac{1}{2}a$

Figure 17-4 Genotype frequencies and allele frequencies. (*a*) Random mating in a population means that the frequency of genotypes does not change from one generation to the next. In this example the allele frequencies are assumed to be equal ($\frac{1}{2}A$ and $\frac{1}{2}a$), but the conclusion that the genotypic frequencies do not change is valid for any ratio of alleles. (*b*) The allele frequencies can be calculated from the genotype frequencies. Note that the allele frequency turns out to be the same as in (*a*).

the population does not change from generation to generation; the frequency of allele *A* remains one-half, as does the frequency for allele *a*. (If we had started with allelic frequencies different from one-half, the frequencies would still remain unchanged in succeeding generations.)

In 1908 an English mathematician, G.H. Hardy, and a German physician, W. Weinberg, independently discovered that the relationship between allele frequencies and genotypes of individuals in a population can be expressed by the simple formula

$$p^2 + 2pq + q^2 = 1$$

which is known as the **Hardy-Weinberg law**. The letters *p* and *q* represent the frequencies of the two alleles in the population; p^2, *pq*, and q^2 represent the genotypes of individuals. Since together the alleles for any genetic locus must equal 100 percent, $p + q = 1$. (In reality, many genetic loci have numerous alleles. The HLA genes, for example, may have as many as forty different alleles in a human population.)

The Hardy-Weinberg law, then, expresses an ideal situation in which matings occur at random among individuals, and it does not take into account other biological mechanisms that cause allele frequencies to change. In populations that are in Hardy-Weinberg equilibrium, allele frequencies remain unchanged from one generation to the next and, consequently, evolution *cannot* occur in such populations. Because the Hardy-Weinberg law demonstrates so clearly how allele frequencies would remain unchanged in a randomly mating population, it provides the means for discovering and analyzing what forces *do* contribute to evolution and how they cause allele frequencies to change in real populations in nature.

Recall the geneticists' definition of evolution: *the cumulative changes in allelic frequencies in populations*. Thus, the situation described by the Hardy-Weinberg law is one in which the population does not evolve, because the alleles are in equilibrium. In reality, no population in nature is in Hardy-Weinberg equilibrium. In different human populations allele frequencies are often quite different. For example, the human M-N blood antigens (another blood group, distinct from the ABO group) are synthesized from a single genetic locus with two alleles. Homozygous (*MM* or *NN*) or heterozygous (*MN*) persons can be distinguished by biochemical blood tests, and the frequencies of the *M* and *N* alleles in any population can be calculated from the Hardy-Weinberg formula. As shown in Table 17-1, the *M* and *N* alleles occur with equal frequency among U.S. Caucasians, whereas the *M* allele is found almost exclusively among Navajo Indians and the *N* allele predominates among Australian Aborigines. Many other examples from natural populations of organisms show that certain forces must change

Table 17-1. Allelic Frequencies for M-N Blood Antigens in Various Human Populations

Population	Number of individuals	Frequencies of genotypes			Allelic frequencies	
		MM	MN	NN	M	N
U.S. Caucasians	6,129	.29	.50	.21	.54	.46
Navajo Indians	361	.85	.14	.01	.92	.08
Australian Aborigines	730	.03	.30	.67	.18	.82

Table 17-2. Decrease in Frequency of Individuals Heterozygous for a Particular Allele as a Result of Self-fertilization

Generation	Frequency of genotypes		
	AA	Aa	aa
1	1/4	1/2	1/4
2	3/8	1/4	3/8
3	7/16	1/8	7/16

gene frequencies to produce the genetic variation that is essential for evolutionary change. We shall consider these forces a little farther on in this chapter.

Why Human Inbreeding Is Undesirable

Most human societies have established taboos (prohibitions) against **inbreeding**, or matings between closely related individuals. Even before the principles of heredity were established, people realized that inbreeding produces more biological problems among progeny than do matings between unrelated individuals. Homozygosity is increased by inbreeding, as is the chance that deleterious recessive alleles will come together in an individual.

It is estimated that each person carries as many as eight lethal recessive alleles in the heterozygous condition and many more sublethal alleles. Using the Hardy-Weinberg formula, we can calculate the increase in the frequency of homozygous individuals in a population that is self-fertilizing—the most extreme form of inbreeding, in which individuals mate with themselves. (This commonly occurs in plants.) The number of heterozygous individuals in self-fertilizing populations decreases 50 percent in each generation, and individuals that are homozygous for recessive alleles increase proportionately (Table 17-2). The extent to which heterozygosity is reduced in a population can be calculated if the amount of inbreeding is known. The calculation of inbreeding values is important for plant and animal breeders, who strive to genetically modify their stocks in order to produce phenotypes they deem desirable. Inbreeding in human populations results in more individuals that are homozygous for recessive alleles, which increases the chance for genetic diseases and defects.

The Forces of Evolutionary Change

The Hardy-Weinberg law ignores the forces that cause nonrandom matings in populations: mating preferences, movement of individuals with different alleles into and out of the mating

 (a)

 (b)

Figure 17-5 Comparison of a wild-type fruit fly and a mutant bithorax fruit fly. Three mutations can be combined in *Drosophila* to produce a bithorax fly with two pairs of wings. It is thought that today's single-winged flies evolved from flies with two pairs of wings. This laboratory-generated mutant demonstrates how mutations create new structures, new organisms, and eventually new species. (Curiously enough, airplanes were first designed with two wings and "evolved" into single-winged aircraft.) (Courtesy of Edward Lewis, California Institute of Technology.)

population, differential mortality of individuals with certain genotypes, and chance deviations from randomness in matings. In addition to the many forces that influence matings, four other important forces that change allele frequencies in populations are *mutation, migration, genetic drift,* and *natural selection.*

Mutation is the ultimate source of all genetic variation and, consequently, the force underlying all evolutionary change (Figure 17-5). Whereas recombination between chromosomes during meiosis can generate new gene combinations, recombination does not produce new *genes*. A mutation is analogous to introducing a joker into a deck of cards. Shuffling a normal deck of fifty two cards rearranges the cards but does not create any new information; inserting a joker does.

If a mutation arises in sperm or eggs, and if that mutation makes an individual better able to survive and reproduce than other individuals in the population, the frequency of the mutant gene in the population will increase. This is what Darwin meant by "survival of the fittest" and what geneticists and evolutionists refer to as **fitness**: the reproductive contribution of an individual to the following generations. A person's fitness has nothing to do with his or her intelligence, health, strength, or looks; rather, fitness is based on the number of progeny produced. The more progeny, the greater the number of that person's genes in the next generation.

Mutations can, of course, have either beneficial or harmful effects in a given environment. If a mutation results in an individual that is less fit for its particular environment, the mutant gene and other genes carried by such individuals will tend to decrease in frequency, because the mutant individuals will not survive and reproduce as well as others. Ultimately, it is the interaction of the individual's genotype with the environment

The Founder Effect: A Gene Spreads Through a Population

BOX 17-3

Most serious hereditary diseases occur infrequently in large human populations because most individuals affected by them do not survive long enough to reproduce. But not all genetic diseases are serious enough to interfere with the individual's ability to reproduce, and some disease-causing genes can increase in frequency in a population. For example, certain genetic diseases can be traced to a *founder effect,* in which a gene (allele) has a frequency markedly higher in a certain population than in the general population from which it came. For instance, the Afrikaner population in South Africa shows a very high frequency of two different inherited diseases that can be traced back to particular individuals among the Dutch settlers who emigrated to South Africa in the seventeenth century.

One of these diseases is *variegate porphyria,* a slightly different form of porphyria from that which afflicted King George III (described in Chapter 13). Persons suffering from this disease excrete porphyrins in their urine, and they are sensitive to sunlight and to certain drugs, including barbiturates. This inherited disease is extremely rare world-wide, but in South Africa more than 10,000 persons suffer from it. Medical geneticists have traced the introduction of the mutant gene into South Africa to a Dutch couple who were married in Cape Town in 1688. All the present-day carriers and affected individuals in South Africa are their descendants and carry the same mutant alleles.

Another inherited disease that occurs with high frequency among both white and colored South Africans (a "colored" is a person whose ancestry includes one or more blacks and one or more whites)—20 per million among these groups as compared to 1 per 10 million among native (black) Africans—is Huntington's chorea. The symptoms of this disease, which (as we saw in Chapter 13) is caused by an autosomal dominant gene, usually begin to manifest in middle age or later, so the disease does not affect the person's fitness. The disease is characterized by a progressive deterioration of the nervous system, which causes convulsive movements, mental deterioration, and eventual death.

All the cases of Huntington's chorea that have been identified in South Africa—almost 500 to date—have been traced to Jan van Riebeeck, who led the first group of Dutch settlers to South Africa in 1658. Over half of all South Africans with Huntington's chorea are direct descendants of van Riebeeck, and the other affected individuals are believed to be distantly related to him.

Founder effects also explain why certain genetic diseases are much more common in small religious communities such as the Amish in Pennsylvania, who tend to intermarry among themselves in order to obtain a spouse who shares the same religious convictions. Unusually high frequencies of certain diseases may occur in all human populations, and the founder effect is only one of a variety of mechanisms that lead to different gene frequencies among populations.

Reading: M. R. Hayden, H. C. Hopkins, M. Macrae, and P. H. Beighton, "The Origin of Huntington's Chorea in the Afrikaner Population in South Africa." *South African Medical Journal,* August 2, 1980.

(which produces the phenotype) that determines reproductive success.

Migration refers to the movement of individuals and their genes into or out of a population. A dramatic example of changes in gene frequency due to migration is the **founder effect** (see Box 17-3). An individual may introduce a new allele into a population, and if that allele does not adversely affect survival and reproduction, its frequency will increase generation after generation. If a group of animals leave a population, travel a long

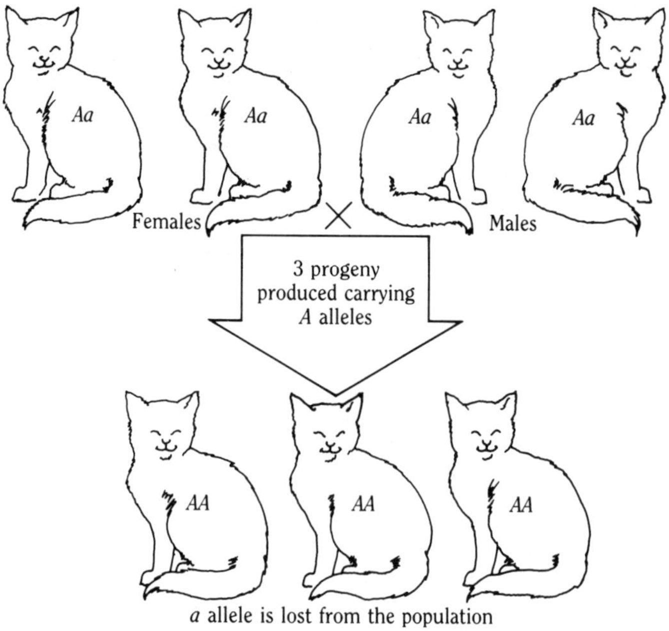

Figure 17-6 Effects of genetic drift. In small populations, an allele may disappear because of random genetic drift. Once the allele has been lost, the chance of it reappearing in the population is negligible, regardless of the size of the population in the future.

distance, and eventually establish a new, isolated population, survival in the new environment may depend on the presence of certain genes that were not subject to strong selection in the original population. As a result of the new environment and natural selection, the frequency of some alleles will increase in the population and the frequency of other alleles will decrease.

Genetic drift refers to the fact that alleles may change in frequency in a population by chance. Chance plays a particularly important role in gene selection in small populations, just as it does in a limited number of coin tosses. If a coin is tossed many times, chance produces, on the average, equal numbers of heads and tails. However, if a coin is tossed just a few times, it is not uncommon to observe a run of all heads or all tails. Allelic frequencies will fluctuate from generation to generation because of random genetic drift.

If the population is small, it is possible for certain alleles to be lost completely. For example, suppose that in a population consisting of just two females and two males, three progeny are produced. Suppose further that by chance only the *A* allele is passed on. In this case the *a* allele is lost forever, even though the three offspring may leave large numbers of progeny and the population might increase in size (Figure 17-6).

A wealth of evidence supports the idea that of the four evolutionary forces that affect evolution by determining whether the frequencies of particular genes increase or decrease in a population, natural selection has the greatest influence. The other three forces—mutation, migration, and genetic drift—do cause gene frequencies to change, but the genetic changes that accumulate over time are principally the result of natural selection. Most biologists believe that Darwin was justified in assigning such great importance to natural selection in the process of speciation and in the overall course of evolution.

Natural selection is a slow process. To appreciate how slowly gene frequencies change, consider our example in which the alleles A and a occur with equal frequencies in a population ($A = .5$, $a = .5$). Let us assume that all homozygous recessive individuals are selected against (die) and so are unable to reproduce or pass on any genes to the next generation. In this extreme case, 25 percent of the population is unable to pass genes on to the next generation. A simple reformulation of the Hardy-Weinberg law can be derived that gives the change in the frequency of the allele being selected against in any future generation:

$$q_n = \frac{q_0}{1 + nq_0}$$

In this formula, q_0 equals the frequency of the a allele (.5 in the first generation) and n equals the number of generations. What will this frequency be in, say, 100 generations—about 2,500 years in terms of human lifetimes? Solving the equation:

$$q_{100} = \frac{.5}{1 + (100 \times .5)} = \frac{.5}{1 + 50} = .0098 = 1\%$$

Thus, even with complete selection against homozygous (aa) individuals, the a allele will still be present in the population at a frequency of 1 percent after 100 generations.

Deleterious genes persist in populations for exceptionally long times because they are carried by individuals in the heterozygous state. Phenylketonuria occurs with a frequency of about 1 per 10,000 in the white population in the United States. Using the Hardy-Weinberg law we find that the frequency of the recessive allele (q) in the population is 0.01 ($q^2 = 1/10,000$; $q = .01$) and that heterozygous carriers ($2pq = 2 \times .99 \times .01 = .0198$) constitute about 2 percent of the entire United States population, or over 4 million persons.

Further calculation shows that matings between heterozygous carriers of PKU produce 98 percent of all the affected individuals in the next generation. The conclusion is inescapable. Preventing individuals affected with PKU from reproducing would have virtually no effect on the number of PKU individuals born in following generations. This disheartening conclusion applies to all recessive hereditary diseases. Elimination of affected individuals scarcely reduces the frequency of the mutant alleles in the population; neither does it reduce the incidence of affected individuals to any significant degree.

This simple application of population genetics has profound social implications that are often unappreciated. Misinformed people still advocate sterilization or euthanasia (mercy killing) for individuals who are affected by serious hereditary diseases, as a means of improving the human gene pool. People who espouse these ideas simply fail or refuse to comprehend the facts of population genetics. Almost all persons who are genetically handicapped as a result of carrying homozygous recessive alleles are the progeny of normal, unaffected heterozygous parents. Eliminating affected individuals who are judged to be genetically undesirable would reduce the number of affected individuals only an insignificant degree, because the population will still contain millions of persons who are phenyotypically normal yet carry the undesirable alleles. (The topic of eugenics is discussed in Chapter 19.)

Evolution by Neutral Mutations

Some biologists are not convinced that all—or even most—evolutionary changes in the past were caused by natural selection. They argue that even if no selection pressure existed gene frequencies would still change significantly in a population; this is referred to as the **neutral mutation theory** of evolution. It is based partly on the fact that bases in DNA can change without changing the amino acid composition of proteins or their cellular functions. Neutral mutations might accumulate for many generations before some environmental change caused them to affect the fitness of the organism.

Yet another theory that some biologists propose is that evolution occurred as a result of abrupt changes in populations—not the gradual, slight changes accumulating over time envisioned by Darwin. According to this **punctuated equilibrium theory** of evolution, little or no change occurs in populations over long periods of time, and then abruptly, as a result of environmental or genetic causes, the population changes and evolves into new species of organisms. The molecular bases underlying these two alternative mechanisms to natural selection are discussed in Chapter 18.

Evolutionists are split into opposing camps over the importance of neutral mutations versus natural selection in past evolutionary change. This disagreement is sometimes interpreted to mean that Darwin's ideas or the facts of evolution are in jeopardy, but that couldn't be farther from the truth. Rather, the controversy is over the relative contributions of various evolutionary forces in the historical development of new organisms and new species. The "neutralists" and the "selectionists" both agree that evolution occurred; their disagreement is over which mechanism has been more significant in producing evolutionary change during the history of life on earth. As is the case with most arguments, both sides are probably partly correct; natural selection and the accumulation of neutral mutations probably both contribute to changes in gene frequencies and, thereby, to evolution.

The Evidence for Natural Selection

In the *Origin of Species*, Darwin established two principles that he believed fundamental to evolution: (1) a certain amount of genetic variation is produced in a population each generation, and (2) this genetic variation is subject to natural selection. Darwin defined natural selection as the "preservation of favorable variations and the rejection of injurious variations." Thus, he perceived evolution in the context of a struggle by individuals for survival.

Natural selection has no purpose. Nor can it see into the future to select for traits that might someday be adaptive or useful. The environment of the moment determines which organisms survive and which genes are passed on to progeny. Francois Jacob, of the Pasteur Institute, has likened natural selection to a tinkerer who "during eons upon eons, would slowly modify his work, unceasingly retouching it, cutting here, lengthening there, seizing the opportunities to adapt it progressively to its new use."

Observations of changes occurring in plants and animals today, as well as the facts of the fossil records, are satisfactorily explained by natural selection. However, as with other scientific hypotheses, natural selection must be confirmed by experiments whenever possible. Among the most convincing experiments to demonstrate natural selection were the **industrial melanism** experiments carried out with the peppered moth in England by the late H. B. D. Kettlewell of Oxford University and his successors.

The peppered moths occur in two distinctive forms: dark-winged (the melanic form) and light-winged (wild-type). On normal, lichen-covered oak trees found in rural areas of England, the light-colored moths are virtually invisible against the trees, whereas the dark-colored moths are clearly visible against the bark's light background (Figure 17-7). In heavily industrialized areas of England, the bark of many oak trees has become covered with soot, a form of pollution that was especially common earlier in the century. Light-colored moths become clearly visible on the soot-covered trees, whereas dark-colored moths remain almost indistinguishable from the background.

The principal predators of these peppered moths are birds. Presumably, the moths' coloration helps protect them from being seen and eaten by the birds. The light-colored moths are

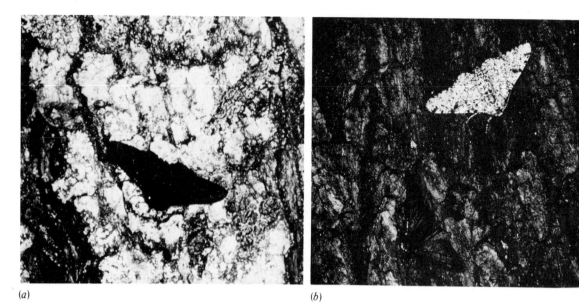

(a)   (b)

Figure 17-7 Industrial melanism. (a) Wild-type moths are nearly invisible on lichen-covered oak trees in rural Wales. The mutant (melanic) form of the moth is quite visible on the same bark. (b) Around English industrial cities such as Liverpool, the oak trees are soot-covered, so the melanic form is better camouflaged on the dark trees. Natural selection results in a higher proportion of melanic moths in industrial areas, whereas the wild-type predominates in rural areas. (Courtesy of L. M. Cook, University of Manchester, England.)

prevalent where they are harder to spot against the lighter (soot-free) trees. If natural selection accounts for this prevalence, then by the same token the dark-colored moths should be prevalent in industrial areas where the trees are covered with soot, because there they would have a selective advantage over the light-colored, wild-type moths. Numerous measurements carried out by Kettlewell and other biologists have indeed showed that the dark-colored moths prevail in sooty environments, just as the light-colored moths prevail in rural, nonindustrial regions.

The observation that wild-type moths are more fit in rural areas and melanic moths more fit in industrial areas is consistent with natural selection. Additional experiments carried out by Kettlewell and his successors have shown that survival of the moths depends on how many of each type fall prey to the birds that eat them. To demonstrate that wild-type moths survive better on lichen-covered trees and melanic moths survive better on soot-covered trees, the researchers pinned many moths of both types to trees in rural and industrial areas and then counted the survivors of each kind after several days. As might be expected, the birds more readily found and ate the melanic moths pinned to lichen-covered trees and the light-colored moths pinned to soot-covered trees.

These ongoing studies of industrial melanism in moths provide strong support for Darwin's hypothesis of natural selection and survival of the fittest. They also show quite dramatically how the frequencies of genes for wing coloration change in the moth population in response to changes in the environment. In the "struggle for survival," the fitter moths—those that avoid being

(a)   (b)

eaten—are able to reproduce and thereby increase their numbers at the expense of those that do not survive to reproduce.

Natural selection can take many interesting twists, one of which is **mimicry**; moths and butterflies provide good examples. Not all moths or butterflies make good meals for birds; some have evolved so as to be able to synthesize substances that birds find unpalatable and even noxious (Figure 17-8). A bird quickly learns which butterflies are tasty and which are not, and the butterfly population changes accordingly. Natural selection will favor moths and butterflies that contain substances that make them less desirable to their predators. And it will also favor any moth or butterfly that looks like (mimics) those that actually contain substances distasteful to the birds. As a result, many species of moths and butterflies closely resemble other, indigestible species but, in fact, are themselves quite palatable. Mimicry (which can entail behavior as well as physical appearance) protects many species of insects, other animals, and plants from predators.

Another fascinating example of mimicry is the light flashes emitted as sexual signals by both male and female fireflies. Male fireflies flash and females flash back in signal patterns that are specific to each firefly species. Females of one firefly species are able to attract—and eat!—males of another species because they are able to mimic the light flashes of females of the other species. So natural selection promotes the survival of individuals or, as in the case of firefly mimicry, may hasten their destruction.

Figure 17-8 A benefit of mimicry. When a bird captures and begins eating a monarch butterfly, it has a severe reaction and learns from the experience not to eat the noxious butterflies. Other butterflies and moths that resemble (mimic) monarchs are also avoided, so their mimicry helps protect them from predators. (Courtesy of Lincoln P. Brower, University of Florida, Gainesville.)

17
EVOLUTION:
NATURAL SELECTION AND POPULATIONS

It is obvious to anyone who studies nature (or people) that certain characteristics have adaptive value in certain environments. The "survival of the fittest" that Darwin described is not a struggle between good and evil. It is simply an inevitable consequence of the organism's genotype interacting with its environment. To repeat, in biology *fitness* is the genetic contribution of an individual organism to succeeding generations. And it is natural selection that largely determines what the frequencies of particular genes will be in the next generation.

Key Terms

natural selection   the differential reproduction of individuals with varying genotypes due to their fitness in a particular environment.

modern synthesis   the unification of evolutionary and genetic theories to explain both the continuity and change of organisms over time.

evolution   the changes in gene frequencies that accumulate over time in populations of organisms.

species   populations of individuals that interbreed with one another in nature but do not interbreed with other populations, from which they are reproductively isolated.

race   an arbitrary subclassification of a species based on physical or genetic differences.

reproductive isolation   the inability of organisms to interbreed because of biological differences.

population genetics   branch of biology dealing with the measurement and calculation of allele frequencies in populations and with how gene frequencies change in successive generations over time.

random mating   mating in which any individual can mate with any other individual in a population.

Hardy-Weinberg law   a mathematical expression of the relationship between allele frequencies and genotype frequencies in an idealized population.

inbreeding   mating between closely related individuals.

fitness   the reproductive contribution of an individual to the following generations.

migration   flow of genes from one population to another due to movement of individuals or gametes into or out of a population.

founder effect   demonstration of a gene whose frequency is much higher in a particular population than it is in other populations.

genetic drift   variation in gene frequencies in a population from one generation to the next as a result of chance.

neutral mutation theory   the idea that mutations arise and accumulate in populations without any selection pressure from the environment.

punctuated equilibrium theory   a model of evolution suggesting that populations remain stable for long periods of time and change abruptly rather than gradually and continuously.

mimicry   a resemblance of one organism to another, which often protects it from predators.

industrial melanism   the natural selection of dark-colored (melanic) moths in industrial areas of England.

Beehler, B. "For Gaudy Display, It's Hard to Beat Birds of Paradise." *Smithsonian*, February 1983.

Cohen, J. E. "Irreproducible Results and the Breeding of Pigs." *BioScience*, June 1976.

Gould, S. J. *The Panda's Thumb*. New York: W. W. Norton, 1980.

Gould, S. J. "The Chance that Shapes Our Ends." *New Scientist*, February 5, 1981.

Gould, S. J. "Evolution as Fact and Theory." *Discover*, May 1981.

Gould, S. J. "Genesis versus Geology." *The Atlantic*, September 1982.

Hardin, G. "Ambivalent Aspects of Evolution." *The American Biology Teacher*, January 1973.

"Hawaii: Showcase of Evolution." *Natural History*, December 1982.

Levinton, J. S. "Charles Darwin and Darwinism." *BioScience*, June 1982.

Lloyd, J. E. "Mimicry in the Sexual Signals of Fireflies." *Scientific American*, July 1981.

Stanley, S. M. "Darwin Done Over: Major Evolutionary Change Is not Gradual but Proceeds by Fits and Starts." *The Sciences*, October 1981.

Review Questions

1. What molecule must change for evolution to occur?

2. How do gene frequencies change in a population that obeys the Hardy-Weinberg equation?

3. If the frequency of one allele in a population is 75%, what is the frequency of the other allele? What percentage of individuals are heterozygous?

4. What organisms were used in the study of industrial melanism?

5. What evolutionary force is involved in the founder effect?

6. What is the difference in chromosome number between chimpanzees and humans?

7. What aspect of a population must change in order for it to evolve?

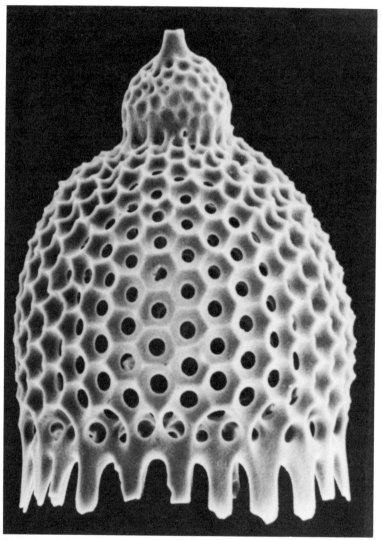

THE FOSSIL SKELETONS OF *RADIOLARIA*, SPECIES OF PROTOZOA, PROVIDE CLUES TO EVOLUTION. (695 ×)

18

EVOLUTION

Biological History in Fossils and Molecules

Everything is the sum of the past and nothing is comprehensible except through its history.

PIERRE TEILHARD DE CHARDIN

What Happened to the Dinosaurs?

BOX 18-1

About 65 million years ago—give or take a few million—something truly catastrophic happened on earth. According to the fossil record, during this period of the earth's history more than half of all the existing genera became extinct. Animals, plants, and even many species of marine life were annihilated over a period that may have encompassed millions of years, although many species, including all of the dinosaurs, vanished more rapidly.

This world-wide biological extinction is evident geologically in the sediments that separate the Cretaceous and Tertiary periods. Fossils of dinosaurs are abundant in the sediments of the early Cretaceous period, but all dinosaur skeletons are absent from the more recent Tertiary rocks. Moreover, many other kinds of fossils found in Cretaceous rocks do not appear in the more recent rock formations, suggesting that some catastrophic event occurred at the end of the Cretaceous.

Various ideas have been proposed to explain this relatively sudden global extinction of so many life forms. A sharp rise in the earth's temperature, dust from numerous volcanic eruptions, a drop in the levels of the oceans, a severe disruption in the food chain—all have been proposed, with very little supporting evidence or general acceptance. Now, however, many scientists are becoming increasingly convinced that the earth was struck about 65 million years ago by an enormous extraterrestrial object, probably a meteorite about 10 miles in diameter, and the physical evidence being uncovered lends strong support to this startling hypothesis.

(Kent and Donna Dannen)

Evolution is the continuous unfolding of new forms of life over time; *evolutionary science* attempts to decipher the earth's biological history from observations, measurements, and experiments. All studies of history, whether biological or social, try to reconstruct past events so that we may come to understand more about the earth's past, including human origins, human evolution, and human behaviors. Even when events in the history of human societies are well documented—for example, World War II or even the American Civil War—they are always subject to conflicting interpretations. By its very nature, biological and social history is imprecise in comparison with mathematics, physics, or chemistry, sciences that generally yield exact solutions to specific problems. The events and consequences of social and biological history are continually being challenged and reinterpreted as new information becomes available.

In 1980 a group of scientists reported an unusual geological finding. They detected large amounts of iridium, a rare platinumlike metal, in the narrow band of clay sediments that characterizes the boundary between Cretaceous and Tertiary geological formations. Rocks from almost any geological formation contain virtually undetectable amounts of iridium because this metal is a rare element in the earth's crust. However, it is known that iridium is abundant in meteorites.

While analyzing the chemical composition of the clay layer separating the Cretaceous and Tertiary formations in Italy, the scientists discovered that the iridium concentration was thirty times greater in this layer than in the rock formations on either side. Since this discovery, unusually high iridium concentrations have been found in geological samples from Denmark, Spain, and New Zealand and in sediment samples from the floors of the Atlantic and Pacific oceans. Most significantly, the high levels of iridium are always found in the layer that separates the Cretaceous and Tertiary rock formations.

The original discoverers of the iridium deposits hypothesized that 65 million years ago earth was struck by an enormous extraterrestrial object that exploded on impact and whose debris fallout around the globe resulted in the iridium deposits that have been detected. Indeed, we know that our planet has been hit by such objects; a few massive craters, dug by the impact of huge meteorites in prehistoric times, are well known to geologists. The well-known Meteor Crater, near Flagstaff, Arizona, is about half a mile in diameter. But it is dwarfed by the two largest meteor craters yet discovered. Each of these—one in Ontario, Canada, the other in South Africa—is more than 80 miles across. It has been estimated that the explosions of the immense objects that dug these craters each released more than a million times the energy yield of the largest hydrogen bombs.

The scientists who proposed that the iridium came from a meteor explosion also suggested

that the environment of the earth was so disrupted by the catastrophic impact that most life forms simply could not survive. The atmospheric ozone layer that protects the earth from intense radiation may have been disrupted temporarily. Or the millions (or billions) of tons of dust ejected into the atmosphere may have obscured the sun for years, causing the earth's temperature to change abruptly and for a long time. Whatever climatic changes were produced by the event, it is apparent from the fossil record that a majority of the species existing at the time, including the dinosaurs, were destroyed.

Small mammals were among the few survivors. What different paths might evolution have followed if this catastrophe had not occurred? What if the dinosaurs had not become extinct? Would mammals still have been able to evolve into primates and humans? It is entirely possible that the human species owes its very existence to a catastrophe that marked the end of the Cretaceous period 65 millions years go.

Reading: Cheryl Simon, "Clues in the Clay." *Science News*, November 14, 1981.

Scientists reconstruct and explain the biological events of the past as accurately as possible given the data at hand, but as new facts are discovered, their hypotheses and interpretations change. The fact that organisms lived and became extinct in the past is as well established as the fact that our present society evolved from past societies and is still changing. The fossil records provide undeniable factual evidence of evolution and of biological change that has occurred over billions of years. For example, we know without a doubt that many species of dinosaurs flourished on earth in the past; we even know that all the species of dinosaurs became extinct during a relatively short period of time about 65 millions years ago. What is less certain are the specific biological processes that produced the dinosaurs in the first place and the specific environmental events that caused their extinction (see Box 18-1).

As a result of recent **hominid** (humanlike) fossil discoveries,

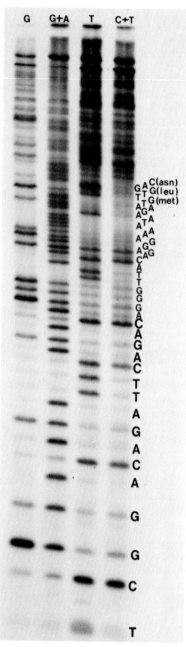

The sequence read from the gel (bottom to top):

T
C
G
G
A
C
A
G
A
T
T
C
A
G
A
C
A
G
A
C
A
G
G
T
T
G
G
T
T
A
C
A
G
A
A
A
T
A
G
A
G
T C(asn)
A G(leu)
G(met)

Figure 18-1 Sequencing a fragment of DNA. The spots on this photograph of a gel represent radioactive fragments of DNA of different sizes. From the presence or absence of the spots in the four columns, the sequence of bases can be directly read by someone familiar with this technique. The actual sequence of bases in this fragment is listed on the right and is read from bottom to top. (Courtesy of R. Rodriguez, University of California, Davis.)

mainly in Africa, it is now possible to trace hominid evolution over a period of 3 to 4 million years. However, modern humans *(Homo sapiens)* appeared quite recently and occupy but a tiny fraction of the overall period of earth's biological history, which spans about 4–5 *billion* years. As recently as 35,000 years ago, hominids were living in Europe who were physically different from us yet similar enough to be classified as a subspecies of modern humans. *Homo sapiens neanderthalensis,* or Neanderthals as they are commonly called, lived in Europe for more than 40,000 years before us, and for reasons unknown they became extinct. Donald C. Johanson, an anthropologist who was the co-discoverer of Lucy, a remarkably complete female skeleton found in Ethiopia and regarded by some anthropologists as the oldest human ancestor found to date, gives this description of the Neanderthal man:

> One hears talk about putting him in a business suit and turning him loose in a subway. It is true, one could do it and he would never be noticed. He was just a little heavier-boned than people of today, more primitive in a few facial features. But he was a man. His brain was as big as a modern man's but shaped in a slightly different way. Could he make change at the subway booth and recognize a token? He certainly could.

The biological history of humans and other organisms can be followed not only from fossil remains but also from measurements of the differences that are observed in the macromolecules of organisms that are alive today. Researchers can determine the similarities between the genes and proteins of organisms belonging to different species, thereby providing insight into evolution. For example, the amino acid sequences of hemoglobin molecules in blood cells of such animals as dogs, pigs, rats, monkeys, apes, and humans have been determined, and these different amino acid sequences have been used to reconstruct the evolutionary relatedness of these organisms, which is referred to as their **phylogeny**, or family tree. And newly developed techniques for determining the sequence of bases in DNA provide evidence of the many genetic changes that have accumulated over time—changes that, as we noted in earlier chapters, arise from mutations and are the basis for evolution (Figure 18-1).

This chapter discusses the various techniques researchers use to study evolution and to reconstruct earth's biological history. To properly evaluate the scientific facts of evolution, we must understand how the ages of fossils and rocks are determined as well as what the differences among biological molecules of different species mean. Only then can we fully appreciate our unique position in the evolutionary process. Darwin expressed his own view of evolution in the statement that concludes his *Origin of Species:*

There is grandeur in this view of life, with its several powers having been originally breathed into a few forms, or into one; and that, whilst this planet has gone cycling on according to the fixed laws of gravity, from so simple a beginning endless forms most beautiful and wonderful have been, and are being, evolved.

The Age of the Earth

Almost all astronomers and physicists agree that the universe came into existence as a result of an explosion of incomprehensible magnitude some 10–20 billion years ago—an event that has popularly become known as the Big Bang (briefly described in Box 1-1). In the instant that began time itself, this cosmic explosion created everything: matter, energy, space, time, and all the laws of physics and chemistry that have governed the properties of matter and energy from that moment until now. The Big Bang is the farthest back in history that science can probe. Before the Big Bang there was nothing—or at any rate, whatever existed must remain a total mystery to the human mind (although a few scientists venture to speculate on what might have been before).

The solar system, which includes our sun and its nine planets, was not created at the beginning of the universe. In fact, compared to the universe, the solar system is quite young; radioisotope dating of earth and moon rocks indicates their age to be about 4.5 billion years. During the early history of the universe, stars and galaxies were just beginning to form from the vast clouds of hydrogen atoms that were slowly being condensed by the gravitational force of their own enormous mass. Not until an enormous number of stars had been born, aged, and then destroyed in gigantic nova or supernova explosions could a solar system such as our own have come into existence.

It is in these stellar explosions that many of the heavier elements that are found so abundantly in the earth's crust were created. No life forms could have existed during the early history of the universe, for cells also require elements that are created only by stellar explosions. Fortunately, nature has supplied us with radioactive elements that provide a means for determining the ages of earth and moon rocks and the ages of fossils that were trapped in various rocks eons ago.

RADIOACTIVE CLOCKS MEASURE TIME

Radioactivity was discovered in 1896 by Antoine Henri Becquerel, a French scientist. Since his discovery, our understand-

Table 18-1. Radioisotopes Used to Determine the Ages of Rocks and Fossils

Material to be dated	Original isotope	Half-life (millions of years)	Decay product
Rocks and older fossils	Rubidium-87	47,000	Strontium-87
	Uranium-238	4,510	Lead-206
	Potassium-40	1,300	Argon-40
	Uranium-235	713	Lead-207
Recent fossils	Carbon-14	5,730 years	Nitrogen-14

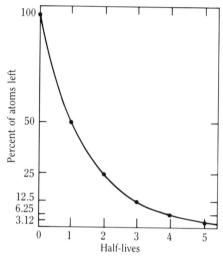

Figure 18-2   Radioisotope dating. There are half as many radioactive atoms in a sample at the end of each half-life interval as there were at the beginning of the interval. If the radioisotope's rate of decay is known, and if the remaining amount of the radioisotope and its decay product can be measured, the age of the sample can be calculated.

ing of radioactive elements and of radioactivity has progressed rapidly, and many applications of that knowledge have been developed. Naturally occurring radioactive isotopes, or radioisotopes (discussed in Chapter 1), provide remarkably accurate "atomic clocks" by which the ages of rocks and fossils can be measured.

Because the nuclei of radioactive elements are inherently unstable, they spontaneously change, or decay, into other elements at a constant rate that can be measured and that never varies even over millions or billions of years. In radioactive decay, atomic nuclei are altered by the emission of energy or particles such as protons and neutrons. It is the emitted particles and energy that are measured as radioactivity. The unchanging rate of nuclear decay provides an exceptionally accurate atomic clock that has been running in rocks and fossils ever since the particular radioactive elements were trapped in them (Figure 18-2).

The **half-life** of any radioisotope is the time—in seconds, weeks, or millions of years—that it takes for half of the nuclei to decay and, thereby, to be converted into some other element. It is impossible to predict which particular atoms will decay in any radioactive sample, but half will always decay in a certain time period, half of the remaining atoms during the next identical time period, and so on. Table 18-1 gives the half-lives of several useful radioisotopes.

Certain stable isotopes are formed only as a result of the decay process of radioactive elements, and these radioisotopes make particularly useful atomic clocks. For example, lead-204 (^{204}Pb) is the most abundant stable isotope of lead, but other stable lead isotopes (^{206}Pb and ^{207}Pb) are formed almost exclusively as the decay products of radioactive isotopes of uranium (^{238}U and ^{235}U). The amounts of various stable isotopes can be determined with great accuracy by an instrument known as a mass spectrograph, which measures the amount of any isotope according to its weight. Because ^{204}Pb can be distinguished from ^{206}Pb in a mass spectrograph, the particular decay products of uranium radioisotopes can be determined with great accuracy.

Thus, atomic clocks are exceptionally reliable because both the half-life of the radioisotope and the amount of the decay product can be measured.

When volcanos spewed forth molten rock during the fiery period of the earth's early history, radioactive uranium, rubidium, and other radioactive elements become trapped in the rocks as the lava cooled and hardened. The atomic clocks in the newly formed rocks began to run. Of course, the radioactive elements had been decaying all along, not just from the time when they became trapped. But they did not begin to function as clocks until they became trapped in rock from which their breakdown products could not escape. The clock works by making available measurable amounts of both the radioactive element and its breakdown products, and it is these *relative amounts* that enable researchers to determine when the clock started "ticking."

By carefully measuring the amount of strontium-87, lead-206, and lead-207 in rock samples and determining the remaining number of radioactive atoms, scientists can calculate the number of half-lives that must have been necessary to produce the relative amounts of the various isotopes and thereby date the rocks. Because many rock samples can be dated by several independent atomic clocks, the age of the earth is now known with a great degree of certainty.

EARTH'S HISTORY IS DIVIDED INTO GEOLOGIC INTERVALS

In 1654 James Ussher, Archbishop of Armagh (in Ireland), argued that the earth was created in the year 4004 B.C.; he arrived at this number by counting the successive "begats" recorded in the Bible. By the nineteenth century, geologists favored an earth age of several million years, based on rates of sedimentation and the degree of salinity of the oceans. In 1846 the British mathematician and physicist Lord Kelvin calculated the age of the earth (incorrectly as it turned out) to be 20 million years, based on what he perceived to be its rate of cooling. Today, atomic clock measurements of the oldest known rocks from both the earth and the moon give an age of 4.6 billion years. To appreciate how long a time this is, consider an evolutionary earth clock in which each hour equals about 187 million years (Figure 18-3). The four geological eras occupy 24 hours on such a clock, but the entire history of hominid evolution takes place in only about 30 seconds.

The geological history of the earth is divided into eras, periods, and epochs. Geologists base these classifications both on the various kinds of rock formations that can be characterized and on the fossil record contained in these rock formations (Table 18-2). Because of the continual evolution of new life forms and the extinction of others during the several billion years of earth history, geological formations can often be identi-

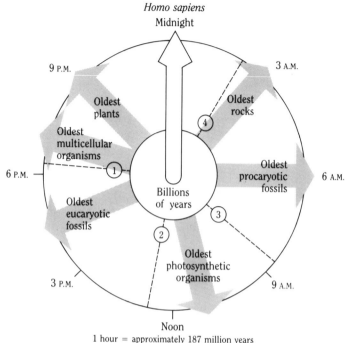

Homo sapiens
Midnight

9 P.M.

3 A.M.

Oldest
plants

Oldest
rocks

④

Oldest
multicellular
organisms

①

6 P.M.

Billions
of years

Oldest
procaryotic
fossils

6 A.M.

Oldest
eucaryotic
fossils

③

②

9 A.M.

3 P.M.

Oldest
photosynthetic
organisms

Noon

1 hour = approximately 187 million years

Figure 18-3 An evolutionary clock. The overall age of the earth is about 4.5 billion years. If this total geological time is set equal to a 24-hour day, the different life forms arose at different times of the day as indicated. In relation to the total age of the earth, modern humans have been around for about 30 seconds of a 24-hour day.

fied by the presence of certain microfossils that have become trapped in the rocks. Geological formations and their fossil remains provide a means of cross-checking the relative ages and events of both geological and biological history. Atomic clocks provide an absolute age.

The lovely Pacific coral islands, or _atolls,_ such as Bikini, Eniwetok, and Kwajalein, were formed by volcanos erupting and afterward sinking beneath the ocean floor. Corals grow on the lava rocks beneath the ocean, and some Pacific coral reefs contain the skeletal remains of the oldest animals on earth. Corals that are still growing today contain fossil remains that are over 100 million years old; thus, these corals began to grow during the Jurassic period.

The coral skeletons on Eniwetok extend almost a mile below the ocean's surface; from changes in the patterns of reef formation, ocean scientists have been able to figure out when the ocean's level rose and fell during millions of years of geological history. At some periods in the past, the ocean level dropped by as much as 180 meters from what it is now, and samples of the coral reef skeletons taken from 80–190 meters below the surface range in age from 160,000 to 700,000 years.

Fossils: Evidence of Biological History

Historians rely on written documents to reconstruct human history and to evaluate the significance of past events. Paleon-

Table 18-2. Biological History on the Geological Time Scale

Era	Period	Epoch	Millions of years before present	Life forms and major events
Cenozoic (Age of Mammals)	Quaternary	Recent	0.01	Modern humans
		Pleistocene	2	Early hominids
	Tertiary	Pliocene	10	Apes
		Miocene	25	Grazing mammals
		Oligocene	35	Flowering plants (angiosperms)
		Eocene	55	
		Paleocene	65	First placental mammals appear
Mesozoic (Age of Reptiles)	Cretaceous		135	First flowering plants
	Jurassic		185	
	Triassic		225	First dinosaurs
Paleozoic	Permian		275	Extinction of much marine life
	Carboniferous (Pennsylvanian, Mississippian)		350	Amphibians: giant swamp trees
	Devonian		400	Age of fishes
	Silurian		430	First land plants
	Ordovician		480	Earliest known fishes
	Cambrian		600	Marine invertebrates; algae
Precambrian			4,600–600	Single and multicellular soft-bodied organisms

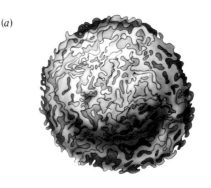

(a)

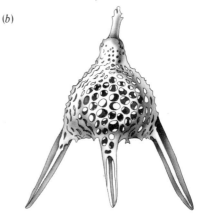

(b)

Figure 18-4 Microscopic fossils. (a) A plant spore from the Mississippian period. (b) A marine protozoan from the late Tertiary. Certain fossils occur only in rocks of particular geological periods. Paleontologists reconstruct biological and geological history from the microfossil record.

tologists, the historians of earth's biological history, study the myriad fossils that have been preserved from the distant past in rocks and sediments and then attempt to reconstruct biological history.

FOSSIL REMAINS

Most people are familiar with the spectacular large-fossil finds: the bones and tracks of such animals as dinosaurs, saber-toothed tigers, mammoths, and mastodons, and the impressions left by the leaves, stems, and other parts of ancient plants. However, important as these large fossils are in understanding biological history, even more valuable information is obtained from microscopic fossil remains such as pollen grains, protozoa, and bacteria (Figure 18-4).

Figure 18-5   Stromatolites in western Australia. These are recent stromatolites, aggregates of microbes cemented together by minerals from the sea. Similar stromatolites have been found that are 2–3 billion years old, indicating the presence of primitive cells that long ago.

(a)

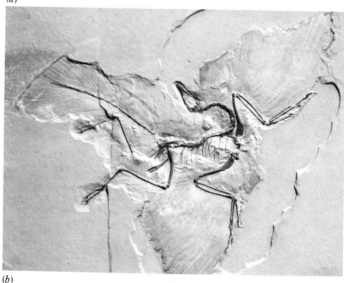

(b)

Figure 18-6   (a) A fossil fish skeleton from Fossil Butte National Monument in Wyoming. This fish lived 40–65 million years ago. (Courtesy Kent and Donna Dannen.) (b) *Archaeopteryx,* one of the transition forms between reptiles and birds. This fossil was found in rocks that are about 150 million years old. (Courtesy American Museum of Natural History.)

Microfossils discovered in rocks from northern Minnesota are believed to be about 2 billion years old. In western Australia, bacterial fossils have been identified in rocks that radioisotopic dating shows to be 3.5 billion years old. Some of the evidence for the existence of single-celled organisms on earth billions of years ago comes from fossils called **stromatolites**: large rocklike structures that were formed by the continuous layering of bacteria (or procaryotic algae) and minerals (Figure 18-5). The stromatolites and their fossil microorganisms support the idea that single-cell organisms probably arose within a billion years or so of the earth's formation. The existence of procaryotic cells billions of years ago is also consistent with the idea that evolution has been modifying and selecting new organisms ever since.

One of the most remarkable aspects of biological history is that such an abundant and detailed fossil record has been preserved. Replicas, almost photographic in their detail, of fish that swam in the oceans about 50 million years ago have been found (Figure 18-6a), as have early reptilian birdlike creatures such as *Archaeopteryx* (Figure 18-6b).

CARBON-14 DATING OF FOSSILS

The biological and geological history we have been discussing goes back hundreds of millions, even billions, of years. What kind of information has been preserved regarding more recent biological events—say, within the past 100,000 years, as human civilizations have developed? The carbon-14 (^{14}C) clock is the most valuable technique for dating organic materials such as bones, mummies, plants, seeds, and so on of more recent vintage. While they are alive and growing, all organisms must assimilate carbon atoms from the environment. The most abundant isotope of carbon is ^{12}C, but the environment also contains a small amount of the radioactive isotope ^{14}C. Plants assimilate both carbon isotopes as they grow, since they cannot distinguish between them in most cellular chemical reactions—nor can the animals that eat the plants.

The carbon gases in the air contain a fixed ratio of $^{12}C/^{14}C$ atoms, and it can be assumed that this ratio was much the same in the recent past as it is today. When an organism dies, the carbon atoms in its tissues remain there, and the ^{14}C atomic clock begins to run. Every 5,730 years half of the carbon-14 atoms decay into stable isotopes of nitrogen; the older the organism, the fewer carbon-14 atoms remain, and so the $^{12}C/^{14}C$ ratio in the sample increases with age. The ^{14}C clock is accurate for ages up to about 40,000 years; beyond that, too few ^{14}C atoms remain to be counted accurately.

The accuracy of the ^{14}C clock has been verified by calibrating it with ring samples from the bristlecone pines of California, some of which are over 5,000 years old (Figure 18-7). Most of the wood in the bristlecone pines, (like wood in other trees) is dead; only a thread of living tissue in the trunk keeps them alive. Small core samples removed from the trunk do not destroy the tree but do allow its age to be determined from the faint but

Figure 18-7 A bristlecone pine. These trees are used to calibrate carbon-14 atomic clocks. Some bristlecones are 5000 years old. The carbon-12/carbon-14 isotopic ratio changes in sections of trees that can be dated by counting rings of yearly growth. (Courtesy Janis Hansen, Kaaawa, Hawaii.)

visible growth rings. The ages of sections of the wood are then used to calibrate the carbon-14 clock.

Molecular Clocks as Measures of Evolution

In addition to fossil evidence, similarities between genes and proteins in different species also provide strong support for evolution. If organisms from widely different species are indeed related, no matter how distantly, by descent, similarities in their genes and other molecules should be apparent. By the 1960s, techniques for determining the amino acid sequences in proteins had advanced to the point where functionally identical, yet structurally different, proteins from different organisms could be analyzed.

CHANGES IN AMINO ACIDS

Cytochrome c is a protein that is involved in the biochemical production of ATP, the main energy-storing compound in every living cell. Cytochrome c proteins have been purified from dozens of different species and their complete amino acid sequences have been determined. If amino acid differences between two organisms are few, it is deduced that they are closely related and that the proteins have diverged in amino acid sequence quite recently in a common ancestor. If amino acid differences are extensive, a longer period of evolutionary divergence is assumed. For example, cytochrome c proteins in humans differ by only one amino acid from cytochrome c proteins in Rhesus monkeys; eleven amino acids separate humans from dogs, and forty-five separate humans from wheat.

A phylogenetic (family) tree can be constructed using the amino acid differences in cytochrome proteins to show the relative relatedness among different species (Figure 18-8). By estimating the mutation rates in DNA that must have caused the amino acid changes in the first place, biologists can use the number of amino acid differences to calculate the age of divergence between various organisms. The ancestral ages that are determined from amino acid differences are then compared with family trees that have been constructed by other physical and physiological measurements—often with remarkably good agreement.

Although it is unlikely that amino acid substitutions have occurred at a uniform rate in organisms over millions of years, it is nevertheless possible to construct a "molecular clock" that provides some estimate of the rate of evolution. Measurements of amino acid changes in seven different proteins from many different species of organisms indicate that the molecular clock runs at an average rate of about 100 amino acid changes per 240 million years. Not all proteins nor all organisms obey this clock, however. Moreover, many assumptions are involved in making these calculations, and molecular clocks are not thought to be as accurate as atomic clocks. Most researchers agree, however, that the more similar the amino acid sequences in functionally comparable proteins from different organisms, the more closely related the organisms are likely to be.

Certain amino acid changes in a protein may alter or destroy its function; in fact, this is one of the most demonstrable consequences of mutations. Yet many amino acid changes can occur in proteins without destroying or changing their function, as is shown by the structurally different cytochrome c proteins that perform identical functions in various species. In fact, analysis of the sequences of bases in genes that code for functionally identical proteins in different organisms gives a surprising result: As many as half of the bases in a gene can be changed without changing the biochemical properties of the protein! Such changes are referred to as **neutral mutations** since they do not affect protein function. It is possible that many neutral mutations accumulate in DNA until one occurs that abruptly causes a change in the function, a change in the phenotype, and eventually an evolutionary change.

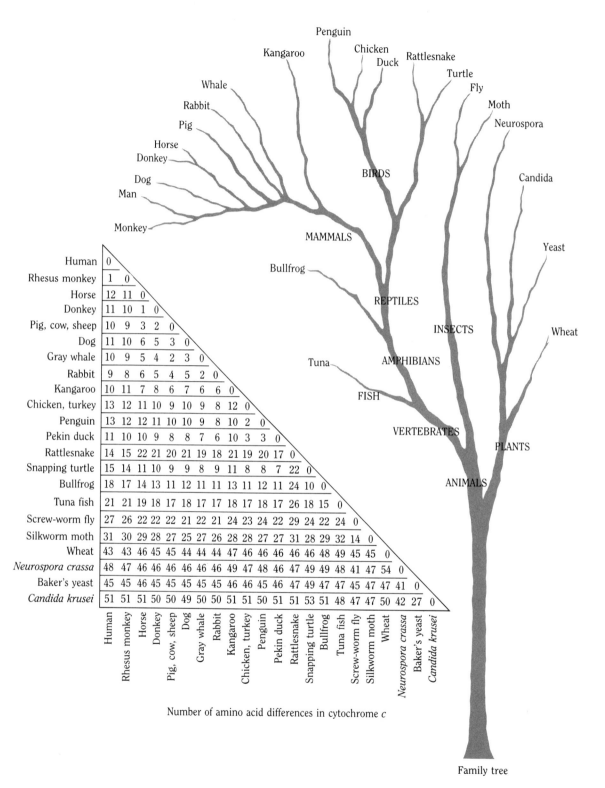

Number of amino acid differences in cytochrome *c*

Family tree

Gradual Evolution or Punctuated Equilibrium—That Is the Question

BOX 18-2

According to the classical Darwinian view that was embraced by most evolutionary scientists until recently, evolution proceeds by a gradual series of changes from one form of organism into another until a new species finally evolves. The mechanism underlying the gradual unfolding of new species was assumed to be natural selection acting on the novel individual phenotypes that stemmed from the genetic variability in the population.

However, some evolutionary scientists now argue that major evolutionary changes, especially speciation, occur abruptly, after long intervals of relatively little change. The hypothesis that new species arise abruptly has come to be called *punctuated equilibrium.* It postulates that species persist virtually unchanged for long periods of time (millions of years) and then suddenly evolve into new species in a relatively short period of time (possibly thousands of years). Proponents of the punctuated equilibrium model of evolution say that the fossil record shows little evidence of gradual changes but does provide numerous examples of abrupt changes in species. The new model also explains why transitional forms are rarely found in the fossil record: there simply aren't very many.

For years the fossil record of the horse was cited as a classic example of gradual evolution: the gradual emergence of a single hoof from a three-toed ancestor, and the gradual increase in size. Now the punctualists reinterpret the fossil record and argue that the modern hoofed horse appeared quite abruptly about 3 million years ago and evolved from an animal that had toes flanking its hoofs.

Depending on the organism and the environmental conditions, it seems likely that evolution proceeded by both mechanisms. It is important to remember that what is being contested by the opposing biologists is not the fact that life forms have evolved but rather the relative importance of specific mechanisms in producing new organisms and new species.

Readings: Steven M. Stanley, "Darwin Done Over." *The Sciences*, October 1981.
Boyce Rensberger, "The Evolution of Evolution." *Mosaic*, September/October 1981.

An idea that is now being debated heatedly among evolutionary scientists is whether species evolve gradually, as was suggested by Darwin, or whether, after long periods of little or no change, new species suddenly appear (see Box 18-2). The Darwinian idea that species evolve by continuous small changes that accumulate over long periods of time is called *gradualism;* the recent controversial hypothesis that speciation (formation of new species) happens abruptly and rapidly is called *punctuated equilibrium* (discussed in Chapter 17).

HEMOGLOBIN AND HUMAN EVOLUTION

New genotypes arise in individuals not only by changes in single bases in DNA but also by gene duplications (duplications of segments of DNA) that allow an organism to preserve genes essential for its survival while creating new genes that may evolve some useful new function. The best known example of gene duplication and subsequent evolution of new functions is the family of genes that direct the synthesis of hemoglobin. It

Figure 18-8 (opposite page) A phylogenetic tree constructed from amino acid differences in cytochrome *c*. The table shows the number of amino acid differences in cytochrome *c* between any two species listed. Note that the number of amino acid differences between humans and Baker's yeast (45) is about the same as between Baker's yeast and *Neurospora crassa* (41), which are both fungi. In terms of evolutionary divergence, yeast and *Neurospora* are as distantly related to each other as humans are to either fungal species. (Redrawn from Dickerson and Geis, *The Structure and Action of Proteins,* Harper & Row, 1969).

Table 18-3. Amino Acid Differences Between
Adult Human Hemoglobin β-chain and the
Hemoglobins of Other Primates

Human compared with	Amino acid difference
Chimpanzee	0
Gorilla	1
Gibbon	3
Rhesus monkey	8
Squirrel monkey	9

appears that primitive animals possessed a single ancestral globin gene that was duplicated numerous times during the evolution of organisms, eventually giving rise to a modern hemoglobin protein that in humans and other primates consists of two α-globin polypeptide chains and two β-globin chains that can be synthesized from any of about a dozen different hemoglobin genes.

Human hemoglobin genes have been studied extensively by analyzing both the amino acid sequences of the hemoglobin proteins and the base sequences of the family of genes that produce the α- and β-chains. The α-genes in humans are on chromosome 16, and the β-genes are on chromosome 11. Different α- and β-genes are expressed during the stages of human development, producing what are called embryonic, fetal, and adult hemoglobins that differ in their amino acid sequences but that all perform the same function of transporting oxygen. In the past few years, hemoglobin "pseudogenes" have been found—that is, genes that have the capacity to synthesize β-chains but that are never expressed or activated in blood cells. It is possible that these "silent" genes served some cellular function in the past but were permanently switched off in cells and are no longer used.

Amino acid differences among the five different β-chains found in the various classes of human hemoglobins have been determined, and from the number of amino acid differences the approximate ages of the gene duplications and of the original ancestral globin gene have been estimated. The original β-gene arose in vertebrates about 70 million years ago and was duplicated 30–40 million years ago; the most ancestral globin gene appeared 400–500 million years ago.

Hemoglobin proteins and their genes provide a molecular record of human evolution. Likewise, a comparison of the amino acid sequences in hemoglobins from various other organisms provides corroborating evidence for the close or distant relatedness of their species. For example, a comparison of adult human hemoglobin with the hemoglobin of other primates shows that chimpanzee and human hemoglobin are identical in amino acid sequence and that gorilla hemoglobin differs from human hemoglobin by a single amino acid (Table 18-3). These negligible differences between the hemoglobins of human, chimpanzee, and gorilla are virtually impossible to explain except by assuming that all three species evolved from a common ancestral primate, possibly 15–20 million years ago.

This conclusion concerning our relatedness to apes is further substantiated by comparing the genetic organization of the chromosomes in these three primates. As has been shown by chromosome banding patterns, eighteen of the twenty-three human chromosomes are virtually identical in structure and gene organization with those in chimpanzees and gorillas, and the other five human chromosomes are also very similar to those in the apes. As William Winwood Reade expressed it a century ago, "It is a shabby sentiment . . . which makes men prefer to believe that they are degenerated angels rather than elevated apes" (Figure 18-9).

As a matter of fact, much of the public controversy over evolution has to do with *human* evolution: Did humans evolve from apes? This notion was extremely distasteful to the Victorians of Darwin's time. One eminent English lady expressed her view of the idea as follows: "Let us hope that it is not true but if it is let us pray that it will not become generally known." And when, in 1860, Thomas Henry Huxley read Darwin's scientific report before the Royal Society, he was asked sarcastically by Bishop Wilberforce whether he claimed descent from monkeys on his mother's or his father's side of the family. Huxley, a distinguished biologist and a formidable advocate of Darwinism who came to be known as "Darwin's bulldog," turned the tables on his interrogator:

> I asserted—and I repeat—that a man has no reason to be ashamed of having an ape for his grandfather. If there were an ancestor whom I should feel shame in recalling, it would rather be a man, *a man of restless and versatile intellect, who, not content with an equivocal success in his own sphere of activity, plunges into scientific questions with which he has no real acquaintance, only to obscure them by an aimless rhetoric, and distracts the attention of his hearers from the real point at issue by eloquent digressions and skillful appeals to religious prejudice.*

Today we know that human beings are not descended from monkeys or apes but that, rather, all three have some common ancestor. But as a commentary on the larger issue of the use and abuse of evolutionary biology, Huxley's eloquent reply is as much to the point today as it was over a hundred years ago.

DNA SIMILARITIES AMONG PRIMATES

Another molecular technique used to measure the similarity of the genetic information from different organisms is **DNA-DNA hybridization**. This technique involves extracting DNA molecules from cells of different organisms, separating the individual strands of DNA by heating (melting them apart), and measuring the extent to which the separated DNA strands from different organisms come together or rejoin to form a stable double helix (Figure 18-10).

The degree to which the strands of DNA rejoin depends on

Figure 18-9. One of our "close" relatives.
(© Nina Leen, courtesy of Life Picture Service.)

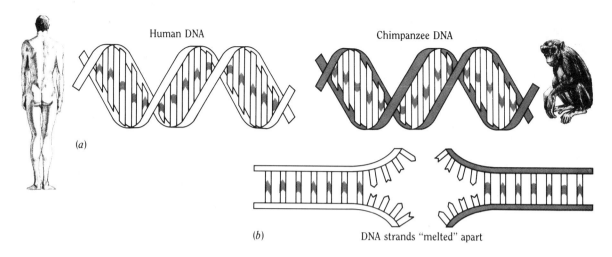

Human DNA

Chimpanzee DNA

(a)

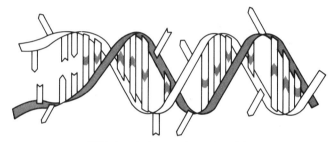

(b)

DNA strands "melted" apart

Figure 18-10   DNA-DNA hybridization, a technique for measuring the degree of homology in the DNAs from different organisms. *(a)* DNA molecules are extracted from cells of humans and chimpanzees. *(b)* The DNA molecules are heated *in vitro* to separate (melt apart) the two strands by breaking the hydrogen bonds between base pairs. *(c)* The separated strands of DNA from the two species are mixed together *in vitro,* and the ability of the different strands to hybridize (rejoin to form a double helix) is measured.

DNA strands from each species rejoin

how similar their base sequences are. If the DNA strands were from the same organism, the base sequences would be virtually identical and the structural correspondence, or *homology,* would be perfect. Thus, the degree of nonhomology between strands is a measure of the genetic dissimilarity between organisms. The degree of nonhomology is measured by the specific **melting temperature** of the hybrid DNA, that is, the temperature at which enough of the hydrogen bonds are broken so that half of the DNA strands in a sample come apart and are unable to stay joined in a stable double helix. If this hybridization technique is applied to DNAs extracted from cells of various primates, human DNA strands hybridize almost as well with chimpanzee DNA strands as they do with human DNA. Human–chimpanzee DNA is the most closely related in base sequence; it exhibits at least 95 percent homology (Figure 18-11).

Why, if humans and chimpanzees are so nearly identical in their proteins, chromosomes, and genes, are they so phenotypically different? No one can mistake a chimpanzee for a person. The answer is that chimpanzee differences must derive from different patterns and regulation of gene expression in the two species rather than from a dramatic difference in the structure of genes themselves. But as yet we know very little about such

Human compared to

	DNA-DNA hybridization (percent)	Time of divergence (millions of years)
Chimpanzee	95	5–15
Gibbon	91	15–25
Rhesus	76	25–40
Capuchin	71	40–55
Galago	47	50–65

Figure 18-11 Degree of genetic and evolutionary relatedness among primates as indicated by DNA hybridization. The primate most closely related to humans is the chimpanzee; other apes and monkeys are more distantly related.

patterns and regulation. One of the more exciting prospects of recombinant DNA technology is that soon we will gain some insight into the mechanisms of gene regulation in human development.

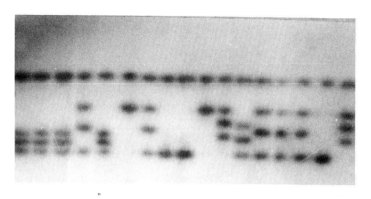

Figure 18-12 Electrophoresis of proteins extracted from plant cells. This photograph of a gel that has been stained for a particular enzyme shows the different structural forms of the enzyme synthesized from different alleles in a population. Each column shows the forms of the protein synthesized in an individual plant. There are more than two forms because the subunit polypeptides synthesized from pairs of different alleles associate in various combinations. (Courtesy Leslie Gottlieb, University of California, Davis.)

Polymorphisms and Genetic Diversity

Individual members of a population of organisms have different phenotypes because of the different alleles that can be carried at any genetic locus. **Polymorphism** (literally, "many forms") refers to the presence of several forms of a trait (or of alleles that determine a trait) in a population. For example, among humans, the various blood types of the ABO, M-N, and Rh blood groups represent polymorphisms, as do differences in skin color, hair texture, height, and so on. Phenotypic differences are usually the result of slight variations in the proteins that are synthesized in different individuals in a population. Because it is the various alleles that determine the kinds of proteins that are synthesized, measuring the frequency of protein polymorphisms provides an indication of the population's genetic diversity.

Polymorphisms are readily analyzed by examining the proteins that are synthesized from the various alleles (Figure 18-12). Because amino acid substitutions usually affect the electrical charge carried on proteins, they can be separated by the process of electrophoresis. Although electrophoresis does not detect all amino acid substitutions (some amino acid changes are electrically neutral and do not change a protein's overall electrical charge), it does provide a minimum estimate of the amount of genetic variation in a population.

The number of polymorphisms varies widely among species. In the many species of the fruit fly *Drosophila* that have been examined, approximately half of all the genes are polymorphic—that is, more than one allele exists at a particular genetic locus. About 30 percent of all genetic loci in amphibians and fishes are polymorphic, while mammals exhibit the fewest polymorphisms: only about 20 percent of the genes of mammals have multiple alleles. Among self-fertilizing plants, polymorphisms account for one-third of all genes.

One of the more unexpected results of studying polymorphisms is that the amount of genetic diversity *within* a population sometimes turns out to be greater than the amount of diversity *between* populations—even between species. For example, the genetic diversity in a population of fruit flies of the same species living in a tropical swamp may be greater than the genetic diversity (as measured by polymorphic proteins) between that population of flies and a different species of flies that lives in an arid desert.

The implication of this finding is that most protein changes may not be particularly important in the evolution of new species. It is more likely that changes in the regulation and interaction of genes are what cause new species to evolve. Even within human populations, genetic variation is nearly as great among individuals who superficially look alike as it is among individuals who look very different. And as pointed out in the previous section, the proteins and genes of humans and chimpanzees are virtually identical, yet their overall phenotypic differences are vast.

Human Polymorphisms and Race

Like the populations of all other species, human populations are highly polymorphic at many gene loci. As discussed in Chapter 16, the human HLA gene loci contain hundreds of different alleles, so no two persons are likely to be identical for these histocompatibility proteins. The particular alleles that produce altered hemoglobins or that cause favism are examples of polymorphisms that have been selected and maintained in populations by natural selection (discussed in Box 7-1). Many other protein polymorphisms in humans are also believed to result from selection of individuals carrying certain alleles in particular environments.

The overall genetic differences between various human populations can be studied by measuring the polymorphisms that are found among individuals. The assignment of people to *races,* a term used to divide a species into subgroups, can be based on morphological differences, on polymorphisms, or on any other characteristics that can be used to distinguish the various subgroups.

Use of the term *race* by politicians, sociologists, biologists, or other persons who seek, for whatever reason, to divide the human species into subgroups has contributed to prejudice in the past, often with disastrous consequences for particular human groups. Harold J. Morowitz, professor of biochemistry at Yale, points out what racial prejudice really signifies: "When we hate an individual for his race, we are hating genes over which he has no control. . . . Prejudice means disliking the genes of a . . . person without paying that much attention to the individual that goes with the genes . . . When you get down to it, what we are hating are DNA molecules."

Human races are arbitrary classes that can be based on geography, nationality, religion, appearance, proteins, or genes.

Traits such as skin color, hair texture, and facial features have traditionally been used to assign people to races. During the past century, the human species has been divided by biologists and anthropologists into as few as two races (the straight-haired race and the woolly-haired race) and as many as thirty races on the basis of other phenotypic differences.

The most important fact that has emerged from studies on human polymorphisms is that most of the genetic variation existing in the entire human species is found within local populations; less than 10 percent of all human genetic variation exists between the so-called major races—Caucasian, Negro, and Asian. As Richard C. Lewontin, professor of population genetics and natural history at Harvard, has explained it:

> Genetic variation between one Spaniard and another, or between one Masai and another, is 85 percent of all human genetic variation. . . . If everyone on earth became extinct except for the Kikuyu of East Africa, about 85 percent of all human variability would still be present in the reconstituted species.

Thus, the term *race* as applied to human populations is a cultural and political concept that, when examined scientifically, is not justified by the overwhelming genetic similarities among people. About 75 percent of all people's genes are identical regardless of their race or nationality.

In all fairness, it should be pointed out that arbitrary selection of certain distinguishing polymorphisms does allow us to assign individuals to previously defined races or ethnic groups. Although one allelic difference is not sufficient to make an unambiguous assignment of an individual to a particular population, selection of several polymorphisms that are present in high frequency in one population but absent or present in low frequency in another does permit assignment of individuals to a particular population with considerable assurance. For example, judicious selection of 5 or 10 specific polymorphisms out of millions of genes would probably allow us to distinguish a Navajo Indian from a Swede or an Ethiopian from a German. The question that must be asked is, why do people insist on establishing differences between populations, given that it has been firmly established that human genetic similarities far outweigh the genetic differences?

Some scientists argue that information about human polymorphisms is essential for understanding the evolution and origin of hominid species, particularly our own. Others are convinced that such studies contribute to ongoing racial prejudice and social inequality. Concern over the motives and goals of scientists who devise ways to distinguish one human racial or ethnic group from another are often raised and may be justified (discussed in Box 5-3). Measurements of human differences, whether they pertain to race, sex, intelligence, or behavior, have generated heated controversy in the past and are certain to continue to do so in the future.

The study of human genetic differences is in itself a valid scientific pursuit and provides useful data on the genetic variability of existing human populations. It is sad indeed that the division of people into racial, ethnic, religious, and national groups has resulted in much human suffering (discussed in Chapter 19). Once again, it is the human use of information that can be dangerous, not the information itself.

Human Origins

In 1859 Charles Darwin suggested that humans are evolved from apes, and the furor stemming from that proposal has scarcely diminished even today. All the molecular evidence accumulated since Darwin's time—the close relatedness of chromosomes, proteins, and genes between humans and apes—supports Darwin's basic hypothesis. The fundamental question raised by these findings of human-ape biological similarities is how such a profound difference in phenotype can occur when there is so slight a difference in the genotypes of the two species. In seeking answers to this question, anthropologists have searched for fossil evidence that might enable them to unravel the course of hominid evolution. In Darwin's time, hominid fossils were virtually unknown; in recent years, however, intensive efforts have uncovered fragments of numerous such fossils and have established the existence of several hominid species in Africa dating back 3–4 million years.

What concerns anthropologists today is not whether hominids evolved from apes but rather the particular path of primate development that culminated in *Homo sapiens;* this is still a controversial subject. **Primates** comprise a mammalian order made up of humans and extinct hominids; apes; monkeys; and smaller more primitive animals such as the lemur. The discovery of Lucy (as mentioned earlier in this chapter) led anthropologists Johanson and White to modify the hominid family tree by proposing that a species (defined by Lucy) called *Australopithecus afarensis* was the common ancestor of two hominid lines, one of which eventually became extinct (Figure 18-13).

Despite the many recent discoveries of ape and hominid fossils, it is not yet possible to reconstruct with assurance the overall biological history of primates and of hominids in particular. However, anthropologists generally agree that the dramatic expansion in the size and complexity of the hominid brain is largely responsible for human evolution.

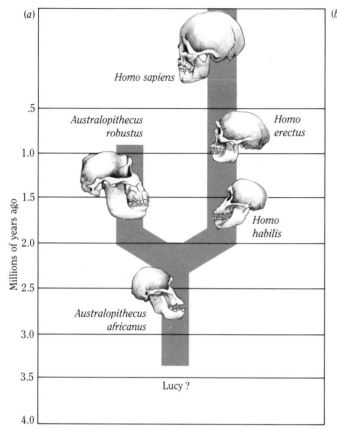

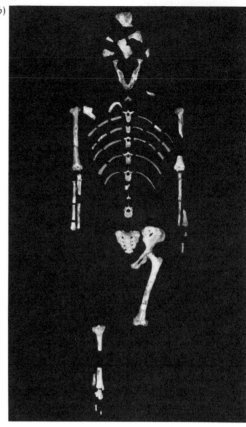

Figure 18-13 *(a)* One of several different hominid family trees proposed from fossil discoveries. The hominid lineage can be traced back about 3–4 million years. *(b)* Lucy is the most complete hominid fossil found to date and is also thought to be the oldest, having been dated at about 3.5 million years. Lucy's place in the hominid family tree is still controversial. (Photo courtesy Donald C. Johanson, Institute of Human Origins.)

The specific genetic changes that led to present-day humans are unknown. However, because many proteins in humans and apes are so similar, it is thought that the significant genetic changes must be regulatory rather than structural. If the expression of blocks of genes was altered, that might have produced changes in development that in turn caused the dramatic alterations in phenotype.

The relative contributions of natural selection, sexual selection, and other evolutionary mechanisms to hominid evolution are still being explored. Anthropologists argue about lineage, evolutionists argue about gradualism versus punctuated equilibrium as the mechanisms of speciation, and geneticists debate whether neutral mutation or natural selection is the most significant mechanism underlying genetic variability.

Creationists point to these scientific controversies as throwing all of evolutionary theory into question. The truth is, however, that these biologists are in complete agreement on the biological relatedness of all organisms, including humans and apes, and on the fundamental facts of evolution. Today's scientists are trying to find out *how* organisms change, *how* one

A Concise History of the Creation-Versus-Evolution Controversy in American Public Education

BOX 18-3

1925   The "Monkey Trial" in Tennessee. John Scopes, a public school teacher, was convicted of teaching evolution to his students in Dayton, Tennessee. He was fined $100.00. The conviction was overturned on appeal.

1942   Mississippi passed a law prohibiting the teaching of evolution in public schools.

1968   The U.S. Supreme Court in the case of *Epperson* v. *Arkansas* ruled that the Rotenberry Act, which prohibited the teaching of evolution in public schools, violated the 1st and 14th amendments to the U.S. Constitution and was unconstitutional.

1969   The California State Board of Education passed a resolution requiring public school teachers to present alternative ideas (creation) in teaching evolution. In 1972 the Board required that all biology texts used in public schools in the State of California have a section on creationism.

1973   Tennessee passed a law requiring a Genesis account of creation in all biology texts used in that state. The law was declared unconstitutional in 1975.

1981   Arkansas Act 590 required that public schools teach "scientific creationism." The act was declared unconstitutional by federal judge William R. Overton in 1982.

1981–1982   Laws mandating the teaching of scientific creationism were passed and subsequently struck down in Mississippi and Louisiana. Similar laws are pending in eighteen other states.

The Creationists have not fared well recently. But the controversy is not over, and it is still worth remembering the closing remarks of Clarence Darrow, the lawyer who defended John Scopes in the famed "Monkey Trial":

If today you can take a thing like evolution and make it a crime to teach it in the public schools, tomorrow you can make it a crime to teach it in the private schools, and next year you can make it a crime to teach it in the hustings or in the church. At the next session you may ban books and the newspapers. Soon you may set Catholic against Protestant, and Protestant against Protestant, and try to foist your own religion upon the minds of men.

After a while, your Honor, it is the setting of man against man and creed against creed, until with flying banners and beating drums we are marching backward to the glorious ages of the sixteenth century when bigots lighted fagots to burn the men who dared to bring any intelligence and enlightenment and culture to the human mind.

Reading: Irving M. Klotz, "Why Not Teach Creationism in the Schools?" *BioScience*, May 1982.

---

species evolves into another, and *how* humans evolved from ancestral primates—not *whether* they changed, for that is clear.

The Confrontation Between Creationists and Evolutionists

From the information presented in this chapter and in Chapter 17, it should be apparent why nearly all scientists consider creationism an invalid alternative to evolution. Not that most scientists are irreligious; on the contrary, most see no conflict in accepting evolution while maintaining their religious beliefs. For most scientists, the distinction between their religious beliefs and the observable, reproducible phenomena they study scientifically causes no conflict or concern.

However, a large group of American Christians who adhere to a fundamentalist (that is, literal) biblical interpretation of biological and geological history cannot reconcile their religious convictions with the facts of evolution. In order to protect their religious beliefs, which they feel are threatened by these scientific facts, Christian fundamentalists deny the scientific basis of evolution and at the same time attempt to legitimize their own beliefs with "scientific creationism."

If this were merely a battle of words and definitions, the issue between evolutionists and creationists could be dismissed as trivial. However, with diligence, zeal, and great expenditure of money, creationists have sought to have their ideas inserted into textbooks, to change the teaching of biology in public schools, and even to force museums to change their fossil exhibits. Through various legal maneuvers over the past fifty years, creationists have petitioned to have the biblical view of biological and geological history taught as science along with the theory of evolution (see Box 18-3).

The confrontation between creationists and evolutionists will not be settled by compromise, because both sides are defending what they regard as truth and as matters of principle. Creationists are defending their religious convictions; evolutionists are defending the basic process of scientific inquiry, and even more importantly, are opposing the instruction of students in religious explanations of biological history. That this is an important issue is borne out by the results of a 1982 Gallup poll showing that almost half of all Americans believe that humans were created only a few thousand years ago. Misunderstanding of scientific facts and principles in general and of genetics in particular has resulted in social turmoil, injustice, and individual suffering—topics discussed in the next chapter.

Key Terms

hominid a mammal whose characteristics are more humanlike than apelike.

phylogeny the evolution of a genetically related group of organisms.

half-life the time during which one-half of the atoms in a radioactive sample will decay.

paleontology the study of the life forms of past geological periods by means of the fossil record.

stromatolites large, ancient rocklike structures formed by bacteria and the minerals trapped by the bacteria.

neutral mutations mutations in DNA that do not alter the function of a protein and so are not directly selected for or against by natural selection.

DNA-DNA hybridization a technique used to measure the similarity in base sequences between DNA molecules from different organisms.

melting temperature in DNA-DNA hybridization, the temperature at which half of the DNA double helices separate into single strands.

polymorphism the presence of several forms of a trait or gene in a population of organisms of the same species.

primates a mammalian order comprising humans, apes, monkeys and certain more primitive animals such as the lemur.

Additional Reading

Futuyma, D. J. "Science on Trial: The Case for Evolution." New York: Pantheon Books, 1983.

Hay, R. L. and M. D. Leakey. "The Fossil Footprints of Laetoli." *Scientific American*, February 1982.

Johanson, D. C. and M. A. Edey. "Lucy." *Science 81*, March/April 1981.

Kitcher, P. *Abusing Science: The Case Against Creationism*. Cambridge, Mass.: MIT Press, 1982.

Lack, D. "Darwin's Finches." *Scientific American*, April 1953.

Landon, M., D. Pilbeam, and A. Richard. "Human Origins a Century After Darwin." *BioScience*, June 1982.

Latter, B. D. H. "Genetic Differences Within and Between Populations of the Major Human Subgroups." *American Naturalist*, August 1980.

Lewontin, R. C. "Are the Races Different?" *Science for the People*, March/April 1982.

Rensberger, B. "The Emergence of *Homo sapiens*." *Mosaic*, November/December 1980.

Rensberger, B. "The Evolution of Evolution." *Mosaic*, September/October 1981.

Ruse, M. "A Philosopher at the Monkey Trial." *New Scientist*, February 4, 1982.

Review Questions

1. Approximately how long ago did dinosaurs become extinct?

2. Approximately how long ago did hominid organisms appear on earth?

3. What natural process allows scientists to determine the ages of rocks and fossils?

4. Are atomic clocks more or less accurate than molecular clocks?

5. What organisms are involved in the formation of stromatolites?

6. Approximately how many genes are the same in any two humans?

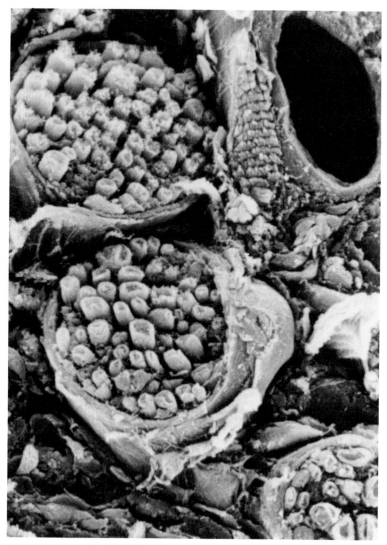

NEURONS (NERVE CELLS) GOVERN THE BEHAVIORS OF ORGANISMS. (1,320 ×) .

19

NATURE VERSUS NURTURE

The Social Consequences of Biological Determinism

Science in general, is not value free—neither in its origins and directions nor, more obviously, in its impact and applications.
JONATHAN BECKWITH

Every individual is different from all other individuals in numerous traits and behaviors. Even identical twins are not phenotypically the same in all respects; a casual observer will detect physical and behavioral differences that distinguish one twin from the other. Similarly, genetically identical plants—those that have been vegetatively propagated or cloned—will not grow at the same rate and will not have the same number of leaves, flowers, stems, or roots. Nor are ants or even bacteria phenotypically identical, although the differences among individuals may not be readily apparent. Taken all together, these observations indicate quite clearly that genes alone do not determine the physical traits and behaviors of organisms; nongenetic factors must also play a role.

Collectively, the nongenetic factors that affect the phenotypes of organisms are called the **environment**. For plants, the environment includes such factors as sunlight, water, nutrients, diseases, and pests. For humans, the environment includes all these things and many more—not only the physical, chemical, and biological components of the environment but also the many intangible social and psychological factors that shape our lives. Love, sorrow, pity, hate, loneliness, and anger can and do have profound effects on our development and behaviors.

An individual's genotype contains all of the genetic instructions essential for human development, growth, and reproduction, but the expression of all genes is affected by environmental factors to a greater or lesser degree. For example, the presence of even small amounts of certain chemicals in the body of a pregnant woman can drastically alter normal gene expression and development in the embryo; small amounts of alcohol at crucial stages of embryonic development, for instance, can adversely affect physical and mental development (discussed in Chapter 13).

Having "normal" genes guarantees neither normal development nor a normal phenotype. A genetic constitution that contains "flawless" information is like a perfectly recorded tape of music: even if the tape is recorded using the most advanced techniques, the quality of the sound will still depend on the quality of the playback equipment. A cheap cassette player will not produce high-fidelity music; weak batteries or a torn speaker cone will distort the sound even though the recorded tape is perfect.

The opposite situation may also occur. The sound on an imperfectly recorded tape or record can be greatly improved by electronic techniques that suppress noise and other unwanted sounds while enhancing the overall musical quality. In an analogous fashion, expression of an "imperfect" genetic constitution may be markedly improved in the appropriate environment. The question, then, is: What are the relative contributions of genotype and environment to an observed human trait? This question is at the center of all controversies about *nature* (heredity) versus *nurture* (environment).

Breeders have known for centuries that the characteristics of plants and animals can be altered by both selectively breeding for desired traits and controlling environmental factors. During the past century, plant and animal breeders have developed methods for quantifying the relative contributions of genes and of the environment to economically important traits in their crops and flocks. The yield of grain from wheat, the amount of milk produced by a cow, and the number of eggs laid by a chicken are traits that can be increased by selective breeding. However, to improve a trait, breeders must know how much of the trait is genetically determined and how much is due to factors in the environment; if, for example, 95 percent of a trait is due to environmental factors, it would be futile to attempt to alter it by breeding.

This chapter describes how geneticists measure the relative contributions of both genes and the environment to plant and animal traits. It also looks at some of the experimental difficulties, social consequences, and human tragedy encountered when an attempt is made to measure the contribution of heredity to variations in human intelligence, criminality, mental illness, and other such traits.

Biological Determinism: Fact or Fiction?

All human traits, including weight, strength, height, sex, skin color, hair texture, fingerprint pattern, blood type, intelligence, and aspects of personality (for example, temperament), are ultimately determined by the information encoded in the DNA. However, environmental factors modify every trait to a lesser or greater degree; for example, the environment plays almost no role in modifying gene expression for blood type or fingerprint pattern, whereas it exerts a strong influence on intelligence and personality. If a person is genetically programmed for type AB blood, environmental factors cannot cause that person to have type O blood; similarly, the environment cannot alter a person's fingerprint pattern after birth. However, we know that factors in the uterine microenvironment affect fingerprints because even identical twins do not have identical ones. And, even though the expression of many genes helps to determine the height, weight, and strength of an individual, all of these traits are greatly affected by environmental factors.

Biological determinism is the idea, advanced by some people, that a society is more of an accurate reflection of its biology than of its cultural history; that is, most variation in human traits and behaviors can be attributed to genes (see Box 19-1). Proponents of biological determinism believe that social and

Pellagra:
Tragic Consequences of Biological Determinism

BOX 19-1

In 1914 the U.S. Public Health Service gave epidemiologist Joseph Goldberger the task of discovering what causes pellagra, a debilitating and often fatal disease that was commonly found among poor and institutionalized people, especially in the South, during the first half of this century. Pellagra, which means "angry skin," is characterized by rough red skin eruptions, and its more serious symptoms include diarrhea, dizziness, lassitude, and, in advanced cases, "feeblemindedness" and death.

Goldberger, an expert on infectious diseases, quickly disproved the idea that pellagra was caused by some infectious agent. He inoculated monkeys with blood and tissues from human victims of pellagra, yet none of the monkeys showed any symptoms of the disease. A few years later, after Goldberger became completely convinced that pellagra was a "disease of poverty" and was simply the result of poor nutrition, he injected blood and skin from pellagra patients into himself, his wife, and other human volunteers to emphasize his claim that pellagra is a vitamin B deficiency disease.

Goldberger's discoveries regarding the dietary cause of pellagra were reported in the leading scientific journals between 1914 and 1917, and he was acclaimed by his medical colleagues for his discoveries. But despite the irrefutable evidence that Goldberger had amassed, nothing was done to eradicate pellagra in the United States for more than twenty-five years. Millions of economically disadvantaged Americans continued to suffer, and hundreds of thousands died from pellagra itself or from their increased susceptibility to infections.

Why did the truth about the causes of pellagra remain hidden from general public view for at least a generation? The explanation involves complex social and economic policies, but a primary cause can be traced to the efforts of a group of American geneticists who believed in *eugenics,* a branch of genetics that advocates the improvement of human traits and behaviors by selective breeding. These eugenicists, whose leader and spokesman was the in-

fluential geneticist Charles Benedict Davenport, believed that most human traits and behaviors are genetically determined.

Davenport headed the Pellagra Commission, an influential body set up by the New York Medical School in 1912 to study the causes of pellagra. In the Commission's report, published in 1917, Davenport and other members all but ignored Goldberger's widely acclaimed discoveries, and as late as 1920 Davenport published an article in the *Journal of the American Psychological Association* claiming that feeblemindedness, criminality, and pellagra were all genetic conditions. Davenport's views, along with those of other eugenicists, greatly influenced public health policies in the United States in the first half of this century and—tragically—prevented the dietary changes that would have alleviated great human suffering and saved the lives of many Americans who were too poor to afford adequate diets.

Readings: Allan Chase, "The Great Pellagra Cover-up." *Psychology Today*, February 1975.
Lori B. Andrews, "Inside the Genius Farm." *Parents Magazine*, October 1980.

intellectual ability and even economic status and differences in sexual preferences among individuals and groups of people are principally due to hereditary differences. It follows from the notion of biological determinism that "genes are destiny"—for each person, his or her genes determine the biological characteristics, the behaviors, and the place of the individual in society. Although even the strictest biological determinists do not deny the effects of environment (nurture), they assert that *most* of the variation in human traits and behaviors (such as intelligence, laziness, homosexuality, selfishness, altruism) is attributable to heredity. The nondeterminist does not deny the role of heredity but asserts that environmental factors are exceedingly important in determining human traits and behaviors.

The crucial problem for geneticists has been to devise experiments that quantify the separate contributions of heredity and environment for a specific trait. With plants and domestic animals, the appropriate genetic experiments are straightforward and have been carried out with many different species. However, the same experiments cannot be performed with people because it would be immoral (and illegal) to cross and cultivate human beings like plants and domesticated animals. To evaluate the notion of biological determinism and to appreciate its social significance, we must first know something about the different kinds of genetic experiments that measure the hereditary and environmental contributions to a trait. After seeing how these experiments apply to plants, it will be possible for us to critically examine the use of alternative genetic techniques in estimating the heritability of human traits.

The Concept of Heritability

The behavioral and physical traits of all persons have a strong hereditary basis; obviously, without the appropriate genes a person would not develop nerve cells, a brain, or a mind. Therefore, intelligence, personality, and even emotions are hereditary in the sense that genes are required for these human traits to exist. These traits vary from one individual to the next; before we can declare that such variation has a genetic basis, we need a method for measuring *how much* of the variation of a trait is determined by hereditary differences and *how much* is determined by environmental differences.

Heritability (which is not to be confused with heredity) is a measure of the amount of phenotypic variation in a population that can be attributed to hereditary differences. Heritability is not a direct measure of the genetic contribution to a trait; *it is a measure of the degee to which phenotypic differences in a population can be accounted for by genetic differences.* Heritability is a quantitative measurement that can vary between zero and one; a value of *zero* heritability means that environmental factors alone cause all the phenotypic variation among individuals, whereas a value of *one* indicates that the variation is totally a result of genetic differences among individuals. Many traits can and do have heritability values approaching 1.0. Human blood group differences, the variation in histocompatibility antigens, and polymorphisms for all kinds of enzymes are almost completely determined by the different alleles that people carry. Because the phenotypic differences in these traits are entirely due to the different alleles, the heritability of these traits has a value of 1.0, which is to say that these traits are essentially unaffected by the environment.

Mathematically, the heritability of a trait is defined as the amount of phenotypic variation in a population resulting from genetic differences divided by the total amount of variation (environmental plus genetic) observed in the population. Heritability is expressed by the formula:

$$H = \frac{V_G}{V_T} = \frac{V_G}{V_G + V_E}$$

in which the symbols are defined as follows:
H = heritability
V_T = total phenotypic variance in the trait
V_G = phenotypic variance due to genetic differences
V_E = phenotypic variance due to environmental differences
(The measure of variation is called the *variance* in statistics.)

While heritability is a well-defined concept, in practice it is often difficult to separate the relative contributions of environmental and genetic variance to the trait. For traits in plants, heritability can generally be measured accurately because inbred lines can be constructed in which the genes governing the trait are all homozygous. In a cross between inbred parental lines of plants, all of the F_1 progeny are genetically identical to each other, so all of the variance in the trait must be environmental (V_E). However, in the F_2 generation, the plants will be genetically heterogeneous because of segregation of the genes, and the variance will be caused by both environmental and genetic differences (V_T). Because V_T and V_E can be measured, V_G can be calculated ($V_G = V_T - V_E$).

It is also important to realize that heritability cannot be determined for traits that do *not* vary among individuals, even though such traits are clearly determined by genes. An obvious example of a genetically determined trait whose heritability cannot be measured is the number of organs in a body. Every person has one nose, and the number is genetically determined. However, since there is absolutely no variation in the number of noses in a population, regardless of differences in the environments, heritability cannot be calculated for this trait. On the other hand, the shape of the nose does vary in human populations, so the phenotypic variation can be measured (at least in principle) and the heritability can be calculated.

HERITABILITY CAN BE MEASURED IN PLANTS
The concept of heritability is best illustrated with plants, where the genotypes and the environments can be manipulated experimentally. Early in this century the Danish plant geneticist Wilhelm Johannsen showed that variation in the weights of

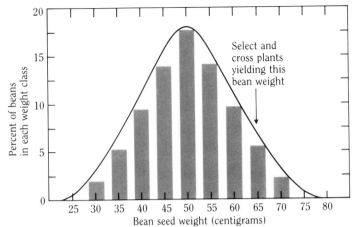

Figure 19-1   Seed weight variance in a population of bean plants. The plants are grown in a uniform environment. If the heritability of seed weight has a large genetic component—that is, if V_G is large—crossing plants from the heavy end of the distribution should increase the average weight of seeds in the next generation.

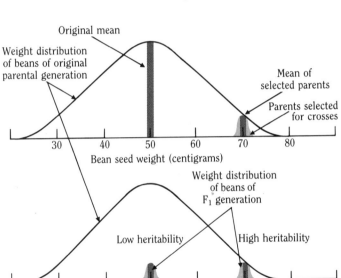

Figure 19-2   How the heritability of bean seed weight is estimated. If the mean seed weight in the F_1 generation of plants is nearly the same as the mean of the selected parents, the heritability of the trait is high. If the mean in the F_1 generation is about the same as the mean of the original parental population, the heritability is low.

beans could be partitioned into genetic and environmental components. Johannsen measured the variance in bean seed weights from bean plants grown in a uniform environment and plotted the weight distribution (Figure 19-1). He realized that if the variation in bean weights among individuals was due partly to particular genes, he should be able to increase the frequency of those genes by selecting and crossing particular plants. Johannsen then went on to selectively breed bean plants that would yield a greater weight of beans per plant. By crossing selected plants that produced heavier beans, he produced strains whose average bean weight was greater than the average weight in the parent population (Figure 19-2).

If the heritability of bean weight were high—that is, if the

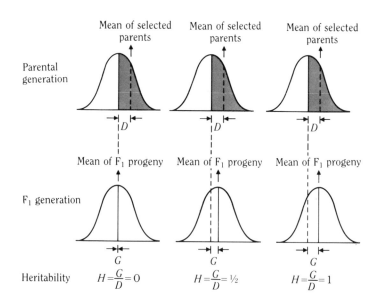

Figure 19-3 Measuring heritability in plants. Plant breeders calculate a value (between 0 and 1) for the heritability of a trait by measuring the differential selection (D) in the parent plants and the selection gain (G) in the F_1 generation. The value for D is determined by measuring the difference between the value in the parent plants selected for the cross and the average value in the population. For example, if the average height of the plants is 12 feet and the plants selected for the cross are 16 feet tall, then $D = 4$ feet. Similarly, the value for G is obtained from the difference in the average height of the F_1 generation plants and the average height of the parental plants. If the mean of the F_1 generation progeny is the same as that of the parents selected for the cross, the heritability is 1. This shows that all of the variation in height is due to genetic differences among plants in the population. If the mean of the F_1 generation plants is unchanged from the mean of the parent population, the heritability is 0. This shows that all of the variation in height is due to environmental factors.

variation were due to different genes in the population—the new statistical mean would be close to the mean of the selected parents; if the heritability were low, the new mean would not be appreciably different from that of the general population. In fact, the heritability of bean weight can be calculated by comparing the statistical means of the parent and offspring populations, as shown in Figure 19-3. The heritability for plants is expressed by a modified formula for heritability, $H = \dfrac{G}{D}$, in which

H = heritability
G = the selection gain
D = the selection differential

(See figure caption for definitions.)

Johannsen showed that many traits in plants have a high heritability and the frequency of particular traits can be increased in a population by crossing individual plants with the desired characteristics. Heritability is a useful measure for plant breeders, who use it in their efforts to improve crops by selective breeding. Heritability is more difficult to measure in animal populations because crosses cannot be performed as easily and environments cannot be as rigidly controlled; furthermore, the generation time for animals is longer than for plants, so acquiring data requires a rather long time. These difficulties are even greater in human populations. Thus, human heritability measurements are unreliable and sometimes result in unwarranted conclusions that appear to support the idea of the biological determinism of human behavior. We shall have more to say about this later in the chapter.

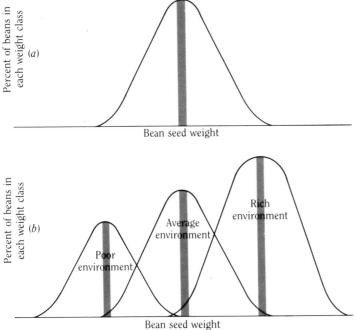

Figure 19-4   Variation in the average distribution of bean weights when plants are grown in different environments. *(a)* A large random population of bean plants grown under average field conditions gives the distribution shown. *(b)* If seeds from the same batch are cultivated under three different environmental conditions, the average distributions of bean weights will be quite different. Three different heritability values would be obtained, even though the original seeds are all from the same batch and, on the average, have the same genes.

HERITABILITY APPLIES TO A SPECIFIC POPULATION

In each experiment that measured the variance of bean weights, heritability was determined for a particular population of plants grown in a uniform environment. (Because heritability is a function of both genes and the environment, changing the environmental variability––the environment in which the plants are grown—would also change the heritability value.) Heritability comparisons are valid only *within* the populations in which the measurements are carried out; it is a fundamental (and frequent) error to apply a heritability value obtained in one population to a different population.

To appreciate why a particular heritability value cannot be assigned to different populations, consider a hypothetical experiment. If a large, random sample of bean seeds is collected from a genetically heterogenous population of bean plants grown in a field with uniform conditions of soil, water, and sunlight and these beans are then planted in the same field, the distribution of bean weights will be about the same in the parent and progeny populations. However, if the same sample of beans is planted in a rich environment—for example, in a field that is heavily fertilized and optimally watered—the average bean weight and the variation in bean weight of the progeny plants will be greater than that of the parents (Figure 19-4). Similarly, if the beans are grown in a poorer environment than the one in which the parental beans were grown, the average weight of the progeny beans and the variation in bean weight will be lower. Thus, if the environment changes, the heritability value will also change, even though the overall genetic constitution of the three populations is identical. Thus, we see that heritability comparisons are valid only *within* populations that are exposed to similar, and if possible identical, environments. This situation is rarely possible to achieve in human heritability studies, so invalid comparisons are often made between different human populations.

This example shows that heritability values must be interpreted cautiously. In many instances—particularly those involving human traits—the genetic and environmental components of phenotypic variation simply cannot be separated from each other. An attempt is made to do so in human studies involving identical twins, but as we will discuss later, these studies also have serious flaws. The genetic manipulation of plants can be controlled and plants can be grown in well-defined uniform environments so that, generally speaking, heritability values are accurate and are useful for improving traits in crops. However, when applied to human populations, heritability measurements are not nearly so reliable. This unreliability makes heritability measurements of human traits a dangerous business, particularly since the values obtained (which may be incorrect) can have serious consequences for individuals and for society if such data are used to influence social attitudes and policies.

Is Variation in Intelligence Inherited?

There is no doubt that heredity determines intelligence to some degree; as we have seen, mutations in any one of hundreds of human genes can result in severe mental retardation and lowering of intelligence. However, environmental factors play an important part in determining intelligence, too. Some examples of environmental factors that affect intelligence are nutritional deficiencies, elevated levels of heavy metals such as lead or mercury in the body, and educational deficiencies. Thus, the question again is, "To what degree is the *variation* in intelligence in a human population the result of genetic or environmental differences?"

In all of the discussion that follows, we accept the fact that intelligence is a trait that shows phenotypic variation and we assume that it can be measured by an intelligence (IQ) test, although this assumption itself is hotly disputed, particularly among educational psychologists. So we see that the question "Is intelligence inherited?" becomes "Can the genetic component of intelligence be separated from other positive or negative environmental factors that also affect measurement of IQ?"

The continuing controversy over IQ differences measured in

414

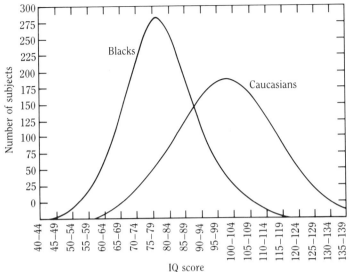

Figure 19-5 A typical distribution of IQ scores obtained from tests administered to black and Caucasian school children. The curves are based on scores from a number of studies. The average difference in IQ score distribution is about 20 points between the two groups. Note, however, the considerable overlap in the two distributions: many blacks scored higher than many whites. Moreover, these data provide no indication as to how much of the difference in the distributions is due to genetic factors and how much to environmental factors.

different human populations, especially differences between black and white Americans, could be resolved if an accurate value for the heritability of the trait of intelligence were available. Virtually all of the IQ controversy in the United States derives from studies in which the average IQ of black populations has been found to be consistently lower than the average IQ of white populations (Figure 19-5). Despite claims that the IQ tests were given to persons who were raised in similar environments, it is impossible to define what would constitute a "similar" environment for the development of a human trait as complex as intelligence. Moreover, we have seen why comparisons of heritability values between populations are not regarded as valid by geneticists.

The heritability of human intelligence has been estimated by some psychologists to be as high as 0.8, on the basis of studies of the IQs of identical twins who were raised separately from each other after birth. Acceptance of such a high estimate for IQ heritability would inescapably lead to the conclusion that differences in human intelligence are mainly genetically determined; with this line of thinking, one would be forced to conclude that, on the average, blacks have fewer of the genetic determinants involved in intelligence. Such a conclusion would have devastating personal, social, and political implications because, if, in fact, 80 percent of the variation in IQ between black and white populations were due to genetic differences among individuals, as claimed by advocates of biological determinism, then no amount of teaching or educational help could be expected to significantly improve the performance of any individual on IQ tests. It could thus be argued (erroneously) that educational enrichment programs are a waste of time and

Fraudulent Science and the IQ Controversy

BOX 19-2

The trait that most dramatically separates humans from all other animal species is intelligence, the ability to think and reason abstractly. Of all the species of animals that have evolved, only humans have come to understand the fundamental laws of nature and have been able to construct technologies and civilizations. For over two centuries, many people have held the view (some would call it a prejudice) that human intelligence is primarily a hereditary trait.

In the nineteenth century, before the discoveries of Darwin and Mendel, an American physician named Samuel George Morton devoted much of his life to proving that people differ in intelligence according to their race. Morton, like many others, equated human brain size with level of intelligence (an idea that seems reasonable but that is unsupported by any evidence—Albert Einstein's brain was quite average in size, for example). Morton measured the brain sizes of human skulls that he collected from a variety of racial groups over a period of twenty years. He, or his assistants, would carefully fill the brain area of a skull with lead shot and afterward weigh the shot to determine the brain size.

Morton concluded from his data that the intellectual superiority of Caucasians was well substantiated by their larger skulls and brains. American Indians and African blacks had the smallest skulls, which, according to Mor-

ton, would account for their inferior social status and supposed lower intelligence. Morton's results and opinions were highly regarded and widely accepted by whites in the nineteenth century. The New York *Tribune* wrote in his obituary that "probably no scientific man in America enjoyed a higher reputation among scholars throughout the world than Dr. Morton."

Morton's summary table of cranial capacity by race showed that the average Caucasian had a brain capacity 9 cubic inches greater than the average brain capacity of an Ethiopian and 5 cubic inches greater than the average American Indian. In 1978, Stephen Jay Gould, a Harvard biologist and historian of science, carefully reexamined all of Morton's cranial measurements and calculations. Gould was able to show that Morton had, either inadvertently or deliberately, manipulated his data to conform to his prejudices. When Gould recalculated all of Morton's original data, the corrected values for cranial capacity showed essentially no differences among races (see table).

Was Samuel George Morton prejudiced or simply careless? No one can say for sure, but an indication of Morton's racial views is given by his description of the Shoshonee Indians:

Heads of such small capacity and ill-balanced proportions could only have belonged to savages; and it is interesting to observe such remarkable accordance between the cranial developments, and mental and moral faculties. Perhaps we could nowhere find humanity in a more debased form than among these very Shoshonees, for they possess the vices without the redeeming qualities of the surrounding Indian tribes; and even their cruelty is not combined with courage. . . . A head that is defective in all its proportions must be almost inevitably associated with low and brutal propensities, and corresponding degradation of mind.

The disturbing saga of "scientific racism" continues into the twentieth century. An English psychologist, Cyril Burt, accumulated vast amounts of data that consistently supported the prevailing view of upper-class English society that intelligence is determined by heredity. Burt was the foremost expert on educational psychology in Britain in the period from about 1913 to 1932, and he administered IQ tests to thousands of English children. It was from his extensive IQ data, particularly the data obtained from twin studies, that American hereditarians such as Jensen,

money, because each person's intelligence and learning capacity would be biologically determined at birth and the environment could do little to change it.

It is important to place the IQ controversy in historical

Shockley, and Herrnstein concluded that human intelligence and social status are governed primarily by people's genes. The data collected by Sir Cyril Burt (he was knighted in 1946 for his "contributions") was cited by scholars for over three decades as the bulwark of the scientific proof that human intelligence is at least 80 percent determined by heredity.

Burt died in 1971; by 1974 doubts were being raised about the validity and accuracy of his twin study data. Much of the initial reinvestigation of the Burt data was carried out by an American psychologist, Leon Kamin. When Kamin discovered systematic flaws and discrepancies in the Burt studies and announced his findings, both Arthur Jensen and Richard Herrnstein came to Burt's defense. Jensen stated in 1977: "The central fact is that absolutely no evidential support for these trumped-up charges of fakery and dishonesty on the part of Burt has been presented by his accusers. The charges, as they presently stand, must be judged as the sheer surmise and conjecture, and perhaps wishful thinking, of a few intensely ideological psychologists." Herrnstein regarded the suggestion of fraud as "so outrageous I find it hard to stay in my chair. Burt was a towering figure of twentieth-century psychology. I think it is a crime to cast such doubt over a man's career."

By 1978 the "towering figure" had been toppled and the only "crimes" involved were those committed by Burt himself, who had fabricated the data that had been used to prove that human intelligence is hereditary. In the definitive biography of Burt published in 1979, L. S. Henshaw confirmed that Burt fabricated the data from IQ scores in twins, was involved in deliberate deception, and even invented collaborators who never existed to support his own biased ideas about intelligence. All scholars and psychologists, including Jensen, Herrnstein, and other former supporters, now agree that the Burt data are fraudulent. Ironically, Burt's biographer attributes his need to falsify the data to a difficult childhood, a failed marriage, and paranoia—shortcomings that are not hereditary, but environmental.

These two well-documented examples of scientific fraud have had profound educational and social consequences. (Burt, for instance, set up the tier system of education in Britain.) Modern hereditarians, while admitting to these past scientific embarrassments, argue that new, uncontaminated twin studies still support the hypothesis that human intelligence is primarily hereditary. And today's eugenicists, such as William Shockley, still believe in human breeding programs to improve intelligence.

Readings: Stephen Jay Gould, "The Brain Appraisers." *Science Digest,* September 1981.
Stephen Jay Gould, *The Mismeasure of Man.* New York: W. W. Norton, 1981.
Stephen Jay Gould, "Morton's Ranking of Races by Cranial Capacity." *Science,* May 5, 1978.
Nigel Hawkes, "Tracing Burt's Descent to Scientific Fraud," *Science,* August 17, 1979.

Cranial Capacities of Skulls as Calculated by Morton and Recalculated by Gould

Morton's values			Gould's values	
Race	Number of skulls	Cranial capacity (in.^{3})	Population	Cranial capacity (in.^{3})
Caucasian	52	87	Native American	86
Mongolian	10	83	Mongolian	85
American Indian	147	82	Modern Caucasian	85
Malay	18	81	Malay	85
Ethiopian	29	78	Ancient Caucasian	84
			African	83

context, since the story of IQ and brain measurements is an extraordinary—though bizarre—chapter in the history of science (see Box 19-2). Foremost among those who have argued for high IQ heritability is Arthur Jensen, a psychologist at the University of California, Berkeley. He is one of those who are convinced that the differences measured between black and white populations are largely due to genetic differences between the races. Jensen's views have been the focus of intense and often acrimonious debate since his article "How Much Can We Boost IQ and Scholastic Achievement?" appeared in the *Harvard Educational Review* in 1969.

Although Jensen misapplied the concept of heritability in that article by comparing IQ differences *between* black and white populations that were exposed to different environments, and despite the inherent unreliability of human heritability values and the methodological inappropriateness of comparing different populations, he has steadfastly stuck to his original conclusion. For example, in 1972 Jensen was still espousing a heritability value of 0.7 to 0.8 for IQ based on MZ twin studies (later shown to be fraudulent). Moreover, he still argued that valid conclusions could be drawn from comparisons of IQ measurements between different populations:

> . . . *high heritability of a trait* within *populations that differ in the trait does, however, increase the* a priori *likelihood of a genetic difference between the populations. The fact of the high heritability of IQ, therefore, makes it very reasonable and likely that genetic factors are involved in the Negro-white IQ difference. No geneticist to my knowledge has argued otherwise.**

In fact geneticists had published numerous criticisms pointing out this fundamental error since Jensen's article appeared in 1969. Yet in both scientific and popular articles Jensen persisted in his claim that black Americans were genetically inferior in intelligence. In 1973 Jensen reiterated his view in somewhat more moderate language:

> *Thus, the substantial heritability of IQ within the Caucasian and probably black populations makes it likely (but does not prove) that the black population's lower average IQ is caused at least in part by a genetic difference.*†

Jensen is well aware of the serious social and educational consequences of his controversial views and conclusions on IQ and race. In a book titled *Educability and Group Differences* he stated

---

*Jensen, A. R. *Phi Delta Kappan*. March 1972, p. 421.
†Jensen, A. R. "The Differences Are Real." *Psychology Today*. December 1973, p. 81.

. . . improving the benefits of education to the majority of Negro children, however, may depend in part upon eventual recognition that racial differences in the distribution of educationally relevant abilities are not mainly the result of discrimination and unequal environmental conditions. *

And finally, in 1981 Jensen authored a book called *Straight Talk About Mental Tests,* written for the general public. In this book he went to great lengths to prove that his views on the genetic basis for IQ differences are correct and that he is not a racist. Yet he persisted in claiming that

. . . something between one-half and three-fourths of the average IQ difference is attributable to genetic factors, and the remainder to environmental factors and their interaction with genetic differences. †

The majority of American geneticists and educational psychologists disagree with Jensen's conclusions. In 1970, a year after Jensen's controversial article appeared, two highly regarded population geneticists, Walter F. Bodmer and Luigi L. Cavalli-Sforza, reviewed all of the genetic evidence on race and intelligence and concluded:

In the present racial climate of the U.S. studies on racial differences in IQ, however well intentioned, could easily be misinterpreted as a form of racism and lead to an unnecessary accentuation of racial tensions. Since we believe that, for the present at least, no good case can be made for such studies on either scientific or practical grounds, we do not see any point in particularly encouraging the use of public funds for their support. ‡

The issue of race and intelligence was of such importance in the United States in the 1970s that members of the Genetics Society of America took a position on the issue and endorsed a statement on race and intelligence, a part of which stated:

It is particularly important to note that a genetic component for IQ score differences within *a racial group does not necessarily imply the existence of a significant genetic component in IQ differences* between *racial*

---

*Jensen, A. R. *Educability and Group Differences*. New York: Harper & Row, 1973, p. 364.
†Jensen, A. R. *Straight Talk About Mental Tests*. New York: The Free Press, 1981, p. 227.
‡Bodmer, W. F. and L. L. Cavalli-Sforza. *Scientific American*, October 1970, p. 13.

groups; an average difference can be generated solely by differences in their environments. The distributions of IQ scores for populations of whites and of blacks show a great deal of overlap between the races, even in those studies showing differences in average values. Similar although less severe complexities arise in consideration of differences in IQ between social classes. It is quite clear that in our society environments of the rich and the poor and of the whites and the blacks, even where socioeconomic status appears to be similar, are considerably different. In our views, there is no convincing evidence as to whether there is or is not an appreciable genetic difference in intelligence between races.

All human populations have a vast store of genes in common; yet within populations, individuals differ in genes affecting many characters. Each population contains individuals with abilities far above and below the average of the group. Social policies, including those affecting educational practice, should recognize human diversity by providing the maximum opportunity for all persons to realize their potential, not as members of races or classes but as individuals. We deplore racism and discrimination, not because of any special expertise but because they are contrary to our respect for each human individual. Whether or not there are significant genetic inequalities in no way alters our ideal of political equality, nor justifies racism or discrimination in any form.

It's our obligation as geneticists to speak out on the state of current knowledge on genetics, race, and intelligence. Although the application of the techniques of quantitative genetics to the analysis of human behavior is fraught with complications and potential biases, well-designed research on the genetic and environmental components of human psychological traits may yield valid and socially useful results, and should not be discouraged. We feel that geneticists can and must also speak out against the misuse of genetics for political purposes, and the drawing of social conclusions from inadeqate data. (Genetics Society of America, Supplement, July 1976, p. 100)

To most scientists, the experimental evidence that is advanced to support a biological basis for differences in intelligence between races is either scientifically invalid or so unconvincing that it cannot be taken seriously. However, the idea continues to find a receptive audience and is believed true by many people (even a few biologists!). As will be discussed in a later section, a similar scientific illogic has been applied to other

Figure 19-6 Identical twins have identical genes yet may have similar or different appearances. (Photo by Steve Dunn.)

complex human traits, such as criminality and homosexuality, to show that these behaviors are also biologically determined. Before discussing these issues we need to examine precisely how human heritability values are obtained.

Twin Studies and Human Heritability

As we have pointed out, the heritability of plant traits can be determined by appropriate genetic crosses in controlled environments. Although we do not cross or raise people like plants, nature does provide "experimental" human material in the form of twins. *Monozygotic* (MZ) *twins* develop from a single fertilized egg that divides very early in development to produce two genetically identical individuals. *Dizygotic* (DZ) *twins* develop from two separate eggs that have been fertilized by different sperm, and DZ (fraternal) twins are genetically different from each other. Everyone knows that identical twins are physically very similar—so much so that people sometimes have trouble telling them apart (Figure 19-6). If traits such as intelligence, personality, and susceptibility to certain diseases are primarily determined by genes, then a comparison of MZ and DZ twins reared in similar or different environments *might* provide a means to measure heritability, just as similar comparisons do for plants and animals.

Heritability studies on intelligence are carried out by com-

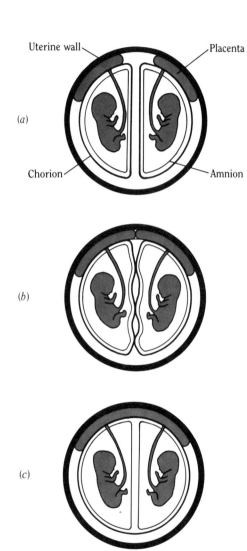

Uterine wall

Placenta

(a)

Chorion

Amnion

(b)

(c)

Figure 19-7 Variation in the uterine development of twins. Variation occurs because of the different ways twins are connected to the uterus. *(a)* Twins with separate placentas and membranes. This arrangement is typical for DZ twins but rare for MZ twins. *(b)* Placentas and membranes fused. This arrangement occurs in both MZ and DZ twins. *(c)* A common placenta and chorion but separate amniotic sacs. This arrangement occurs only in MZ twins.

paring the IQ scores of identical (MZ) twins reared in the same environment (that is, together in the same family) with the IQ scores of MZ twins separated as infants or while quite young and subsequently reared in different families. Different families are assumed to represent different environments, whereas the same family is assumed to represent the same environment. In early studies of this kind it was further assumed that any variation between the IQs of twins reared together and those reared apart must reflect environmental effects on intelligence.

The degree to which a trait is similar in identical twins is called **concordance**. For most traits, MZ twins show a higher concordance for a particular trait than DZ twins, and these findings are interpreted to mean that variation in these traits has some genetic basis. For example, the physical traits of MZ twins have a higher concordance than the same physical traits in DZ twins. However, as emphasized earlier, to obtain a heritability value that can be interpreted with confidence, the genetic contribution to the trait must be separated unambiguously from the environmental contribution. The example of bean-weight heritabilities shows how readily different heritability values may be obtained under different environmental conditions.

With twins it is impossible to reliably separate the influence of hereditary and environmental factors. One source of error is introduced even during the development of MZ twins *in utero*. Each individual of the pair may be exposed to different nutrients, chemicals, or experiences prior to birth and develop differently (Figure 19-7). Identical twins usually have somewhat different birth weights, which shows that the environment has already affected their development. And from birth on, each twin will have countless experiences that are not shared by the other twin, even if they are reared in the same family environment. For example one twin of a pair may nearly drown, may be hit by a car, may contract a serious disease, or may consistently be treated differently by one or both parents.

Psychologists attempt to circumvent these environmental variables by studying pairs of twins who are separated from each other and reared separately in what are expected to be different environments. However, it cannot be established how similar or different various environments really are. Thus, many psychologists and scientists believe twin studies are misleading as a measure of heritability for human traits. In the case of intelligence, for example, nobody knows what IQ tests really measure (though they are called intelligence tests) or what environmental factors specifically affect the performance of people on IQ tests. Thus, even ignoring the scandalous history of deliberately falsified twin data (discussed in Box 19-2), it does not seem likely that twin studies can prove how much of the variation in human intelligence is due to different genes and how much is due to different environments.

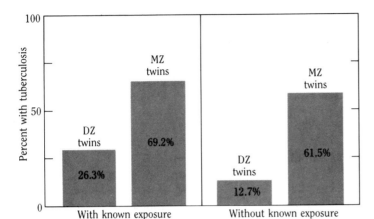

Figure 19-8 Concordance in identical (MZ) and fraternal (DZ) twins for tuberculosis. At the time these studies were performed, the higher concordance among identical twins was interpreted to mean that TB was a hereditary disease.

TWINS AND TUBERCULOSIS

Twin studies have been employed not only in efforts to prove that variation in intelligence is largely biologically determined but also in attempts to prove the genetic basis of other behaviors, ranging from criminality to spelling ability, and of other traits, such as disease susceptibility to tuberculosis (TB). For example, in the 1940s the high concordance of tuberculosis in identical twins was cited by some people as proof that TB susceptibility was genetically determined (Figure 19-8).

Once again, the issue is not whether particular genes may predispose certain individuals to contract tuberculosis but rather whether the relative amounts of genetic and environmental contributions can be measured by twin studies. The increased concordance of TB for identical twins as compared to fraternal twins might be explained by differences in uterine development or by the more closely shared environment of MZ twins, which exposes them to infection more often than that of DZ twins. Deaths from TB have declined dramatically in the United States not by changing peoples' genes but by eliminating or controlling the environmental factors that actually cause TB. Overall, genes in the U.S. population have not changed in the past eighty years, nor has the causative agent of TB, *Mycobacterium tuberculosis,* disappeared from the environment. What *has* changed are peoples' diets, sanitary habits, and living conditions. The high heritability seemingly demonstrated for TB from the degree of concordance in twin studies could have led to the conclusion that the incidence of TB could not be changed by changing the environment—a conclusion that obviously would have been false.

TWINS AND OTHER DISEASES

Belief is widespread today (even among physicians) that there is a genetic basis for schizophrenia and for insulin-dependent diabetes, a metabolic disorder characterized by abnormal insulin production. In these diseases, too, the issue is not whether genes are involved or whether certain human genotypes may be more susceptible than others but whether the relative genetic and environmental contributions can be measured. For diabetes and schizophrenia, twin studies are again used in an attempt to obtain heritability values and to establish the degree to which these diseases are caused by genetic differences among people. A recent analysis of the genetic basis of schizophrenia, written for the medical profession, concludes that:

> *Twin studies provide further support, but not conclusive proof, for genetic transmission of schizophrenia. Ironically, the only definitive conclusion the twin data allow is that environmental variables are important causal factors in schizophrenia—a conclusion demanded by the fact that far less than 100 percent of MZ twin pairs are concordant for schizophrenia.* *

The hereditary basis for insulin-dependent diabetes is also in doubt, based on the very twin studies that are often cited as establishing its genetic basis. In the largest study of diabetes in identical twins, published in 1976 in *The Genetics of Diabetes Mellitus,* the concordance was only 50 percent—thirty-two individuals out of thirty-two pairs of identical twins did *not* have diabetes, even though their co-twin had been diabetic for many years. Despite the research efforts over the years to establish a genetic basis for schizophrenia and diabetes, no defective genes, gene products, or chromosomal abnormality have been shown to cause either of these diseases.

The most significant fact to emerge from twin studies over the years is that some people expend enormous effort in attempts to determine whether human behaviors and diseases are genetically determined. Examination of all the scientific evidence indicates unambiguously that the environment is responsible for TB and for criminality, yet these traits also show a high concordance in MZ twin studies. Thus, the degree of concordance does not reliably indicate the degree of genetic involvement. The environment clearly is important in the variation in intelligence and in susceptibility to diabetes and schizophrenia. It is important to realize that while we can change environments, we cannot change our genes. Twin studies continue to produce misleading results (and unfortunate social consequences) because the relative contributions of genes and the environment simply cannot be unambiguously separated and measured for complex human behaviors.

---

*Kinney, D. K. and S. Matthysse. *Genetic Transmission of Schizophrenia.* Palo Alto: Annual Review of Medicine, 1978, p. 461.

The Race Betterment Movement Aims

To Create a New and Superior Race thru EUTHENICS, or Personal and Public Hygiene and EUGENICS, or Race Hygeine.

A thoroughgoing application of PUBLIC AND PERSONAL HYGIENE will save our nation annually:

1,000,000 premature deaths.

2,000,000 lives rendered perpetually useless by sickness.

200,000 infant lives (two-thirds of the baby crop)

The science of EUGENICS intelligently and universally applied would in a few centuries practically

WIPE OUT

Idiocy   Insanity   Imbecility   Epilepsy

and a score of other hereditary disorders, and create a race of HUMAN THOROUGHBREDS such as the world, has never seen.

Evidences of Race Degeneracy

Increase of Degenerative Diseases - - - { Cancer / Insanity / Diseases of Heart and Blood Vessels / Diseases of Kidneys / Most Chronic Diseases / Diabetes

Increase of Defectives { Idiots / Imbeciles / Morons / Criminals / Inebriates / Paupers

Diminishing Individual Longevity

Diminished Birth Rate

Disappearance, Complete or Partial, of Various Bodily Organs - - - { According to Wiedersheim there are more than two hundred such changes in the structures of the body

Figure 19-9   Examples of eugenics propaganda in the United States in the early years of this century. (From *Official Proceeding of the Second National Conference on Race Betterment,* August 4–8, 1915, Battle Creek, Michigan. Published by the Race Betterment Foundation, p. 147.)

Eugenics: The Science of Breeding Humans

The eugenics movement began in England in 1883 with the publication of a book by its founder, Francis Galton, titled *Inquiries into Human Faculty.* This was followed several years later by his book *Hereditary Genius.* In the first book Galton defined **eugenics** (from the Greek *eu,* "good" or "well," plus *genes,* "born") as "The science which deals with all influences that improve inborn qualities of a race." The American geneticist Charles B. Davenport was a powerful advocate of eugenics in the United States, and in his book *Heredity in Relation to Eugenics* he gave this definition: "Eugenics is the science of improvement of the human race by better breeding."

From around 1870 to 1905, poverty, "feeblemindedness," and insanity were all commonly regarded as resulting from hereditary defects, and between 1905 and 1930 the idea that *all* human weaknesses are the result of defective genes was widely believed (Figure 19-9). The eugenics movement was particularly strong and influential in the United States during the early 1900s. Eugenics ideas, promulgated by some geneticists and other scientists, were readily embraced by politicians, aristocrats, and successful businessmen who were quite ready to be convinced that their superior socioeconomic status was due largely to their superior genetic endowment. It also seemed quite logical to most people then, and to some even now, that the way to improve the human species was to eliminate the "defectives"—or at least prevent them from breeding—while at the same time to encourage breeding among "superior" individuals.

Eugenics ideas are not supported by any scientific or sociological evidence, yet these notions, which are abhorrent to many people, are not easily laid to rest. Even today, similar views rephrased in modern pseudoscientific terms are espoused by modern eugenicists. William Shockley, a Nobel Laureate in physics, advocates improving human stock by breeding and has donated his own sperm to further these efforts. Shockley believes that

*Nature has colour-coded groups of individuals so that statistically reliable predictions of their adaptability to intellectually rewarding and effective lives can easily be made and profitably be used by the pragmatic man in the street.**

What Shockley means by this statement is that dark-skinned individuals are intellectually inferior to light-skinned individuals and that the "man in the street" recognizes this fact.

Eugenics policies based on a self-serving misunderstanding of available scientific evidence have caused a great deal of individual suffering and have had serious political and social consequences over the years. For example, the passage of the Johnson-Lodge Immigration Restriction Act in 1924 was a major accomplishment of the American eugenics movement. Eminent geneticists and psychologists testified that American racial stock was deteriorating because of the influx of immigrants from southern and eastern Europe. Much of the testimony that helped passage of the Act centered on data collected by a fervent American eugenicist, Henry Goddard. In 1912 Goddard was paid to administer IQ tests to immigrants and he found the following frequencies of feeblemindedness: Italians, Jews, and Hungarians, 83 percent; Russians, 87 percent! Based on these figures, immigration quotas for these four groups were lowered, and the Johnson-Lodge Act remained in effect until 1962.

In the 1930s Nazi Germany used the ideas of the eugenics movement to justify the incarceration and extermination of "undesirables" to prevent what the Nazis referred to as "the menace of race deterioration." This genocidal policy focused on Jews, but other European ethnic groups—gypsies, for example—were its victims, too. Before and during World War II the Nazis carried out genocide on an incomprehensible scale, exterminating millions of people on the basis of their religion, nationality, or supposed mental inferiority.

The Nazi leaders also found support for their policies among influential scientists, including Konrad Lorenz, the founder of **ethology** (the study of the biological basis of animal behaviors). Lorenz had correctly shown that many animal behaviors, partic-

---

*Shockley, W. *Phi Delta Kappan*, January 1972, p. 307.

ularly among birds, are biologically determined, the most famous example being the capacity for *imprinting* in geese. During a critical period after hatching, baby geese follow whatever animal is present—be it a mother goose or a person—and from that moment on treat the individual as a parent. Lorenz felt justified in extrapolating his animal findings to human behaviors. Lorenz's statements such as the following were embraced by the Nazis and used to support their genocidal acts:

> *The selection for toughness, heroism, social utility . . .*
> *must be accomplished by some social institution if*
> *mankind, in default of selective factors, is not to be*
> *ruined by domestication-induced degeneracy. The racial*
> *idea as the basis of our state has already accomplished*
> *much in this respect. . . . The most effective*
> *race-preserving measure is that which gives the greatest*
> *support to the natural defenses. . . . We may—and we*
> *must—rely on the healthy instincts of the best of our*
> *people . . . for the extermination of elements of the*
> *population loaded with dregs. Otherwise, these*
> *deleterious mutations will permeate the body of the*
> *people like the cells of a cancer.**

The eugenics movement was powerful enough in the United States to force enactment of laws even more discriminatory than the Immigration Act of 1924. To prevent the infiltration of "undesirable genes" into the Caucasian population (and possibly of "desirable genes" into black populations), thirty-four states passed **miscegenation** laws, making marriage illegal between blacks and whites and, in some states, between Orientals and whites, too. These miscegenation laws remained in effect in many states until 1967, when they were declared unconstitutional by the U.S. Supreme Court.

About the same number of states also passed laws authorizing sterilization for a variety of supposedly inherited conditions and behaviors, including feeblemindedness, tendency to commit rape, alcoholism, and criminality. Even the brilliant and liberal Supreme Court Justice Oliver Wendell Holmes was influenced by the eugenics movement and wrote in his decision upholding the compulsory sterilization laws: "Experience has shown that heredity plays an important part in the transmission of insanity and imbecility."

No doubt heredity can play a role in these traits in some cases, as we have pointed out, but defective genes are not what caused the "feeblemindedness" in the thousands of pellagra vic-

---

*Chorover, S. L. *From Genetics to Genocide*. Cambridge: Massachusetts Institute of Technology Press, 1979, p. 105.

tims described in Box 19-1. Human experience is not a particularly good tool for measuring the hereditary contribution to a trait. In the course of human history, experience has told people that the earth is flat, that small persons are inside sperm, and that the sun revolves around the earth. Experience may "tell" a banker that he gained his wealth because of his superior genes, or it may "tell" a family of musicians that their talent was inherited, but these beliefs are not supported by genetic or other scientific evidence. Despite the injustices that have occurred in the past and the genocidal acts that have been carried out, the idea of improvement of human qualities by selective breeding continues to hold appeal for some people.

The ideas of the eugenics movement that were so scientifically and politically influential during the first half of this century have now been discredited and discarded for the most part, at least in their previous form. But in the past ten years the potentially dangerous ideas of biological determinism of human behaviors have been revived in a new scientific form known as sociobiology.

What Is Sociobiology?

The term *sociobiology* was introduced as early as 1946, but it attracted little notice until a book entitled *Sociobiology: the New Synthesis* by Harvard biologist Edward O. Wilson appeared in 1975. Wilson, who is generally regarded as the founder and most influential spokesman of sociobiology, defines it as "the scientific study of the biological basis of all forms of social behavior in all kinds of organisms, including man." Most of the book is devoted to descriptions of the biology, evolution, and social organization of insects and other animals. It is only in the final chapter that Wilson extrapolates his observations from animal studies to human behaviors and societies and postulates that such human behaviors as aggression, homosexuality, criminality, and morality are biologically (genetically) determined and are a consequence of evolution and natural selection.

The central thesis of **sociobiology** is that the behaviors of animals and societies are determined to a large degree by genes whose existence in present-day populations has resulted from evolutionary mechanisms, principally natural selection. Sociobiologists are ardent evolutionists and apply the Darwinian ideas of adaptation, fitness, and natural selection to explain animal behaviors whenever possible. One of the goals of sociobiology is to extend the principles of evolutionary and population genetics to human behaviors and cultures and to reorient such disciplines as anthropology, sociology, psychology, philosophy, and economics toward biological and evolutionary explanations.

From its very inception, sociobiology has been sharply criticized as yet another attempt to mask society's ills with genetic explanations. When *Sociobiology: the New Synthesis* appeared in 1975, critics attacked Wilson for lending support to the biological determinists who sought to explain existing social evils by the presence of "defective" genes in human populations. Wilson's critics not only denounced the scientific flaws in his arguments—which they claimed were serious—but also pointed out the potentially dangerous social consequences of sociobiology. Sociobiologists, it was argued, were continuing to advocate the pseudoscientific ideas of the eugenics movement.

Wilson argued in response that he and other sociobiologists were concerned only with discovering the scientific truths underlying animal behaviors and that establishing the genetic basis of human behavior was not their primary goal. However, in retrospect, the critics' warnings appear to have been justified, since three years later another book by Wilson appeared—*On Human Nature*—in which he did attempt to establish the biological basis for human social behaviors. The purpose of *On Human Nature*, Wilson asserted, "is simply the extension of population biology and evolutionary theory to social organization." By 1981 Wilson had undeniably joined the ranks of the biological determinists and with C. J. Lumsden published another book, *Genes, Mind, and Culture*, which presented a theoretical model showing how human genes determine human behaviors. And in a lecture presented at the American Genetic Association, Wilson asserted: "Natural selection acting on behavior within particular cultures alters the frequencies of the genes underlying the developmental processes of cognition and behavior."*

SOCIOBIOLOGY'S EXPLANATION OF ALTRUISM

Altruism is a behavior that benefits another individual at some expense to the doer. Altruistic behaviors can reduce the individual's chances of survival; hence, the individual's genes will not be passed on if reproduction has not yet occurred. For example, providing food for others while going hungry oneself is altruistic behavior; saving another person from death at the risk (or cost) of one's own life is an extreme form of altruism.

W. D. Hamilton, a British scientist, showed in an important series of articles, beginning with "The Genetical Theory of Social Behavior" (1964), that altruism can be explained genetically and, at least in theory, could have been advantageous to animals during the course of evolution. Hamilton showed that sister female ants are genetically more related to each other than to their brother ants or even to their parents, for the following reasons (Figure 19-10). Female worker ants are sterile and expend most of their time and energy feeding the queen (altruistic

---

*Wilson, E. O. "Epigenesis and the Evolution of Social Systems." *The Journal of Heredity*, 72(1980), p. 70.

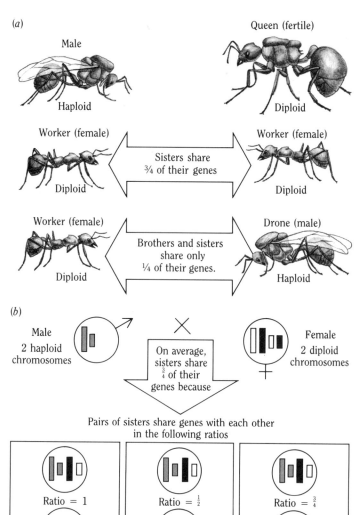

Figure 19-10 Insects such as ants exhibit altruistic behaviors. (a) Because males are haploid and females are diploid, sterile female worker ants share, on average, three-fourths of their genes with one another. They share only one-fourth of their genes with their brother drones. By caring for the queen, female worker ants pass on more of their genes to the next generation than if the workers themselves had progeny. (b) The diagram shows how sister ants share their genes. Only two chromosomes are shown. All female worker ants inherit the father's chromosomes, and on average, they inherit half of the mother's. Thus, a population of sister ants have three-fourths of their genes in common.

behavior) and helping her to produce more progeny ants. Since the female workers are sterile, they produce no offspring themselves. How, then, are their genes transmitted to future generations? The answer is that their altruistic behavior in supplying the queen with food and caring for the nest ensures that their genes are passed on to succeeding generations through the vehicle of the queen. This is because all sterile females produced by the queen share three genes out of four with one another, so the female ants' altruistic behavior is, genetically speaking, advantageous to them. If worker ants were not sterile and produced their own progeny, they would share only half of their genes with the next generation instead of three-fourths.

The evolutionary mechanism that has been advanced to

explain altruistic insect behavior is called **kin selection**, a process that operates differently from natural selection. Darwin showed that new traits arise by natural selection and give an advantage to individuals who are reproductively more fit as a consequence of those traits. In kin selection, however, it is the genetic relatedness of the entire family group that is subject to selection, not the individual. Individuals per se are not important in kin selection; in fact, an individual's death may ensure that more of that individual's genes are actually passed on by its surviving relatives. For example, a male lion that is killed protecting his pride of lionesses and cubs may have ensured that his genes survive and are passed on by his numerous offspring.

Kin selection provides an evolutionary mechanism that can account for the presence of genes that govern altruistic behavior in insects. Whether this mechanism also accounts for altruistic behaviors in other species of animals remains an open question. In kin selection the reproductive fitness of the individual is no longer of paramount importance; rather, it is the **inclusive fitness** of the entire genetically related group that must be considered.

Years ago, J. B. S. Haldane, a famous British biologist and geneticist, quipped that he would give up his life for three brothers or nine cousins. The mathematical logic of his statement derived from the number of genes that he shared with his brothers or cousins. The number of genes shared by various family members is expressed by a **coefficient of relatedness** (see Table 19-1). If Haldane's relatives were to produce more progeny than Haldane himself, then more of "his" genes would actually be present in the next generation (though they would be distributed among numerous individuals).

To sociobiologists, it is the evolutionary mechanisms of kin and group selection that are thought to be important in changing the frequency of genes in populations—particularly genes that affect behavioral traits. Individuals no longer need to be regarded as the units on which selection acts, as envisioned by Darwin. Genes themselves become the units of selection; hence the term "selfish gene" has come into widespread use. The British ethologist Richard Dawkins popularized the concept in his book *The Selfish Gene,* published in 1977. In Dawkins' view,

> . . . *a predominant quality to be expected in a successful gene is ruthless selfishness. This gene selfishness will usually give rise to selfishness in individual behavior. However, . . . there are special circumstances in which a gene can achieve its own selfish goals best by fostering a limited form of altruism at the level of individual animals.*

Despite Dawkins' assertion, the fact remains that genes reside in the cells of animals; because only animals mate and

Table 19-1. Coefficients of Relatedness (r) Between a Person and Specified Relatives*

Relative	r
Parent	1/2
Offspring	1/2
Full sibling	1/2
Half sibling	1/4
Grandparent	1/4
Grandchild	1/4
First cousin	1/8
Nephew or niece	1/4
Uncle or aunt	1/4
Second cousin	1/32

*Coefficients of relatedness express the mathematical probabilities of shared genes among close relatives. However, the calculations do not take into account the biological fact—derived from measurements of the number of polymorphic gene loci in human populations—that 70 to 80 percent of all genes in unrelated persons are identical. Totally unrelated persons—even those of different races—have identical alleles at the majority of the genetic loci. While coefficients of relatedness are useful in population genetics and for genetic counseling, the values are misleading. Close relatives are actually only a few percent more similar in their genetic constitution than are unrelated people.

reproduce and pass on genes to future generations, it can be argued that the organism is still the unit of selection. The genetic explanation for the altruistic behaviors of the social insects such as ants, bees, and termites may turn out to be correct. However, it also is the case that many insect species with social behaviors as complex as those of the ants and bees do not exhibit any altruistic behaviors. This implies that altruism (regardless of whether it has a genetic basis) is not an essential evolutionary mechanism for development of complex animal behaviors and societies. Moreover, humans are not insects, nor are human cultures necessarily a reflection of the biology that governs insect societies. This fact was recognized by Wilson in an article in which he described "slavery" among ants: "The slave-making ants offer a clear and interesting case of behavioral evolution, but the analogies with human behavior are much too remote to allow us to find in them any moral or political lesson."* In his more recent statements, however, Wilson has drawn parallels between insect behavior and human behaviors and has ignored his own earlier admonition.

DOES SOCIOBIOLOGY REPRESENT A NEW FORM OF BIOLOGICAL DETERMINISM?

The "new synthesis" that is alluded to in the title of E. O. Wilson's first influential book and that sociobiology seeks to construct is the extension of genetic and evolutionary principles to all aspects of animal behavior, animal social oganization, and human cultural differences. Although most sociobiologists do not dismiss the contributions of environment to animal behaviors, their main focus is on demonstrating the biological basis of all behaviors. When applied to organisms that can be experimentally manipulated both genetically and environmentally in such a way as to give clear-cut results, such as insects or slime molds (microorganisms that aggregate and exhibit social behavior), sociobiology may provide new information and insight about the biological basis of behavior. However, sociobiologists discuss human behavior as if it is a slightly modified extension of the behavior of other organisms. This line of reasoning can produce erroneous conclusions that may have unfortunate social consequences.

For example, Wilson and other sociobiologists argue that male homosexuality in humans is genetically determined to a significant degree and can be explained in evolutionary terms by altruism and kin selection. According to this argument, in early human societies homosexuals may have helped raise their brothers and sisters, and even if the homosexual males did not leave progeny, their homosexual "genes," carried in a heterozy-

---

*Wilson, E. O. "Slavery in Ants." *Scientific American*, June 1975, p. 36.

Are XYY Males Overly Aggressive?

BOX 19-3

The addition or deletion of any autosome or X chromosome in the genome of an individual can have significant mental and physical effects. An extra autosome is usually lethal; those few trisomies that aren't lethal produce disorders such as Down's syndrome, which is characterized by severe mental retardation. An extra X chromosome in males or the loss of an X chromosome in females causes sterility and usually some physical and mental abnormalities. The Y chromosome, however, is unique in that, while it is essential for male sexual differentiation, it is the only chromosome on which no other essential genes have been shown to be carried.

When chromosome studies were carried out on inmates of penal institutions in Scotland in 1965, it was reported that 3.6 percent of the males examined had an XYY chromosome constitution; that is, they carried an extra Y chromosome in all of their cells. This study led to the speculation (by people other than the original investigators) that criminally deviant behavior in males might be caused by an extra Y chromosome. The popular media picked up the idea and fanned the speculation until, in 1968, it became sensationalized. In that year Richard Speck murdered several nurses in Chicago, and it was widely reported, based on a faulty karyotype analysis, that Speck was an XYY male. Subsequent karyotype analysis revealed that Speck's sex chromosomes were normal—that is, he was XY—but the retraction received little attention.

We now know that about 1 in every 1,000 males has an XYY chromosome constitution. Subsequent chromosome studies of inmates in penal and mental institutions have confirmed that about 2 percent of imprisoned males have an XYY genotype—a figure higher than the proportion in the general population. Does this mean that an extra Y chromosome in males causes, or even increases the likelihood of, criminal behavior? The answer should be an unequivocal *no* (recall the discussion in Box 2-1 on cause-effect relationships; correlation does not indicate causation).

Despite their overrepresentation in penal institutions, only about 1 percent of all XYY males acquire a criminal record. Since 99 out of every 100 XYY males are living normal, socially acceptable lives, it certainly is not fair to conclude that XYY males are "genetically" prone to crime. Yet in the 1970s various medical experts, including a prestigious group at the Harvard Medical School, wanted to carry out genetic screening of newborn males and psychological studies of those having an XYY chromosome constitution. Quite clearly, parents told that their infant is XYY and hence at risk for aggressive, criminal behavior would be affected psychologically, and such information would be bound to have a serious effect on how the parents raised and disciplined their child. Such a state of affairs might well lead to a self-fulfilling prophecy of behavioral problems and would certainly alter the parent-child relationship. Even if the dubious ethics of such a program were disregarded, it would be impossible to separate the genetic and environmental contributions to the behavior of the children studied.

Reading: Jonathan Beckwith and Jonathan King, "The XYY Syndrome: A Dangerous Myth." *New Scientist*, November 14, 1974.

gous state by brothers, sisters, and other relatives, would have been transmitted to subsequent generations. In Wilson's view: "Genes favoring homosexuality could then be sustained at a high equilibrium level by kin selection alone." Psychologists and sociologists, on the other hand, view homosexuality as a social and cultural phenomenon and hold that evidence for a biological basis for it is nonexistent at present.

Some sociobiologists argue, on the basis of observations of animal behavior, that aggression in human societies is genetically determined and that human "aggression genes" are the result of natural selection. Once one accepts the idea that human aggression is mainly the result of heredity, it then becomes easier to argue that criminality is genetically determined (see Box 19-3). And proof that criminal behavior is genetically based, according to proponents of this view, is the high concordance of criminality in identical twins. The opposing view to this biological determinism is that identical twins generally share close emotional and psychological experiences and are probably exposed to the same home or social situations that foster criminality. It seems clear that individuals and society are endangered by the loose logic of biological determinism and by application of twin studies to conclusions about human behaviors.

As the anthropologist S. L. Washburn has observed:

The basic problem with postulating genes for social behaviors is that it shifts the nature of the explanation from the logic of social facts to the logic of genetics. Even though the environment may be important in both cases, it makes a great deal of practical difference if the cause of crime is in large part genetic, or if crime is a word standing for a wide variety of behaviors learned under varying circumstances and defined differently in different cultures. *

The success or failure of sociobiology as applied to human social behavior will ultimately be decided after accumulation of considerably more evidence, discussion, and time. Whether sociobiology is a valid and scientific approach to understanding human behavior is not known; at present the reader must decide based on the existing evidence. For this reason it seems appropriate to quote E. O. Wilson, who, more than any other scientist, is the leading authority on sociobiology:

A key question of human biology is whether there exists a genetic predisposition to enter certain classes and to play certain roles. Circumstances can be easily conceived in which such genetic differentiation might occur. †

The question of interest is no longer whether human social behavior is genetically determined, it is to what extent. The accumulated evidence for a large hereditary component is more detailed and compelling than most persons, including even geneticists, realize. I will go further: it already is decisive. ‡

---

*Washburn, S. L. *Sociobiology*. September/October 1978, p. 39.
†*Sociobiology: The New Synthesis*. Cambridge, Mass.: Harvard University Press, 1975, p. 554.
‡*On Human Nature*. Cambridge, Mass.: Harvard University Press, 1978, p. 19.

Moderately high heritability has been documented in introversion-extroversion measures, personal tempo, psychomotor and sports activities, neuroticism, dominance, depression and the tendency toward certain forms of mental illness such as schizophrenia. [*]

Scientists and humanists should consider the possibility that the time has come for ethics to be removed temporarily from the hands of the philosophers and biologicized . . . [†]

The hereditary factors of human success are strongly polygenic and form a long list, only a few of which have been measured. IQ constitutes only one subset of the components of intelligence. Less tangible but equally important qualities are creativity, entrepreneurship, drive, and mental stamina. Let us assume that the genes contributing to these qualities are scattered over many chromosomes. [‡]

Sociobiologists and others seem to *assume* that genes determine behaviors, but as indicated in our discussion of heritability, such claims are not well substantiated by experiment—certainly not for humans. The nature-versus-nurture controversy about human behaviors will not be resolved, nor will it disappear, in the near future. All of us are inherently curious about why we behave as we do, and speculations about the causes of human behavior will continue. Despite the assurances of sociobiologists that genes and evolutionary mechanisms largely determine the behaviors of individuals and the customs of societies, no scientific evidence allows us to draw these conclusions yet.

Experiments can be conducted that tell us how DNA replicates, how it changes from generation to generation, and how it is transmitted from one organism to another. However, people cannot be manipulated like molecules of DNA, so less edifying experiments must be devised. Only by understanding the limitations of human experiments and the potential misuse or misapplication of heritability measurements can we properly evaluate the results and claims of both sides of the controversy.

Too often nature-versus-nurture arguments are based on what appear to be reasonable inferences that may, however, be misleading. For example, Wilson and other sociobiologists argue that because many people in many cultures have phobias (intense fears) of snakes or spiders, these phobias must have a genetic basis and phobic genes must have been selected for

---

[*]*Sociobiology*, p. 550.
[†]Ibid., p. 562.
[‡]Ibid., p. 555.

during evolution. Yet such arguments ignore the fact that other phobias are caused by very recent inventions and modern social situations. For example, many people are phobic about flying in airplanes—a fear that cannot have been selected for during millions of years of evolution.

The Future

If the facts of heredity and evolution have anything to tell us regarding the human condition, it is that all organisms on earth are biologically related. However, the fact that all organisms are biologically related does not mean that the behaviors of organisms are biologically determined. Nor does it mean that insect genes for altruism have been preserved in human chromosomes. We have all observed how different the members of families are, even though parents and children are closely related genetically and biologically. Observation tells us how important the environment is in determining human behavior, even though we cannot quantify the contribution of nurture or of nature. It is difficult, perhaps impossible, to design experiments that will unambiguously evaluate the contribution of genes and the environment to human traits. Because history has shown us how dangerous the personal and social consequences of biological determinism can be, we need to be exceptionally cautious in accepting assertions by sociobiologists or others that human intelligence, homosexuality, and aggression are determined primarily by heredity. People need to rely on clearly established scientific information and principles if individual and social injustices are to be reduced and prevented in the future.

Key Terms

environment all nongenetic factors that affect the phenotypes of organisms.

biological determinism the idea that most of the variation in human traits and behaviors is genetically determined.

heritability the amount of phenotypic variation in a population that is caused by genetic differences among individuals.

concordance the degree to which identical (MZ) twins are alike with respect to a particular trait.

eugenics the science of improving people by breeding.

ethology the study of the biological and evolutionary bases of animal behaviors.

miscegenation interracial marriage or cohabitation.

sociobiology the study of the biological (genetic) basis of all forms of social behaviors in all organisms, including humans.

altruism self-endangering behavior that benefits other individuals.

kin selection selection for altruistic behavior in a group of individuals that are genetically related.

inclusive fitness the sum of the fitness of an individual plus its influence on the fitness of relatives; the total effect of kin selection.

coefficient of relatedness the degree to which family members are genetically related.

Additional Reading

Block, N. J. and G. Dworkin (eds.). *The IQ Controversy*. New York: Pantheon Press, 1976.

Chase, A. *The Legacy of Malthus: The Social Costs of the New Scientific Racism*. New York: Alfred A. Knopf, 1977.

Chorover, S. L. *From Genesis to Genocide: The Meaning of Human Nature and the Power of Behavior Control*. Cambridge, Mass.: MIT Press, 1979.

Dawkins, R. *The Selfish Gene*. New York: Oxford University Press, 1976.

Ehrlich, P. R. "Human Carrying Capacity, Extinctions and Nature Reserves."*BioScience*, May 1982.

Gould, S. J. *Ever Since Darwin*. New York: W. W. Norton, 1979.

Mehler, B. "The New Eugenics." *Science for the People*, May/June 1983.

Montagu, A. (ed.). *Sociobiology Examined*. New York: Oxford University Press, 1980.

Rensberger, B. "The Nature-Nurture Debate: On Becoming Human." *Science 83*, April 1983.

Shannon, T. A. (ed.). *Bioethics*. Ramsey, N.J.: Paulist Press, 1982.

Wallace, B. "Genetics and the Great IQ Controversy." *American Biology Teacher*, January 1975.

Review Questions

1. What does it mean for a trait to have a heritability of 1.0?

2. Why is it inappropriate to compare heritability values between different populations?

3. What percentage of genes are shared by identical twins? by fraternal twins?

4. What kind of evolutionary mechanism can account for altruistic behavior in insects?

5. Which human chromosome is involved in criminal behavior?

Chapter 1

1. By weight.
2. Covalent, ionic, and hydrogen bonds.
3. See Table 1-1.
4. In stellar nuclear reactions.
5. See Table 1-3.
6. The explosion that created all the matter and energy in the universe.
7. Alpha particles (2P, 2N), beta particles (electrons), gamma rays (photons).
8. Electrons.

Chapter 2

1. All organisms are composed of cells; the chemical composition of all cells is basically similar; all organisms have evolved from a common ancestor.
2. RNA.
3. DNA is the hereditary material in cells and carries genetic information.
4. Phages are much smaller (invisible under the light microscope), lack organelles, and cannot grow or reproduce outside living cells.
5. ^{35}S to label proteins; ^{32}P to label DNA.
6. 46; 46; 46; 23.
7. Mitosis (M).

Chapter 3

1. Bases (A, T, G, C), deoxyribose sugars, and phosphates (PO_4).
2. Four; one.
3. Replication, mutation, and recombination.
4. DNA; DNA.
5. DNA; either DNA or RNA; only RNA.

Chapter 4

1. Males (F^+, F', Hfr); females (F^-).
2. Transduction, transformation, and conjugation.
3. By regulation of gene expression.
4. See Table 4-1. A human cell is about a half million times larger than a tumor virus.
5. Adenosine triphosphate (ATP).
6. To synthesize ATP and chemical energy.
7. To convert light energy to chemical energy.

Answers to Review Questions

Chapter 5

1. Translation.
2. Transfer RNA (tRNA).
3. At a promoter site.
4. At an AUG (start) codon.

Chapter 6

1. 64.
2. A frameshift mutation.
3. To bind and transport oxygen to cells.
4. Transfer RNA.
5. UAG, UAA, UGA.
6. Transfer RNAs.

Chapter 7

1. All four.
2. One base pair; two or more base pairs.
3. Dominant.
4. Germinal mutations arise in sperm and eggs and can cause hereditary changes.
5. Ultraviolet (UV) light.
6. Repair of damaged DNA by enzymes.
7. Cigarettes.

Chapter 8

1. Ionizing radiation, carcinogenic chemicals, tumor viruses, radioactivity.
2. Unregulated cell growth and division.
3. Any agent that causes mutations is a mutagen; if it causes cancer in animals, it is also a carcinogen.
4. Cigarette smoke.
5. Fewer than 5%.
6. DNA is synthesized from RNA.

Chapter 9

1. At least five: RNA polymerase, repressor, lactose, cyclic AMP, and CAP protein.
2. RNA polymerase.
3. Eucaryotic cells.
4. They lack an enzyme able to break down the sugar lactose.
5. Yes—identical twins are clones.
6. The base sequence of each promoter site, which determines the attachment of RNA polymerases to that promoter.

Chapter 10

1. Restriction enzymes that break DNA specific sites.
2. To attach methyl groups to DNA at specific sites.
3. Plasmids.
4. Three (see text).
5. —GAATTC—
   —CTTAAG—
6. DNA.

Chapter 11

1. It is determined by a single gene consisting of dominant and recessive alleles.
2. Seven traits.
3. Meiosis.
4. Because fruit flies have a haploid set of 4 chromosomes, $2^4 = 64$ different gametes.
5. No.
6. One gene with many different alleles.

Chapter 12

1. Gametes: sperm and egg.
2. Sexual selection.
3. Hormones.
4. Klinefelter's syndrome, Turner's syndrome, Down's syndrome.
5. X chromosome.
6. Initiating development of males.

Chapter 13

1. Yes. Cancer is probably initiated by genetic changes in somatic cells.
2. A point mutation that causes an enzyme to become inactive.
3. They cause abnormal development in embryos.
4. Hemophilia.
5. Men.
6. One-fourth.

Chapter 14

1. About 5%.
2. Spina bifida, anencephaly, neural tube defects.
3. More than 100.
4. Hemoglobin.
5. Phenylketonuria (PKU).
6. No.

Chapter 15

1. Mouse chromosomes.
2. No—red blood cells lack chromosomes.
3. By characteristic bands produced by specific dyes.
4. Four.
5. Yes (see Box 15-1).
6. Yes (somatic hybrids).

Chapter 16

1. Yes, in sterile environments.
2. IgE.
3. Only one.
4. Recombination.
5. B lymphocytes or plasma cells.
6. Four genes, each consisting of many alleles; one gene with three alleles.

Chapter 17

1. DNA.
2. Gene frequencies do not change.
3. 25%; 37.5% are heterozygous.
4. Moths.
5. Migration.
6. In humans, N = 23; in chimpanzees, N = 24.
7. Gene frequencies must change.

Chapter 18

1. About 65 million years.
2. About 3-4 million years.
3. Decay of radioisotopes.
4. Much more accurate.
5. Bacteria and blue-green algae.
6. About 75% are identical.

Chapter 19

1. It is determined only by genetic factors; the environment plays no role.
2. Because the environments are different, the measurement of heritability is affected.
3. 100%; 50%.
4. Kin selection.
5. None. The biological consequences of an extra Y chromosome in XYY males are unknown.

Essay Topics

Chapter 1
1. Why do you think all cells and organisms are composed of carbon-containing organic molecules?
2. Is it possible that human physiology and behaviors are governed by forces other than the four fundamental forces of nature listed in Table 1-4?
3. Can the property of organisms we call "life" be explained solely by the chemistry of molecules and cells? Why or why not?
4. Why do both procaryotic and eucaryotic organisms exist in nature?
5. Describe some of the essential functions that enzymes perform in the human body.

Chapter 2
1. What do you regard as the most important properties of living organisms, particularly humans?
2. Describe some of the different kinds of environments that bacteria inhabit.
3. Discuss the strengths and weaknesses of the early experiments showing that DNA carried genetic information.
4. Discuss why some scientific discoveries are premature—that is, are not accepted until many years afterward.

Chapter 3
1. What is the difference between conservative and semiconservative DNA replication? How would you test for this difference experimentally?
2. Describe the mechanism that allows chromosomes to exchange segments of their DNA.
3. Discuss the essential difference between Lamarck's and Darwin's ideas of how organisms evolve.

Chapter 4
1. Discuss the similarities and differences between plasmids, transposons, and insertion sequences.
2. How should antibiotics be used?
3. Why are antibiotics used to treat bacterial infections but not viral infections?
4. Discuss how bacterial plasmids can cause some kinds of diarrhea.
5. What is meant by a genetic map? What kind of information does it contain?

Chapter 5
1. Describe the process that is used to synthesize protein molecules in cells.
2. What is meant by the quaternary structure of proteins?
3. What kind of molecules might have carried genetic information before cells arose on earth?

Chapter 6
1. Discuss why the genetic code had to be at least a triplet code.
2. Why is the genetic code universal in all organisms with only a couple of exceptions?
3. Why do most mutations that occur in human cells have harmful effects rather than beneficial ones?
4. If living organisms were found on a planet in a remote galaxy, would you expect them to have the same genetic code found on earth?

Chapter 7

1. All humans carry several potentially lethal mutations in their DNA. Why is this the case?

2. Describe how bacteria are used to detect carcinogenic substances in the environment.

3. What is the difference between missense and nonsense mutations?

4. Describe how ionizing radiation and radioactive substances can cause mutations.

5. Why is the use of sodium nitrite to preserve meats such as bacon and hot dogs controversial?

Chapter 8

1. Why do epidemiologists believe that most cancers are preventable?

2. How does sunlight cause skin cancer in some sensitive persons? Why can excess exposure cause cancer in anyone?

3. Is it all right for pregnant women to be x-rayed? Why or why not?

4. What is the oncogene hypothesis?

5. How might genetic screening of workers prevent some cancers?

Chapter 9

1. How is development of an organism from a zygote determined by differential gene expression?

2. Describe the various regulatory sites in the *lac* operon and their functions.

3. Explain how totipotency was demonstrated in experiments using the toad *Xenopus Laevis*.

4. How might split genes have arisen during evolution?

5. Describe how it was shown that the *lac* repressor is a protein.

Chapter 10

1. Describe how recombinant DNA experiments are performed.

2. Has an adult human ever been cloned? Should such experiments be attempted?

3. Why are R-plasmids important factors in some human diseases?

4. How would you go about constructing a library of human genes?

5. What do you consider to be the most important benefits and risks of recombinant DNA technology?

Chapter 11

1. What is the law of segregation? the law of independent assortment?

2. Why is recombination an important genetic mechanism for evolution?

3. Why are genetic maps of human chromosomes useful?

4. What is a pleiotropic gene? Give several examples in humans.

5. Why did the ideas of pangenesis last for so many centuries?

Chapter 12

1. How are primary and secondary sex characteristics determined in humans?

2. Is homosexuality in human populations determined primarily by genetic or by environmental factors? Support your answer.

3. Describe how sex chromosomes were first discovered and identified.

4. Describe some of the effects in persons who have one X chromosome too few or too many.

Chapter 13

1. List the three criteria used to establish the hereditary basis of a disease or abnormality.

2. Why are pregnant women advised not to take drugs of any kind?

3. What is achieved in carrying out a pedigree analysis?

4. What is polygenic inheritance? Give several examples.

5. Why was the term "mongolism" applied to individuals with trisomy 21?

Chapter 14

1. Who do you think should receive genetic counseling in our society? Who should receive amniocentesis?

2. Discuss the relative risks and benefits of such prenatal tests as amniocentesis, ultrasound scanning, and fetoscopy.

3. How can society benefit from prenatal tests and genetic counseling?

4. What is the difference between genetic screening and genetic counseling?

5. What are the risks and benefits of widespread genetic screening for a large number of hereditary defects?

6. When, in your opinion, does life begin? Support your answer.

Chapter 15

1. Discuss some of the important scientific uses of the technique of somatic cell hybridization.

2. Explain how genetic complementation occurs and how it is used in experiments.

3. What kind of information is obtained by electrophoresis of proteins?

4. How are human chromosomes visualized and identified in a karyotype analysis?

5. What is immunological tolerance? Describe the cellular mechanisms that determine tolerance.

Chapter 16

1. Describe the various functions of the human immune system.

2. What is immune surveillance? What function does it serve in the body?

3. Explain the clonal selection model. How does it account for the origin of antibody diversity?

4. Why is it that skin generally cannot be grafted from one person to another?

5. How are monoclonal antibodies produced? List some of their important applications.

Chapter 17

1. Explain why natural selection is regarded as the fundamental mechanism for biological evolution.

2. Why are islands such as Hawaii and the Galapagos so important in the study of evolution?

3. Describe the various forces in nature that change the frequency of genes in populations.

4. Why do most human cultures have taboos that prohibit matings between close relatives?

5. Eugenicists advocate sterilization of persons carrying deleterious recessive alleles. Why would this practice *not* reduce significantly the number of infants born with hereditary diseases caused by such alleles?

Chapter 18

1. Describe how radioisotopes are used to determine the age of rocks and fossils.

2. What kind of evolutionary information is provided by the fossil record?

3. Should creationism be taught in public schools? Why or why not?

Chapter 19

1. What is meant by "nature versus nurture" with respect to human behavior?

2. Explain in detail the differences between hereditary and heritability.

3. Discuss the flaws and problems associated with human heritability measurements.

4. Describe some of the historical consequences of biological determinism in human societies.

5. Do you think there is a genetic difference between black and white populations that determines performance in sports? Support your answer.

A

ABO blood group three alleles that direct the synthesis of red blood cell antigens, which determine whether transfused blood will be accepted or rejected by a recipient.

achondroplasia an autosomal dominant defect that affects cartilage growth, causing dwarfism.

acrosomal cap proteins surrounding the head of sperm that facilitate attachment to the egg.

activating enzymes the twenty different enzymes responsible for attaching tRNAs to their corresponding amino acids.

alleles alternative functional states of the same gene occurring on homologous chromosomes.

allophenic mice mice that are formed by mixing embryonic somatic cells from two genetically different embryos and implanting the composite embryo into the uterus of a female mouse.

alpha-fetoprotein (AFP) a protein produced during embryonic development, the level of which in the blood can be used to detect neural tube defects.

altruism self-endangering behavior that benefits other individuals.

Ames test a test for the mutagenic and carcinogenic potential of chemicals that involves measuring the reversion of histidine-requiring mutants of *Salmonella* bacteria.

amino acids the twenty different molecules that are found in proteins.

amniocentesis a procedure in which a sample of amniotic fluid and fetal cells is removed from the uterus and examined for genetic or biochemical defects.

androgen insensitivity syndrome a syndrome characteristic of women who have an XY chromosome constitution.

androgens a group of male hormones (the major one is testosterone) that are produced early in development and result in formation of testes.

aneuploidy the characteristic of having one chromosome more or less than the set normally found in an organism.

antibody an immunoglobulin molecule capable of combining with specific antigens and thereby inactivating them.

anticodon the particular three bases in a tRNA molecule that recognize the codon corresponding to the amino acid attached to the end of the tRNA.

Glossary

antigen any foreign substance that stimulates an antibody response in an animal.

atoms the smallest units of any substance.

ATP adenosine triphosphate, the molecule in all cells that supplies the energy for most chemical reactions.

autoimmune disease a disease such as some forms of rheumatoid arthritis or lupus erythematosus, in which the immune system reacts to normal body cells and damages or destroys normal tissues.

autoradiography visualization of DNA by incorporation of radioactive chemicals that afterward expose a photographic film.

autosomes all chromosomes in eucaryotes excluding the sex chromosomes.

B

β-galactosidase an enzyme that breaks apart lactose into the sugars glucose and galactose.

B lymphocyte a white blood cell that carries the genetic information for synthesis of particular kinds of antibodies.

bacterial colony a visible spot of bacterial growth consisting of millions of cells.

bacterial strains bacteria of the same species that share many phenotypic characteristics but differ from one another in some respect.

bacteriophages infectious particles containing DNA or RNA that are able to grow and reproduce in bacteria. Phages generally destroy the bacteria they infect.

Barr bodies   dark staining spots in cell nuclei that contain a condensed, inactive X chromosome.

bases   the chemical entities in DNA whose sequence encodes the genetic information: adenine (A), thymine (T), cytosine (C), and guanine (G).

biological determinism   the idea that most of the variation in human traits and behaviors is genetically determined.

biosynthetic pathway   a series of enzymes that act sequentially to synthesize new molecules.

break-rejoin model   model explaining how genetic recombination results from the physical exchange of segments of DNA.

C

cancer   unregulated growth of plant or animal cells. The various diseases that result from the growth of masses of cells.

cancer-promoting agents   chemicals that, while not themselves carcinogens, promote formation of tumors by cells previously exposed to carcinogens.

capacitation   chemical changes produced in sperm as they swim through the vagina and uterus.

carcinogen   any agent that can cause cancer.

carcinomas   cancers of skin and membrane cells.

carriers   individuals who carry a mutant gene on one chromosome but who are themselves unaffected because the allele on their other chromosome is normal.

catabolite activator protein   the positive controlling element for glucose-sensitive operons in bacteria.

cell cycle   the four stages of growth and reproduction of eucaryotic cells.

cells   the fundamental units of life. All cells are capable of reproducing themselves.

central dogma   the rules that govern the exchange of information among DNA, RNA, and protein molecules.

centromere   a region on chromosomes that partially determines their structure and that assists in segregating them during mitosis and meiosis.

chiasma (pl. chiasmata)   the visible physical crossing over between the nonsister chromatids of homologous chromosomes at meiosis.

chromatid   one of the two sisters formed after a chromosome divides.

chromosome banding   the patterns of bands that become visible when chromosomes are stained with dyes.

chromosomes   structures in the nuclei of eucaryotic cells that carry the genetic information in the form of DNA molecules.

clonal selection model   a theory explaining how antibody diversity is generated during embryonic development and how the immune system distinguishes "self" from "nonself."

clone   individuals with identical genetic information. Identical human twins are clones since both developed from a single fertilized egg that split into two early in development.

cloning vehicles (vectors)   small DNA molecules— usually plasmids or viruses—into which foreign DNA (genes) is inserted. When the cloning vectors are reintroduced into microorganisms, the plasmids or viruses replicate their own DNA as well as the foreign DNA. Thus, billions of identical copies of the foreign genes are produced (cloned).

code degeneracy   the property of the genetic code by which some amino acids are coded for by more than one codon.

code universality   the fact that in every kind of organism each codon specifies the same amino acid.

codon   a group of three bases in mRNA that specifies one of the twenty different amino acids.

coefficient of relatedness   the degree to which family members are genetically related.

concordance   the degree to which identical (MZ) twins are alike with respect to a particular trait.

congenital defect   any physical abnormality that is detectable at birth.

conjugation   mating between a male and female bacterium in which DNA is transferred from the male and incorporated into the DNA of the female by recombination.

covalent bond   a chemical bond created by the equal sharing of electrons by two atoms.

crossing over   the exchange of segments between pairs of homologous chromosomes during meiosis.

cytoplasm   the part of the cell that lies outside the nucleus.

D

degradative pathway   a series of enzymes that act sequentially to break down small molecules.

deletion   loss of one or many base pairs in DNA.

density-dependent growth   growth of somatic cells *in vitro* that stops when a certain cell density is reached or when neighboring cells come into contact with each other.

differentiation   the process by which cells become increasingly (and irreversibly) specialized in their functions in tissues and organisms during development.

diploid cells   cells that contain two copies of each different chromosome.

disulfide bridges   a covalent bond between two sulfur atoms in different cysteine amino acids in a protein.

DNA   deoxyribonucleic acid, the macromolecule carrying the hereditary information in cells.

DNA cloning   the production of billions of copies of a piece of DNA as part of a plasmid introduced into a microorganism.

DNA-DNA hybridization   a technique used to measure the similarity in base sequences between DNA molecules from different organisms.

dominant allele   the allele in a heterozygous individual that determines the phenotype.

Down's syndrome   an inherited human disease caused by an extra chromosome number 21 in all of the individual's cells.

duplication   duplication and insertion of a group of base pairs into DNA.

E

electrons   negatively charged particles that orbit the nuclei of atoms.

electrophoresis   a technique used for the separation of nearly identical proteins based on differences in their electrical charge.

element   a substance that cannot be broken down further by ordinary chemical or physical means.

endonuclease   an enzyme that breaks and removes bases from DNA (or RNA) at specific sites within the molecule.

environment   all nongenetic factors that affect the phenotypes of organisms.

enzymes   proteins that catalyze chemical reactions.

epistasis   the masking of the expression of a gene by a gene located elsewhere in the genome.

equivalence rule   the rule that the number of adenine bases in all DNA molecules equals the thymine bases and the number of cytosine bases equals the guanine bases.

ethology   the study of the biological and evolutionary bases of animal behaviors.

eucaryotes   complex unicellular or multicellular organisms.

eugenics   the science of improving people by breeding.

evolution   the changes in gene frequencies that accumulate over time in populations of organisms.

exons   the discontinuous segments of DNA in eucaryotic genes that carry information that is translated into the sequence of amino acids in proteins.

exonuclease   an enzyme that removes bases from the ends of DNA (or RNA) molecules.

F

F plasmids   small, circular DNA molecules in bacteria that can be transferred to other bacteria. Occasionally, an F plasmid facilitates transfer of bacterial chromosomal genes in addition to its own.

family pedigree   the pattern of a trait's occurrence in family members from generation to generation.

feedback inhibition   a mechanism by which the final product of a biosynthetic pathway (say an amino acid) inhibits the activity of the first enzyme in that pathway.

female gametes   eggs in all sexually reproducing organisms.

fertilization   the fusion of a sperm and an egg to form a zygote.

fetoscopy   visual observation of a fetus by inserting an optical instrument into the uterus.

fitness   the reproductive contribution of an individual to the following generations.

forward mutation   a change from the wild-type to a mutant organism.

founder effect   demonstration of a gene whose frequency is much higher in a particular population than it is in other populations.

frameshift mutation   a mutation that results from the insertion or deletion of one or two base pairs in DNA.

G

gene   a sequence of bases in DNA that codes for a functional cellular product, usually a protein.

gene dosage   having more or less than the normal number of genes.

gene expression   the flow of information from genes in DNA into RNA and ultimately into proteins.

gene libraries   millions of bacteria containing all the genetic information from an organism whose DNA has been cloned in the bacteria.

gene linkage   a measure of the degree to which a pair of genes assort independently at meiosis or when crosses are performed between genetically different individuals. Also, a measure of the relative distance between genes on chromosomes.

genetic locus   the specific location of a gene on a chromosome.

genetic map   the imputed arrangement of genes (alleles) on a chromosome according to the observed frequency of recombination between them in genetic crosses.

gene pool   all of the alleles in all the individuals in a population.

generation time   the time required for growing cells to double their mass or to divide into two cells.

genetic complementation   in somatic cell hybrids, the provision of essential functions by genes on different chromosomes from different cell lines. More generally, a test for whether mutations occur in different functional genes or on two different chromosomes.

genetic counseling   a process of communicating the risks of having a defective child to prospective parents.

genetic drift   variation in gene frequencies in a population from one generation to the next as a result of chance.

genetic load   the decrease in fitness in a population due to the number of carriers of deleterious genes.

genetic map   the locations of genes with respect to one another in chromosomes.

genetic screening   a diagnostic procedure that detects individuals in a population who carry defective genes.

genome   the total amount of genetic information contained in an organism.

genotype   the particular set of genes present in the DNA of organisms.

germinal mutation   a mutation that arises in the chromosomes of sex cells and that is inherited.

H

half-life   the time it takes for one-half of the nuclei of a radioactive isotope to decay.

haploid cells   cells in which each different chromosome is present in single copy. Sperm and eggs are haploid cells of animals.

Hardy-Weinberg law   a mathematical expression of the relationship between allele frequencies and genotype frequencies in an idealized population.

hemophilia   a recessive sex-linked (X-linked) disease that causes afflicted persons to be "bleeders" because of the lack of a blood clotting factor.

hereditary (genetic) diseases   diseases resulting from abnormal chromosomes or mutant genes that are transmitted from one or both parents to offspring.

heritability   the amount of phenotypic variation in a population that is caused by genetic differences among individuals.

hermaphrodites   individuals in which both the male and female reproductive organs are present.

heterogametic sex   the sex of a species that produces gametes with two types of sex chromosomes.

heterokaryon   a single cell with separate nuclei, formed by the fusion of two cells *in vitro.*

heterozygous   having different alleles of the same gene at the same locus on homologous chromosomes.

Hfr bacteria   male bacteria that transfer their chromosomes with high frequency to female bacteria.

histocompatibility   the condition that determines whether tissue transplanted from one animal to another will be accepted or rejected.

histones   a small group of proteins that are thought to regulate gene expression and chromosome structure in animal cells.

hominid a mammal whose characteristics are more humanlike than apelike.

homogametic sex the sex of a species that produces gametes containing only one type of sex chromosome.

homologous chromosomes the two members of a chromosome pair that were inherited from each parent and that pair up (synapse) at meiosis.

homozygous having identical alleles of the same gene at the same locus on homologous chromosomes.

homunculus a little individual that early Greeks believed was contained in the head of sperm.

hormones a large class of different molecules in plants and animals that regulate gene expression and other physiological processes.

Huntington's disease (chorea) an autosomal dominant disease causing neurological abnormalities; the gene is not expressed until late in life, after an individual may already have produced progeny.

H-Y antigen a male-specific protein that initiates development of male sexual organs.

hybrid plant a progeny plant with a different combination of traits than those observable in either parent.

hybridoma a somatic cell hybrid between a B lymphocyte and a cancer cell; it produces monoclonal antibodies.

hydrocephalus an abnormal accumulation of fluids in the brain that causes mental retardation if it is not drained.

hydrogen bond the sharing of a hydrogen atom between two other atoms. Weaker than a covalent or ionic bond.

hypermutability a rate of mutation much higher than is normally observed.

I

immune system the numerous mechanisms in vertebrates that recognize and eliminate foreign substances and alien cells.

immunoglobulin the scientific term for antibody.

in vitro **fertilization** fertilization of one or several eggs by sperm outside of the body, such as in a test tube.

inbreeding mating between closely related individuals.

inclusive fitness the sum of the fitness of an individual plus its influence on the fitness of relatives; the total effect of kin selection.

inducer any small molecule (for example, lactose) that is able to switch on the expression of one or more genes.

industrial melanism the natural selection of dark-colored (melanic) moths in industrial areas of England.

insertion sequence (IS) small pieces of DNA at the ends of transposons that have identical base sequences. These sequences pair with identical sequences in other DNAs and are responsible for transposon movement.

introns the segments of DNA in eucaryotic genes that separate exons and that are untranslated.

inversion clockwise or counterclockwise rotation of a group of base pairs so that the original sequence of bases is changed.

ion a positively or negatively charged atom.

ionic bond a bond caused by the electrical attraction between two atoms. Weaker than a covalent bond.

ionizing radiation electromagnetic radiation with energy sufficient to strip electrons from atoms, thus creating ions (electrically charged atoms).

isotopes atoms with identical chemical properties but with different weights because of differences in the number of neutrons in the nucleus.

K

karyotype visual arrangement of all of the chromosomes from a single cell so that they can be identified and counted.

kin selection selection for altruistic behavior in a group of individuals that are genetically related.

Klinefelter's syndrome a syndrome characteristic of men who have an XXY chromosome constitution.

L

Lamarckianism the inheritance of acquired traits; an idea named after the French scientist Jean Baptiste Lamarck.

law of independent assortment Mendel's explanation of the failure of two pairs of genes to segregate together.

law of segregation Mendel's law explaining the separation of homologous chromosomes or genes into separate gametes during meiosis.

lethal gene   a gene that causes the death of the individual (before or after birth) if it is expressed.

leukemias   cancers of blood cells.

leukocyte   any kind of white blood cell.

lymphatic system   a complex organ system consisting of the lymph nodes, tonsils, spleen, and thymus; it performs many functions, one of which is immunity.

lymphocyte   a particular type of white blood cell that is associated with the immune system.

Lyon hypothesis   the suggestion that only one X chromosome is expressed in each female cell and that the other X chromosome in that cell is inactive.

lysis   the bursting open of a cell due to disruption of the membrane that surrounds it.

lysogenic bacteria   bacteria that carry a prophage in their DNA.

M

macrophages   large phagocytic white blood cells that engulf and destroy foreign particles.

male gametes   sperm (pollen in plants).

meiosis   process by which the haploid set of chromosomes winds up in gametes (sex cells).

melting temperature   in DNA-DNA hybridization, the temperature at which half of the DNA double helices separate into single strands.

messenger RNA (mRNA)   an RNA molecule whose sequence of bases is translated into a specific sequence of amino acids (a polypeptide).

metabolism   the sum of all the chemical reactions in a cell or organism.

metastasis   the process by which tumor cells are carried to different parts of the body, where they grow into new tumors.

migration   flow of genes from one population to another due to movement of individuals or gametes into or out of a population.

mimicry   a resemblance of one organism to another, which often protects it from predators.

miscegenation   interracial marriage or cohabitation.

missense mutation   a single base-pair change that causes the substitution of one amino acid for another in the protein.

mitochondria   organelles in the cytoplasm of eucaryotic cells whose special function is to produce chemical energy in the form of ATP.

mitosis   process of chromosome segregation and cell division.

modern synthesis   the unification of evolutionary and genetic theories to explain both the continuity and change of organisms over time.

modification enzymes   enzymes that recognize particular sequences of bases in DNA and attach methyl (CH_3) groups to one of the bases in the recognized site. The pattern of methylation (modification) of the DNA protects it from attack by the restriction enzymes of that strain.

monoclonal antibody   an antibody of a single type that is produced by genetically identical somatic hybrid cells (clones).

mutagen   any environmental agent that increases the frequency of mutations.

mutation   any change in one or more base pairs in DNA; a heritable change in the genetic information.

N

natural selection   the differential reproduction of individuals with varying genotypes due to their fitness in a particular environment.

neural tube defects   defects in the formation of the brain or spinal cord in developing embryos.

neutral mutation   a point mutation that does not change the amino acid in a protein and, hence, does not change the protein's function or the organism's phenotype.

neutral mutation theory   the idea that mutations arise and accumulate in populations without any selection pressure from the environment.

neutrons   particles in the nuclei of atoms that are identical to protons but do not carry an electrical charge.

nondisjunction   failure of homologous chromosomes (or chromatids) to separate properly at meiosis, resulting in gametes with too few or too many chromosomes.

nonsense mutation   a single base-pair change that generates a stop codon and terminates the polypeptide chain.

oncogenic viruses   viruses capable of causing cancers in animals when the viral genes are expressed.

O

oncogene hypothesis   the idea that viral genes became integrated into the chromosomes of animal cells millions of years ago. These genes are normally unexpressed but if induced may convert normal cells into tumor cells.

operator (*o* site)   a sequence of base pairs in DNA to which repressor proteins attach, preventing transcription of structural genes of the operon.

operon   a segment of DNA consisting of two or more adjacent genes capable of synthesizing polypeptides, along with the regulatory sites that govern their expression. The genes in an operon are transcribed together into continuous mRNA molecules.

origin of replication   site or sites in DNA where synthesis of a new molecule is initiated.

P

paleontology   the study of the life forms of past geological periods by means of the fossil record.

pangenesis   the concept that each part of an adult organism produces a tiny replica of itself that is collected in the "seed" of the organism and then transmitted to offspring.

parthenogenesis   reproduction by females without fertilization by male gametes. This form of reproduction produces only female offspring.

penetrance   the proportion of individuals carrying a particular allele that actually express the trait.

peptide bond   the specific covalent bond that joins amino acids together in polypeptide chains.

phenotype   the observable characteristics (traits) of an organism that result from the interaction of its genotype with the environment.

phenylketonuria (PKU)   an inherited human disease caused by a defect in the gene that converts phenylalanine to tyrosine.

photosynthesis   the cellular process of converting light energy to chemical energy.

phylogeny   the evolution of a genetically related group of organisms.

pilus   the tube through which DNA is transferred from one bacterium to another.

placenta   the organ that connects a pregnant woman's blood supply with that of the embryo.

plaque assay   a technique for counting viruses.

plasma cell   a specialized cell derived from B lymphocytes that synthesizes antibodies.

pleiotropy   the ability of a gene to affect several unrelated traits in an individual.

point mutation   any change in one DNA base pair.

polygenic inheritance   inheritance in which traits or diseases are caused by more than one gene and the number and chromosomal locations of the genes are undetermined.

polygenic mRNA   an mRNA molecule that carries information from more than one gene and that is translated into different polypeptides.

polymorphism   the presence of several forms of a trait or gene in a population of organisms of the same species.

polypeptide   a chain of amino acids joined together; one or several polypeptide chains make up a protein.

polyploid cells   cells that contain more than two copies of each different chromosome.

population genetics   branch of biology dealing with the measurement and calculation of allele frequencies in populations and how gene frequencies change in successive generations over time.

porphyria   an autosomal dominant disease causing physical and mental symptoms and characterized by variable penetrance, which means that the mutant allele is not expressed in all individuals.

primary sex differentiation   the constitution of the sex chromosomes in the zygote.

primates   a mammalian order comprising humans, apes, monkeys and certain more primitive animals such as the lemur.

procaryotes   simple single-celled organisms such as bacteria.

promoter (*p* site)   a sequence of base pairs in DNA to which RNA polymerase enzymes attach to initiate transcription of structural genes.

prophage   an unexpressed phage genome carried in the DNA of bacteria.

protein   a macromolecule consisting of one or more chains of amino acids that perform catalytic or structural functions in cells.

protons   positively charged particles in the nuclei of atoms.

provirus   an unexpressed viral genome carried in one of the chromosomes of animal cells.

pseudorevertant   a bacterium that grows in the same conditions as wild-type but actually contains two (or more) mutations.

punctuated equilibrium theory   a model of evolution suggesting that populations remain stable for long periods of time and change abruptly rather than gradually and continuously.

R

R plasmid   a small circular DNA molecule that carries genes whose products make bacteria resistant to antibiotics. R plasmids can be transferred to other bacteria by conjugation.

race   an arbitrary subclassification of a species based on physical or genetic differences.

radioactivity   the release of energy from an atom's nucleus in the form of particles or radiation.

radioisotope   an isotope whose nucleus is unstable and decays spontaneously, emitting energy.

random mating   mating in which any individual can mate with any other individual in a population.

recessive allele   the allele in a heterozygous individual that is not observable in the phenotype.

recombinant DNA (genetic engineering; gene splicing)   DNA created by the joining together of segments of DNA from different organisms *in vitro*

recombination   the breaking and rejoining of genetically different DNA molecules; the appearance of traits in progeny that were not observed in parents.

red blood cell   a cell in the blood that contains hemoglobin and that transports oxygen.

regulation of transcription   a process that prevents or alters synthesis of RNA generally and mRNA in particular.

replica plating   a technique in which bacteria are transferred to a series of Petri plates from an original plate by means of a velvet pad.

replication   the duplication of DNA molecules.

replication fork   the region in DNA molecules in which synthesis of new DNA is occurring. The replication fork moves continuously until replication is completed.

repressor   a protein that "turns off" or prevents a gene or a group of genes in DNA from being expressed.

reproductive isolation   the inability of organisms to interbreed because of biological differences.

restriction enzymes   enzymes that recognize particular sequences of bases in DNA and cleave the DNA at or near that site. If the DNA is modified by a particular pattern of methyl groups, the DNA is protected from cleavage by particular restriction enzymes.

reverse mutation (revertant)   the restoration of an original base pair in DNA by a second mutation arising at the same site as the first mutation. A change from mutant to wild-type.

reverse transcriptase   an enzyme present in certain RNA viruses that transcribes the information in the RNA molecule into DNA after the RNA has infected an animal cell.

Rh factor   a red blood cell antigen that can cause anemia in a fetus.

ribosomal RNA (rRNA)   the RNA molecule that is the main structural component of ribosomes.

ribosomes   structures in cells on which protein synthesis occurs. Ribosomes contain three types of rRNA molecules as well as about fifty-five different proteins.

RNA   a nucleic acid composed of a chain of ribonucleotides.

RNA polymerase   the enzyme used to synthesize rRNAs, tRNAs, and mRNAs from genes in DNA.

RNA splicing enzyme   an enzyme that removes bases from eucaryotic mRNA corresponding to introns and that splices the exons together in mRNA before translation occurs.

S

sarcomas   cancers of bone and muscle cells.

secondary and tertiary sex differentiation   the development of male or female sex organs and masculine or feminine physical characteristics.

sex chromosomes   the chromosomes that determine the sex of many organisms.

sexual dimorphism   the distinctly different physical forms of males and females in a species; for example, male birds are generally larger and more colorful than their female counterparts.

sexism   prejudice based on a person's sex; discrimination against women.

sickle-cell anemia   a serious blood disease caused by a mutant hemoglobin. One amino acid is changed at position 6 of the β-chain as a result of a mutation.

sociobiology   the study of the biological (genetic) basis of all forms of social behaviors in all organisms, including humans.

somatic cell hybrid   a somatic cell with a single nucleus containing chromosomes from two genetically different cells that fused to form a single one.

somatic cells   all cells other than the sex cells in plants and animals.

species   populations of individuals that interbreed with one another in nature but do not interbreed with other populations, from which they are reproductively isolated.

split genes   eucaryotic genes in which genetic information is encoded in the DNA in discontinuous segments.

stromatolites   large, ancient rocklike structures formed by bacteria and the minerals trapped by the bacteria.

structural gene   a gene that codes for the synthesis of a polypeptide or protein.

suppressor mutation   a mutation that suppresses the effects of another mutation. The two mutations usually occur in different genes, although this is not always the case.

synapsis   the side-by-side pairing of homologous chromosomes during the prophase stage of meiosis.

T

T lymphocyte   a white blood cell that regulates responses by other cells of the immune system.

teratogen   an environmental agent that adversely affects the development of an embryo.

terminator site   a sequence of bases in DNA where transcription stops; the RNA polymerase detaches from DNA at this site.

terminus of replication   site or sites in DNA where replication stops.

tetrad   the two pairs of sister chromatids formed from a pair of homologous chromosomes.

thalidomide   a teratogenic drug (formerly used as a tranquilizer) that interferes with limb development in human embryos.

thymine dimer   adjacent thymine bases in DNA that are covalently linked by UV light.

topoisomerases   enzymes that change the form, shape, or structure of DNA molecules.

totipotency   the ability of a cell to proceed through all the stages of development, producing a normal adult organism. The nucleus from a single cell, in principle, contains all of the information for reconstructing the complete organism.

transcription   the process of synthesizing RNA molecules from specific segments of DNA molecules.

transduction   transfer of pieces of DNA from one bacterium to another by means of a phage.

transfer RNA (tRNA)   a special type of small RNA molecule that helps line up amino acids in the proper sequence by serving as a link between an mRNA codon and the amino acid it codes for.

transformation   the transfer of DNA from one bacterial strain into a genetically different strain, thereby transforming the genotype and phenotype of the recipient.

transformed cells   a line of established cells that can be cultured indefinitely *in vitro* and whose properties are quite different from those of the normal cells from which the established line was derived.

transition mutation   single base-pair substitutions resulting from adenine ↔ guanine or cytosine ↔ thymine exchanges.

translation   the process of converting the information in the sequence of bases in a messenger RNA molecule into a sequence of amino acids in a polypeptide.

translocation   movement of a group of base pairs from one location to another.

transposon   a group of genes, usually ones that confer antibiotic resistance, that move as a unit from one DNA molecule to another.

tumor   a mass of cells that accumulates at a particular site. If the cells spread to other parts of the body causing disease or death, the tumor is *malignant* Benign tumor cells remain at the original site and do not usually cause disease.

Turner's syndrome   a syndrome characteristic of women who have an XO chromosome constitution.

U

ultrasound scanning production of a visible image of a fetus in the uterus by using sound waves.

V

variable expressivity a characteristic of genes that are expressed in a dominant or recessive manner but whose expression is affected by other genes or by the environment.

viruses particles containing DNA or RNA that infect living cells.

W

wild-type alleles alleles normally found in nature, as opposed to mutant alleles.

Z

zona pellucida the gelatinous outer covering of an egg that allows sperm of the same species to attach and that also facilitates fertilization by a single sperm.

zygote a fertilized egg formed by the fusion of male and female gametes; the first cell of a new individual.

Index

M

macromolecules, 6, 12–15
macrophages, 337, *337*
malaria, 129, 145
male-female differences, 259, 260–261
malignant tumors, 161. *See also* cancer
mammoths, 319
mass spectrograph, 382
mating
    nonrandom, 365
    random, 363, 364
mating behavior, 360–361, 373
matter, creation of, 5
Matthei, Heinrich, 124
Mayr, Ernst, 20
McClintock, Barbara, 87
meiosis, 45, *232*, 233–236, *234*, 245
    recombination in, 65, 235–237
melanin, 130
melanism, industrial, 371–372
Mendel, Gregor, 224–232, *224*
mental retardation, 278, 279, 294, 302, 326. *See also* Down's syndrome
Meselson, Matthew, 52, 63–64, 165
Meselson-Stahl experiment, 52–54, *52*
mesothelioma, 166
messenger RNA (mRNA), 102, 103
    artificial, 124
    discovery of, 118
    genetic code of, 119–127
    inhibition of, 182
    isolation of, 214
    polygenic, 122, *122*
    splicing of, 105
    stability of, 105
    translation of, 105–110, *108, 109*
metabolic diseases, 130, 303
metabolism, 27
    inborn errors of, 130
metacentric chromosomes, 325
metastasis, 162
methionine, *13*
methyl groups, 209–210
mice
    allophenic, 329–330, *329*
    cloning of, 193, *195*, 331
microfossils, 385–386
migration, 367–368
milk, digestion of, 187
Milstein, Cesar, 349
mimicry, 373
miscegenation laws, 427

missense mutation, 140, *140*
mitochondria, 18, 30, 78, 81–82, 121, 266
mitosis, 45, 46
M-N blood antigens, 364–365
modification enzymes, 209–210
molecular clocks, 388–394
"Molecular Structure of Nucleic Acids" (Watson and Crick), 50
molecules, 5, 6, 12–15
mongolism, 277. *See also* Down's syndrome
monoclonal antibodies, 348–352
Monod, Jacques, 118, 183, 184–191, 223
mononucleosis, 174
monozygotic twins, 421–424
Morgan, T. H., 241
Morowitz, Harold J., 397
Morton, Samuel George, 416–417
mosaicism, 261–262
movable genetic elements, 87–88
mRNA. *See* messenger RNA
mule, 363
Müller, H. J., 146
multiple alleles, 61, 143, 345, 396
Mulvihill, John J., 167
muscular dystrophy, 55, 306
mutagenicity, 149
mutagens, 138, 144–149, 150–153
mutations, 27, 31, 54–55, 61, 134, 138
    in bacteria, 56–59
    in cancer, 160, 161, 162, 165
    causes of, 144–149, 151–154
    chromosomal, 141–142
    in E. coli, 186–188
    evolution and, 366
    forward, 149
    frameshift, 123, *123*
    frequency of, 139
    germinal, 274
    in hemoglobin gene, 127–128, 129
    missense, 140
    neutral, 141, 370, 389
    nonsense, 140
    phenotypic effects of, 129, 143–144
    point, 139–140
    random, 59–60
    reverse, 150

somatic, 155–156, 161, 162, 165, 274
spontaneous, 138–139
suppressor, 131
transition, 140
mutation rates, 55, 146–147
Mycobacterium tuberculosis, 423
myelomas, 349

N

National Institute of Health, 218–219
natural selection, 26, 56, 60, 129, 356, 359, 368–369, 371–373, 428, 429
nature vs. nurture, 405–436
Nazi Germany, 426–427
Neanderthals, 380
neo-Darwinism, 357
nerve gas, 114
neural tube defects, 295–297
neutralism vs. selectionism, 370–371, 400
neutral mutations, 141, 370–371, 389
neutrons, 7
Nirenberg, Marshall, 118, 124
nitrosamines, 149
nitrous acid, 149, 155
nondisjunction, 244–245, 275
    of X chromosomes, 255–256
nonrandom mating, 365
nonsense mutations, 140, *140*
nosocomial infections, 83
nucleic acids, 12, *13*. *See also* DNA; RNA
nucleus, 17, 18

O

occupation, and cancer, 164, 166, 168
oncogene hypothesis, 171–172
oncogenic viruses, 169–172, *169*
On Human Nature (Wilson), 429
On the Origin of Species by Means of Natural Selection (Darwin), 26, 356, 371, 381
operator site, 185–190
operon, 184–191
orbital shells, 11
organelles, 17–18, *17*

INDEX